Albert Cheung-Hoi Yu • Kai Gao • Jiangshan Zhan
Editors

Systems Neuroscience

Second Edition

Springer

Editors
Albert Cheung-Hoi Yu
Hai Kang Life Corporation Ltd
Hong Kong, China

Kai Gao
Children's Medical Center
Peking University First Hospital
Beijing, China

Jiangshan Zhan
Pediatrics-neurology
Baylor College of Medicine
Houston, TX, USA

ISSN 2190-5215 ISSN 2190-5223 (electronic)
Advances in Neurobiology
ISBN 978-3-031-69187-4 ISBN 978-3-031-69188-1 (eBook)
https://doi.org/10.1007/978-3-031-69188-1

This Springer imprint is published by the registered company Springer Nature Switzerland AG
The registered company address is: Gewerbestrasse 11, 6330 Cham, Switzerland

Contents

Contributors

Yuri I. Alexandrov Shvyrkov's Lab. Neural Bases of Mind, Institute of Psychology, Russian Academy of Sciences, Moscow, Russia

K. R. Arutyunova Shvyrkov's Lab. Neural Bases of Mind, Institute of Psychology, Russian Academy of Sciences, Moscow, Russia

Fengfeng Bei Department of Neurosurgery, Brigham and Women's Hospital, Harvard Medical School, Boston, MA, USA

Fatma Betül Bozkurt Department of Bioengineering, Gebze Technical University, Gebze, Kocaeli, Turkey

Tunahan Çakır Department of Bioengineering, Gebze Technical University, Gebze, Kocaeli, Turkey

Russell W. Chan Hai Kang Life Corporation Ltd, Hong Kong, China
Hong Kong Centre for Cerebro-cardiovascular Health Engineering (COCHE), Hong Kong, China

Yanxiang Deng Department of Pathology and Laboratory Medicine, Perelman School of Medicine, University of Pennsylvania, Philadelphia, PA, USA
Epigenetics Institute, Perelman School of Medicine, University of Pennsylvania, Philadelphia, PA, USA

Elif Düz Department of Bioengineering, Gebze Technical University, Gebze, Kocaeli, Turkey

Bradley Jay Edelman Brain-Wide Circuits for Behavior Research Group, Max Planck Institute of Biological Intelligence, Planegg, Germany
Emotion Research Department, Max Planck Institute of Psychiatry, Munich, Germany

Kai Gao Children's Medical Center, Peking University First Hospital, Beijing, China

V. V. Gavrilov Shvyrkov's Lab. Neural Bases of Mind, Institute of Psychology, Russian Academy of Sciences, Moscow, Russia

A. G. Gorkin Shvyrkov's Lab. Neural Bases of Mind, Institute of Psychology, Russian Academy of Sciences, Moscow, Russia

Pengfei Guo Department of Pathology and Laboratory Medicine, Perelman School of Medicine, University of Pennsylvania, Philadelphia, PA, USA
Epigenetics Institute, Perelman School of Medicine, University of Pennsylvania, Philadelphia, PA, USA

Kaoru Ide Graduate School of Brain Science, Doshisha University, Kyoto, Japan

Atılay İlgün Department of Bioengineering, Gebze Technical University, Gebze, Kocaeli, Turkey

E. A. Kuzina Shvyrkov's Lab. Neural Bases of Mind, Institute of Psychology, Russian Academy of Sciences, Moscow, Russia

Minshu Li Department of Neurology, Tianjin Neurological Institute, Tianjin Medical University General Hospital, Tianjin, China

Zhichao Liang Department of Biomedical Engineering, Southern University of Science and Technology, Shenzhen, GD, P.R. China

Qiang Liu Department of Neurology, Tianjin Neurological Institute, Tianjin Medical University General Hospital, Tianjin, China

Qing Liu The Brain Cognition and Brain Disease Institute (BCBDI), Shenzhen Institute of Advanced Technology, Chinese Academy of Sciences, Shenzhen-Hong Kong Institute of Brain Science-Shenzhen Fundamental Research Institutions, Shenzhen, Guangdong, China

Quanying Liu Department of Biomedical Engineering, Southern University of Science and Technology, Shenzhen, GD, P.R. China

Youzhi Qu Department of Biomedical Engineering, Southern University of Science and Technology, Shenzhen, GD, P.R. China

Fumiya Sawatani Graduate School of Brain Science, Doshisha University, Kyoto, Japan

A. A. Sozinov Shvyrkov's Lab. Neural Bases of Mind, Institute of Psychology, Russian Academy of Sciences, Moscow, Russia
Faculty of Psychology, National Academic University of Humanities, Moscow, Russia

Xiaoyan Sun CAS Key Laboratory of Genomic and Precision Medicine, Beijing Institute of Genomics, Beijing, China
China National Center for Bioinformation, Beijing, China
University of Chinese Academy of Sciences, Beijing, China

O. E. Svarnik Shvyrkov's Lab. Neural Bases of Mind, Institute of Psychology, Russian Academy of Sciences, Moscow, Russia

Susumu Takahashi Graduate School of Brain Science, Doshisha University, Kyoto, Japan

Shui Ying Tsang Hai Kang Life Corporation Ltd, Hong Kong, China

Dilara Uzuner Department of Bioengineering, Gebze Technical University, Gebze, Kocaeli, Turkey

Chen Wei Department of Biomedical Engineering, Southern University of Science and Technology, Shenzhen, GD, P.R. China

Fuqiang Xu The Brain Cognition and Brain Disease Institute (BCBDI), Shenzhen Institute of Advanced Technology, Chinese Academy of Sciences, Shenzhen-Hong Kong Institute of Brain Science-Shenzhen Fundamental Research Institutions, Shenzhen, Guangdong, China

Yizheng Yao Jiangsu Key Laboratory of Neuropsychiatric Diseases and Institute of Neuroscience, Soochow University, Suzhou, China
Clinical Research Center of Neurological Disease, The Second Affiliated Hospital of Soochow University, Suzhou, China
Department of Neurosurgery, Brigham and Women's Hospital, Harvard Medical School, Boston, MA, USA

Albert Cheung-Hoi Yu Hai Kang Life Corporation Ltd, Hong Kong, China
Neuroscience Research Institute, Peking University, Beijing, China

Weiqi Zhang CAS Key Laboratory of Genomic and Precision Medicine, Beijing Institute of Genomics, Beijing, China
China National Center for Bioinformation, Beijing, China
University of Chinese Academy of Sciences, Beijing, China
Institute for Stem Cell and Regeneration, Chinese Academy of Sciences, Beijing, China

About the Editors

Albert Cheung-Hoi Yu PhD, JP received his PhD from the University of Saskatchewan, Canada, and spent the subsequent years at various institutions, including University of California San Francisco (UCSF), Stanford, The Hong Kong University of Science and Technology (HKUST), and the Peking University. Undoubtedly, he possesses a broad cross-cultural experience base and is actively involved in both the academic and the business environment for over 40 years. He is now a world-renowned bio-entrepreneur and a scientist with over 20 years of experience in this field. Moreover, he has published over 170 peer-reviewed scientific papers in reputed international journals, in addition to publishing several commentaries in journals like *Science* and *Nature Biotechnology*. Apart from his aforementioned designations, Prof. Yu is also the founder and the president of Guangdong-Hong Kong-Macau Greater Bay Area Biotechnology Alliance, on the board of trustees of the Gordan Research Conference, Justice of the Peace and a former member of the board of directors of the Hong Kong Science and Technology Park (HKSTP). With a global network and an expanding base of established contacts in China, Prof. Yu has been a significant facilitator of dialogue regarding the development of the biotech sector in Asia.

Kai Gao PhD is an Assistant Professor at Children's Medical Center, Peking University First Hospital, Beijing, China. He received his PhD from Neuroscience Research Institute, Peking University. He received postdoctoral training at Peking University First Hospital. He is mainly engaged in the research of pathogenic genes, pathogenesis, and intervention strategies of neurodevelopmental disorders in children. He has hosted or is hosting two projects of National Natural Science Foundation and one project of Beijing Natural Science Foundation. He has published 42 peer-reviewed scientific papers in *Brain*, *Science Bulletin (Beijing)*, *World Journal of Pediatrics*, *Movement Disorders*, *Glia*, and *Epilepsia*.

Jiangshan Zhan PhD received his BSc in medicine from Peking University, China, and PhD in medical research from University of Munich (Ludwig Maximilian University of Munich), Germany, with highest degree *Summa* cum *laude*.

He received postdoctoral training at Neuroscience Research Institute, Peking University, China, and Jan and Dan Duncan Neurological Research Institute, Baylor College of Medicine, USA. He is currently a review editor and youth editorial member for prestigious journals such as *Neural Regeneration Research, Innovation Discovery, Frontiers in Molecular Neuroscience, Neurochemical Research* and a committee member of the Hong Kong Biotechnology Organization. He has published 3 academic books and 23 journal papers in *Cell Genomics*, *Cell reports*, *Glia*, *Neurobiology of Disease*, and *Immunology*.

Part I
Principle and Molecular Foundations

Determination of Neuronal Activity and Its Meaning for the Processes of Learning and Memory

Yuri I. Alexandrov, A. A. Sozinov, O. E. Svarnik, A. G. Gorkin, E. A. Kuzina, V. V. Gavrilov, and K. R. Arutyunova

Abstract Despite the years of studies in the field of systems neuroscience, the functions of neural circuits and behavior-related systems are still not entirely understood. The systems description of brain activity has recently been associated with cognitive concepts, e.g., a cognitive map, reconstructed via place-cell activity analysis and the like, and a cognitive schema, modeled in consolidation research. The issue we find of importance is that cognitive concepts reconstructed in neuroscience research are mainly formulated in terms of the environment. In this chapter, we present the idea of an element of individual experience that serves as a model of behavioral interaction with the environment, rather than a model of the environment itself. This intangible difference entails the need for substantial revision of several well-known phenomena, including long-term potentiation. The principal questions we address are: how do elements of experience appear and change during learning and performance; and how do the links between these elements create the whole structure of individual experience? We argue that learning and memory research need a clear distinction between processes that provide the emergence of new elements of experience (functional systems) and processes underlying the retrieval and/or changes in the existing experience. We propose to view the activity of a neuron as an "action" directed to the future adaptive "microresult," essential for meeting its metabolic needs. This anticipatory neuronal activity is coordinated with the activity of many other cells of the body in the organism-wide functional system, ensuring the achievement of an adaptive "macroresult" at the behavioral level. From

Y. I. Alexandrov (✉)
Shvyrkov's Lab. Neural Bases of Mind, Institute of Psychology, Russian Academy of Sciences, Moscow, Russia

A. A. Sozinov
Shvyrkov's Lab. Neural Bases of Mind, Institute of Psychology, Russian Academy of Sciences, Moscow, Russia

Faculty of Psychology, National Academic University of Humanities, Moscow, Russia

O. E. Svarnik · A. G. Gorkin · E. A. Kuzina · V. V. Gavrilov · K. R. Arutyunova
Shvyrkov's Lab. Neural Bases of Mind, Institute of Psychology, Russian Academy of Sciences, Moscow, Russia

A. C.-H. Yu et al. (eds.), *Systems Neuroscience*, Advances in Neurobiology 41,
https://doi.org/10.1007/978-3-031-69188-1_1

this perspective, the mechanisms of learning are considered as the formation of functional systems, and memory is considered as a dynamic structure constituted by systems formed at different stages of individual development.

Keywords Memory · Learning · Memory consolidation · Memory reconsolidation · Apoptosis · Neurogenesis · Specialization of neurons · Systemogenesis · Gene expression · Long-term potentiation

Abbreviations

RSC	Retrosplenial cortex
IEG	Immediate early gene
TFS	Theory of Functional Systems
LTP	Long-term potentiation

1 The Systems View of Neuroscience

1.1 Goal-Directed Behavior and the Concept of a Result

Contemporary neuroscience research often adopts the tradition of isolating meaningful segments of behavior and underlying brain activity from the continuum of behavior. For example, brain activity is often averaged over the periods of time following the presentation of stimuli (tasks) and responses to them. This means that all predictive activity that occurs prior to the stimulus is completely lost. If we consider all behavior as goal-directed, i.e., aimed at achieving a desired outcome rather than just being a set of reactions to incoming stimuli, then behavior and brain activity should be viewed in relation to future events, not just events in the past. This perspective was suggested by Anokhin (1974) in the Theory of Functional Systems (TFS) and has been developed further within the System-Evolutionary Theory (Shvyrkov, 1986).

The central proposition of the TFS is the concept of the system-creating factor that unites sparse brain and body elements into a functional system. The system-creating factor is the *result* of each particular system, i.e., an adaptive effect in the organism-environment interaction achieved when the elements of this system are activated. The concept of a result is isomorphic and can be applied to any system for various research objects and behavioral situations. When applied to behavior, the result is an event in the future, not in the past, where the stimulus is. The system is understood as a dynamic organization of activity of components with different anatomical localizations, aimed at achieving an adaptive outcome, the result (see also Alexandrov et al., 2000). A result is specific for each system because it is the central factor in its creation. In turn, a system can only be identified through the particular result it was formed to achieve.

How can the result, an event in the future, be a reason for current activity? In the TFS, this "time paradox" is explained via the concept of a goal. A goal is a model of the result, it's a predicted outcome of the activity. Thus, the TFS combines the principles of causality and teleology in the description of behavior in a form acceptable for those dedicated to causality as a necessary principle of scientific analysis (Bunge, 1963).

The TFS enables a holistic view of behavior through studying the result-driven organization of the entire organism-environment interactions (Anokhin, 1974). In contrast, the traditional approach views functions as direct effects of a certain substrate, e.g., motor, sensory, emotional, and motivational. A similar view has also been expressed by other authors, suggesting that the activity of any brain area should be studied in relation to behavior, viewed as the process of implementation and selection of systems (Cisek & Kalaska, 2010). The "task space" representation, proposed by Weible et al. (2009) for the anterior cingulate cortex, is, to a certain extent, a similar idea if expanded to the whole organism, rather than applied to one brain region. In the TFS, a systemic function is not localized in any brain structures or parts of the body but applied to the whole organism that interacts with the environment (see Westlin et al., 2023 for a similar view). A function is defined through the result that an organism aims to achieve by a particular activity.

Extensive research within the TFS framework has led to the creation of the System-Evolutionary Theory (Shvyrkov, 1986; see also Alexandrov, 2015). One of the central propositions of this theory is the solution of the psychophysiological (mind-body) problem. The psychic processes (mind) describe the whole organism and its behavior, while physiological processes tend to be considered at the level of elements of the body. From the positions of the system-evolutionary theory, the *organization* of physiological processes into a system is based on neither psychic, nor physiological, but specific (informational) systemic processes. Their substrate is physiological activity, whereas their informational content is psychic. Thus, behavioral acts are not only based on localized physiological processes but also on the processes of their organization. In other words, psychical and physiological processes are different aspects of the same systemic processes.

It is important to note that the described solution of the psychophysiological problem avoids theoretical reductionism (see Dudai, 2002 for types of reductionism), i.e., degrading the psychic phenomena to physiological processes. This aspect is especially important within the problem of consciousness (Alexandrov & Sams, 2005): since consciousness cannot be simplified by analytical means, it is often discarded from the scope of scientific investigation (Kandel, 2006).

The focus of systemic psychophysiology is to study the formation and actualization of systems, their taxonomy, and the dynamics of intersystem relations in behavior. Systemic psychophysiology rejects the reactivity paradigm and employs goal-directed principles in the analysis of the activity of individuals and, more importantly, neurons. Therefore, this approach is free from eclectic explanations of goal-directed behavior by reflex-based mechanisms (see Alexandrov, 2015; Alexandrov & Pletnikov, 2022 for more details). We acknowledge that the presented view shares certain ideas existing in cognitive science. For example, U. Neisser's concept of cognitive schema also includes prediction of incoming information, guidance of the exploration, and modification during execution (Neisser, 1976; see also Moscovitch et al., 2016). Moreover, it presumes simultaneous activation of schemas at different levels of hierarchy (also see Sect. 4). Other resemblant views on brain activity are presented by Engel et al. (2001) and von Stein et al. (2000). The concept of cognitive maps (Tolman, 1948) reconstructed via place-cell activity analysis (Burgess & O'Keefe, 2011; Hartley et al., 2013; O'Keefe, 1976) and cognitive schemas (Bartlett, 1995) modeled in consolidation research (Hennies et al., 2016; Tse et al., 2007, 2011) also share some resemblance with the ideas and interpretations discussed here (see also Collins & Shenhav, 2022; Dudai et al., 2015). However, a clear definition of a system and the system-creating factor (the result) are of critical importance for shifting toward a deeper understanding of cognitive phenomena as the product of an individual's active interaction with the environment, rather than viewing these cognitive phenomena in terms of the environment itself. This leads to the interpretation of neuronal and other physiological processes within living organisms as a manifestation of their *activity*.

1.2 Activity Paradigm

The view of behavior as aimed at future results presumes that activity is a fundamental feature of life and any living matter. The activity presents in various forms at different levels of matter organization (Anokhin, 1974). The view that behavior is driven by a goal, a model of its result, is in line with the principle of activity.

In contrast, the traditional view considers activity and behavior as being caused by past events, whether they are stimuli, triggers, and other events in the environment, or internal processes, such as "motivation." However, the need for a stimulus-triggering behavior disappears when a behavioral act is considered as part of a behavioral continuum, rather than an isolated event. The behavioral continuum is a succession of behavioral acts performed by an individual during their lifetime. Any given behavioral act within the continuum is deployed after the result of the previous act has been achieved and evaluated. The evaluation of the results is an essential step that needs to be completed before the initiation of the following act begins. These processes serve as a transition from the execution of one act to a subsequent one. There is no room for a stimulus in the continuum. The environmental changes described as stimuli are contained in the model of a preceding result. They are conditions, but not causes of behavior. Unexpected changes either have no effect on the continuum (i.e., "ignored") or serve as a condition for behavior that interrupts the succession. In cases when intended behavior is interrupted by something in the environment (a mismatch, see Sect. 3.1), an individual will either attempt to repeat the interrupted behavioral act, or try another act to overcome the obstacle, or even learn a new one (systemogenesis, see Sect. 2.1) in order to achieve the desired outcome. In any case, all behavioral acts are aimed at the future outcome, the result.

Behavior is a manifestation of activity at the level of the whole organism, and the neuronal processes should also be considered as manifestations of activity but at the cellular level.

1.3 Active Neuron

Neuronal activity is often viewed as the transmission of signals, excitation, or habituation, within circuits, or networks. The function of a neuron, in this view, is to receive and forward such signals sometimes referred to as information. Events recorded from neurons are considered as responses to incoming information, which may travel further along the cell to stimulate other neurons. Thus, according to this view, a neuron, just like an organism, responds to stimuli.

The activity paradigm dictates that the functioning of the whole organism and that of a single cell in a multicellular organism are directed at achieving adaptive relationships with the environment. In this view, events in a neuron, or any other living cell, should be looked at as an execution of a genetic program. To continue functioning, a cell requires receiving metabolites from the environment, including other cells (Shvyrkov, 1986; Alexandrov, 2015; Alexandrov & Pletnikov, 2022). Consequently, neuronal activity can be looked at as the behavior of an organism, which does not just respond to incoming stimuli but actively interacts with the environment in order to achieve adaptive outcomes. Events in the neuron are "actions" that change its microenvironment to satisfy its "needs," causing modifications in blood flow, metabolic inflow from glial cells, and activity of other neurons. Therefore, a neuron is not a conductor or a calculator—it's an organism within an organism.

A neuron can satisfy its metabolic "needs" only by interacting with other elements of an organism forming a functional system. Their joint cooperative activity leads to an adaptive relationship between the whole organism and its environment, as well as between each of the cells and its microenvironment. As soon as the result is achieved by the organism, and metabolites are received by a neuron, the firing of the neuron ceases.

Recently, other authors also expressed ideas, based on their experimental results, that neuronal activity is predictive, and that cell metabolism and impulse activity are clearly connected and timed to behavioral events (Mann et al., 2021; Luczak et al., 2022).

This view of neuronal activity corresponds to the evolutionary perspective that showed similarities between the survival principles of single-cell organisms and neurons within the nervous system. It has been shown that cooperation between cells in the colonies of single-cell organisms and cells in a multicellular organism provides for breathing, nutrition, and other essential functions for these cells within the group—they achieve that by synchronizing their metabolic processes (e.g.,

Weber et al., 2012). Recent evidence shows that these processes are also mediated by bioelectric fields, produced upon joint activation of cooperating neuronal ensembles (Pinotsis & Miller, 2023). An organism meets a variety of metabolic requirements by diverse behavioral acts. It can also be argued that along with regular functioning, a neuron is active during apoptosis or "altruistic suicide" (see Sect. 3.3).

The system-evolutionary view of neuronal activity implies that the process of learning, in normal and pathological states, is not just a formation of connections between neurons but also a formation of new groups of metabolically cooperating cells, which include neurons and other cells of the body (see also Alexandrov, 2008; Alexandrov & Pletnikov, 2022; Venerin et al., 2020).

2 The Formation of Memory during Learning and Systemic Structure of Behavior

2.1 Systemogenesis

One of the foundations of the TFS is its approach to development, described through the concept of systemogenesis. Systemogenesis refers to the idea that the maturation of organs is not homogeneous. The first elements to mature during early ontogenesis are those parts of organs and tissues that are essential for achieving the results of the systems to ensure the survival of the organism at these stages of individual development (Anokhin, 1974).

It has been argued that systemogenesis occurs not only in early development but also throughout life and is observed during learning. Learning is viewed as the formation of a new system that provides a new behavioral act or a new way of achieving an adaptive result. It was proposed that due to the successive emergence of new systems during a lifetime, the role of different neurons in the organization of behavior should be considered as based on individual history (Alexandrov & Alexandrov, 1982). This idea was expanded within the system-evolutionary theory, which led to the development of the system-selective concept of learning (Shvyrkov, 1986). Considered in terms of G. Edelman's (1987) view of neuronal ensemble formation during learning, the selection of cells happens based on their features, as opposed to learning as instruction, i.e., modification of neuronal features caused by stimulation. The principle of selection underlies the development of immunity and the processes of evolution in a similar way, although at different timescales.

According to G. Edelman, the selection process starts in the early ontogeny, when a large number of neurons die during brain maturation. The cells that survive this selection were termed a "primary assortment". The "secondary assortment" is formed upon learning via behavioral interaction with the environment at the second stage of the selection.

J.-P. Changeux had also defined two stages of the selection: the "neural Darwinism" in the early ontogeny, including the prenatal stage, is proposed to include a

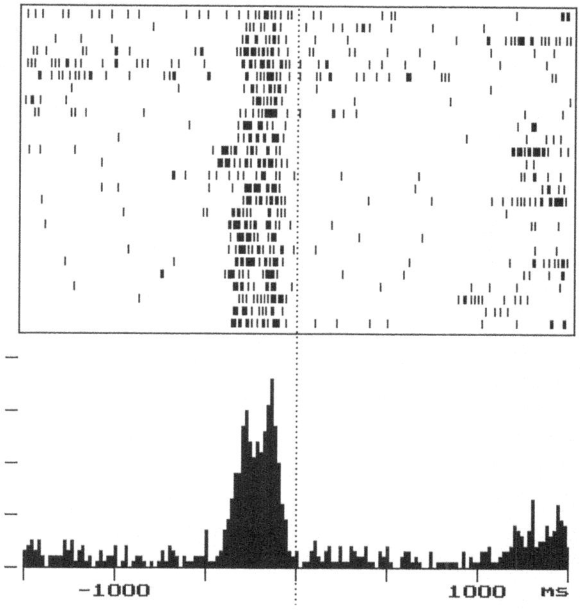

Fig. 1 Activity of a representative specialized neuron #4c20 recorded in rabbit's posterior cingulate cortex. The raster plot shows spikes during successive turns of the animal toward a feeder. Below is the histogram of these spikes (40 ms in a bin, the ticks of the ordinate show tens of spikes) aligned to the mid-wall marker (dashed line). In this experiment (Sozinov et al., 2012), the activity of neurons was recorded with glass electrodes (3–6 MOhms @ 1 KHz) after reaching the learning criterion. For a definition of specialization, see text

selection of effective synapses; and the "mental Darwinism" in adults is considered a modification of the existing synaptic connections. However, the units assumed to be the subject for selection are not connections, but groups of neurons that were selected at the first stage and had cooperative activity (Changeux & Connes, 1999).

The essence of the selection process in both individual development and evolution is the resulting outcome. As R. Wright (1995) puts it, it's not the "truth" that the brain needs, but success. Similarly, in evolution, thrive those who have survived, not those who were right (Cacioppo & Gardner, 1999). Natural selection does not work on independent features, it works on whole organisms, the phenotypes. The basis for the selection is the achievement of results beneficial for the given phenotypic variation. These results are the only objects of the selection (Shvyrkov, 1986; Fodor, 2007). Success in the selection process, defined via the quality of achieved results, includes the formation of "pre-specialized" (see below in this section) and "specialized" neurons (see example in Fig. 1).

In the frame of the system-evolutionary theory, individual development is presented as a sequence of systemogeneses that provide new interactions with the environment. The formation of a new system during systemogenesis is considered as the emergence of a new element of individual experience during learning. Any new system consists of neurons that were selected from the "reserve" cells—presumably, those are low active or silent cells. These cells correspond to the primary assortment (according to Edelman (1987) and are termed here as "pre-specialized." These are the neurons subjected to selection during learning, when some of them become specialized in relation to the new system. This selection process is defined by specific metabolic features of these cells. Accordingly, the group of selected cells corresponds to the secondary assortment. Thus, any new system is an addition to the existing systems; it is "superimposed" on them.

These evolutionary considerations presume that a formed specialization of a neuron remains unchanged for the rest of its lifetime. Learning processes involve new, unspecialized, cells, any previously formed specializations remain the same. The results of experiments support this hypothesis: the number of "silent" neurons in the brain that become active during learning increases, and newly formed specializations remain constant (Schmidt et al., 1976; Thompson & Best, 1990; Wilson & McNaughton, 1993; Swadlow & Hicks, 1997; Williams et al., 1999; Greenberg & Wilson, 2004; Brecht et al., 2005; Chestek et al., 2007; Jackson et al., 2007; Fraser & Schwartz, 2012; Zhang et al., 2022) during the whole period of registration—weeks or even months and years in some instances (see McMahon et al., 2014).

Adult neurogenesis in birds and mammals (Paton & Nottebohm, 1984; Carleton et al., 2003) has been shown to be related to learning. Learning facilitates the survival of newly formed neurons (the "use it or lose it" principle—Kempermann et al., 1998) and their proliferation (Prickaerts et al., 2004). Inhibition of neurogenesis disrupts memory formation (Shors et al., 2001). Based on these and other related data (Anacker & Hen, 2017; Frankland et al., 2013), we hypothesize that adult neurogenesis may support the formation of new systems (see general scheme in Fig. 7). Therefore, both the "reserve" cells and newly formed cells can be specialized in relation to new systems during learning. Adult neurogenesis may also contribute to reinstatement of the primary and secondary assortments of neurons in pathology (Xue, 1998). This compensation of loss of neurons, including the pre-specialized neurons, may also occur in a healthy brain. Since the latter assumption is less grounded with empirical evidence, it is marked with a question mark in Fig. 7.

2.2 The Formation of Neuronal Specializations During Individual Development Continues Phylogenesis

The emergence of the nervous system is a "revolutionary" event in its evolution because it provided a radical increase in complexity and variability of behavior. The complexity of organisms and the size of their genome do not appear to correlate (Gregory, 2001). However, the number of cell types does correspond to the phylogenetic complexity (Bonner, 1988). Importantly, it is the nervous system that contributed the most to this increase in complexity. The cell types in the nervous systems are of great, evidently innumerable, variety (DeFelipe, 2011). Moreover, the combinations of different cell specializations are individual because the specialization is formed in relation to the elements of individual experience—the functional systems. Therefore, the number of unique sets of specializations equals the number of individuals. In other words, every individual has a unique (although culture-specific) composition of systems. The scope of all possible specialization types depends on the species and the subset of neurons pre-specialized during early ontogeny (the primary assortment). Accordingly, the composition of neuronal specializations (the secondary assortment) is individual. Within the presented view of development as a formation of new specializations, the ontogeny appears as phylogeny continued through the increase of the number of cell specialization types.

2.3 The Patterns of Neuronal Specializations in Different Species

The research in our laboratory includes a recording of neuronal spikes from the brain during cyclic operant appetitive behavior in animals. The experimental protocols are in accordance with the Council of the European Communities Directive of November 24, 1986 (86/609 EEC) and the National Institutes of Health "Guidelines for the Care and Use of Animals for Experimental Procedures" and approved by the ethics committee of the Russian Academy of Sciences. The specialization of a neuron in relation to a system is assessed via the probability of its activation in behavioral acts. If this probability reaches 100% in one or more acts, then the neuron is considered specialized, and activations during that act (s) are called "specific" activations (see Alexandrov et al., 2013 for more details). An example of specific activations of a specialized neuron is presented in Fig. 1.

In line with the TFS and the system-evolutionary theory, an individual reflects interactions with the physical world, rather than the world itself. This reflection depends on the individual's goals and existing experience and can be described based on the structure of memory (see paragraph 4). Any individual is essentially a composition of both the phylogenic and ontogenic memory. Thus, we proposed that the structure of memory in different species and even different individuals, who acquire new behavior in the same "resultative milieu" (operant food-acquisition behavior), would have similarities and differences, which can be revealed by comparing patterns of neuronal specializations in various brain regions. Similarities are due to the identical results achieved within the same task, whereas differences are due to the species-specific peculiarities and the history of learning. The patterns of neuronal specializations are defined as relative numbers of neurons specialized in relation to different systems. Therefore, a pattern reveals a particular "kit" of systems.

We have compared the patterns of neuronal specializations in the homological areas of the cingulate (retrosplenial cortex according to Paxinos & Watson, 1997, in rats, and according to Vogt et al., 1986, in rabbits) and motor cortices of rats and rabbits. The animals were trained to perform an operant appetitive behavior in a chamber with two pedals that activated corresponding two feeders. Viewed from the top, the chamber was axially symmetric with a pedal and a feeder in adjacent

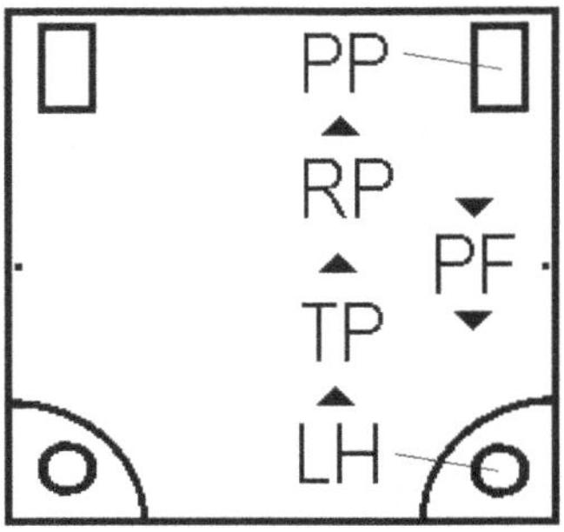

Fig. 2 The schematic view of the experimental chamber. The view from the top includes round feeders and rectangular pedals. The behavioral markers were based on beam crossing or electric contact. The markers were used to identify the following behavioral acts (that corresponded to the stages of learning): lowering the head and taking food from the feeder (LH); lifting the head from the feeder and turning it toward the pedal (TP) until reaching the middle-wall diode; proceeding to pedal corner (RP) until the start of pressing the pedal; pedal pressing (PP); running from pedal to feeder (PF). This sequence of acts was looped (arrows) during neuronal recordings (10+ cycles on each side). The number of acts varied in different experiments

corners of each of the two sides of a square—Fig. 2). The rats' chamber was 1/3 of the size of the rabbits' chamber.

The correct performance was a looped movement from pressing a pedal through turning to the corresponding feeder facing a wall to eating from the feeder and turning back to the same pedal (10–15 repetitions of the whole cycle then switching to the opposite side). The behavioral cycle was divided into several acts (Fig. 2): lowering the head and taking food from the feeder; lifting the head from the feeder and turning it toward the pedal; moving to the pedal corner; pedal pressing; running from the pedal to the feeder. The definition of behavioral acts was based on the stages of learning that had been introduced daily during training. A significant increase of the firing rate above background frequency was considered as an activation. Details of the experimental setup and procedures have been described in more detail elsewhere (e.g., Alexandrov et al., 1990, 2013).

The symmetric arrangement of the chamber allows for the classification of neurons according to how discriminatory its firing is in relation to the behavioral acts. For example, the neuron shown in Fig. 1 has activations in each of the consecutive runs to the pedal on the right side of the chamber. Its specialization would depend on specific activations on the left side. If there's none, the neuron is specialized in relation to a system that subserves approaching the right pedal, which is specific for the new behavior. However, if there is an activation during approaching the right feeder, the firing might be explained by a left turn of the animal's head and/or body (verified in additional tests). In this case, the neuron may belong to a system that subserves (and presumably has subserved) acts of behavior that had been firmed prior to the training in the chamber. Consequently, as opposed to the neurons of the "new" systems, the latter example would be one of the "old" systems.

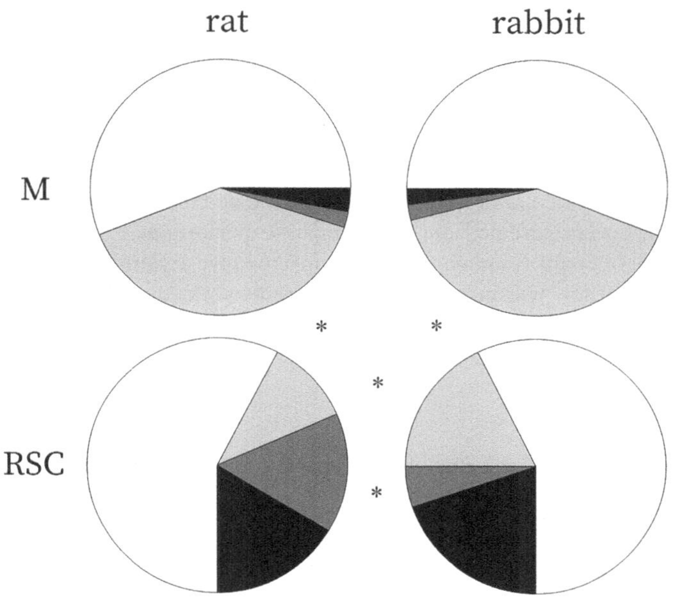

Fig. 3 Percentage of neurons with different specializations in motor and retrosplenial cortical areas in rats and rabbits. White—unidentified neurons, light gray—"old", dark gray—"new" feeder-related, black—"new" pedal-related. Asterisk: significant differences between percentages of neurons. See text for more details

In one of the experiments within this paradigm, activity of single neurons was recorded from retrosplenial (RSC) and motor (M) cortices in rats and rabbits (see Gavrilov et al., 2002, for more details). In the RSC, most of the specialized neurons were classified as belonging to the systems of "new" behavioral acts, whereas most of the M neurons were of "old" systems (e.g., context-independent activations during any particular movement, or during taking any food or non-eatable objects from the feeder or anywhere in the chamber). The percentage of neurons with "new" specializations in the RSC was at least seven times higher than in the M (Fig. 3). The difference between the numbers of neurons with old and new specializations between the two cortical areas was significant and equidirectional in rats and rabbits. The relative content of the old and new systems and specialization patterns in the two cortices were similar in the two species. At the same time, the characteristics of specialization patterns differed—presumably, due to the ethological peculiarities of the species. The rats had relatively more neurons specialized in relation to the new acts of taking food from the feeder. We contrasted the numbers of neurons with feeder-related specializations (acts PF and LH in Fig. 2) and those with pedal-related specializations (acts RP and PP). While these numbers were about the same in rats' RSC, the rabbits had some four times more pedal-related neurons than the feeder-related ones. This difference may be explained by the peculiarities of food taking and manipulation, which is of great variety in rats (Whishaw et al., 1998). Moreover, the greater number of neurons with new specializations in rats can also be explained by their more differentiated and complex behavior.

Thus, the presented data demonstrate both task-related similarities and species-derived differences in the neuronal subserving of similar behavior in the homological cortical areas of rats and rabbits. In this experiment, the recording of neurons was performed after the acquisition of the pedal-pressing behavior, and the specializations were revealed during asymptotic performance. However, a deeper understanding of cognitive phenomena requires the investigation of their emergence. As discussed earlier, within our theoretical framework, the systems that underlie certain behaviors appear via systemogenesis, whereas in conventional terms, new memory undergoes a process of consolidation.

2.4 The Traditional View of Memory Consolidation

The processes of acquisition and consolidation of memory attract the best modern expertise in methods and conceptual schemes (e.g., Feld & Born, 2017; Kitamura et al., 2017; Moscovitch et al., 2016). However, most of the research is still based on the old Descartes concept of memory traces: "The traces are made of the pores that become more permeable as the spirit repeatedly passes through them during the behavior execution."

The issues that follow this idea are those of the mechanisms and limitations of pore enlargement, the brain structures with different numbers of pores, pore permeability duration, etc. These issues, translated from the seventeenth century to modern terms: from pores to synapses and from spirits to neuronal firing, maintain their essence under the concept of engram. Unfortunately, approaches to consolidation (even new ones: Gisquet-Verrier & Riccio, 2022), albeit very distinguished (see Buzsaki et al., 2022; Dudai, 2012 for reviews), rely mainly on long-term increases of conduction effectiveness in neural circuits, networks, etc. "The current central dogma of synaptic consolidation is that it involves stimulus ("teacher")-induced activation of intracellular signaling cascades, resulting in posttranslational modifications, modulation of gene expression, and synthesis of gene products that alter synaptic efficacy" (Dudai, 2012, p. 228).

2.5 The Systems View of Memory Consolidation

From the systems point of view, the neuron is a result achiever, rather than a conductor of excitation. Therefore, the issue of the increase in conduction efficiency is out of the scope of the system-evolutionary approach. The learning process is considered as the formation of a new system of co-active cells (including neurons) of different localization, not necessarily directly connected. This view excludes the concept of "traces," left solely by the instructive input due to plasticity of the nervous system.

The systems view of consolidation was formed based on the system-evolutionary theory. However, the experimental evidence has led other authors to similar conclusions. For example, G. Horn claims that the cross-correlational analysis of neuronal activity in IMHV of domestic chicks does not confirm that the connectivity of "imprint-responsive" neurons is increased during learning, as predicted by the Hebbian rule. "Rather, – the author concludes, – the neurons might form a set of parallel, largely uncoupled elements that are likely to provide a larger storage capacity than a system with tightly coupled elements" (Horn, 2004, p. 121). Although functional connectivity may indeed increase after learning (Abdou et al., 2018), we believe that the connectivity affords synchronization of metabolic activity between structurally connected neurons (see Sect. 1.3), and G. Horn's conclusion remains accurate for the rest of the neurons of the same specialization with no direct connections and even more so for the somatic cells. Different kinds of the network approach in the analysis of synaptic (Hoshiba et al., 2017), cellular (Adams et al., 2017), and whole-brain (Lohmann et al., 2016) processes share some aspects of the systems approach, but they largely retain an instruction-based view on learning and hence, the issue of a unit-of-analysis (Korhonen et al., 2017).

The systems description of the consolidation process necessarily includes two groups of interdependent processes: systemic specialization and accommodative reconsolidation. As described earlier, systemic specialization involves morphological and functional modifications of a neuron that provide its inclusion into a new system. The definition of accommodative reconsolidation requires several preliminary considerations.

Firstly, a new memory is dynamic and adaptive, rather than a stable entity (Bartlett, 1995). The recent progress of memory reconsolidation research shows memory undergoes modifications at the molecular level after its post-consolidation retrieval (Nader, 2015; Sara, 2000). Memory formation and reactivation require protein synthesis, but consolidation and reconsolidation processes are not identical (Anokhin et al., 2002; Dudai & Eisenberg, 2004). Therefore, consolidation, which depends on protein synthesis, reveals a wider range of "active" memory processes (Nader, 2003, 2015), rather than just those of "new" memory.

The idea of reconsolidation does not contradict the idea of permanent specialization aforementioned. Reconsolidation does not undo the changes that had underlied the long-term memory formation (Nader et al., 2000). However, it does constitute another, supposedly less influential, step of the differentiation process for a neuron.

Learning is a specialization of a group of neurons in relation to a new system. The new system is not a substitution, but an addition to the previously formed systems. This addition would require coordination between the new elements and the elements that formed elements. There is evidence suggesting that reconsolidation may indeed be the general mechanism of prior memory reorganization after new learning (see Buzsaki et al., 2022; Dudai et al., 2015; Hupbach et al., 2008; McKenzie & Eichenbaum, 2011).

We suggested earlier that the neurons specialized in relation to a system of one behavioral act may modify their activity and be involved in another behavior without changing their original specialization (Alexandrov, 2008). Acute (Alexandrov et al., 2001) and chronic tetrode (see Alexandrov, 2008) recordings provided more evidence for the reorganization of an existing system upon acquisition of a new behavioral act. Namely, the chronic record of neuronal activity was made during the acquisition of the appetitive operant behavior, described in Sect. 2.3. When the animals reached a good level of performance in the pedal-pressing task on the first side of the chamber, the pedal was turned off to start training on the second side. Once achieved the same level of performance on the second side, the animals were returned to the first side. Consequently, the sides were alternated 10–20 cycles of pedal-pressing each. The activity of several neurons was tracked on the first side of the chamber before and after training on the second side.

Three of these neurons with activations specific to acts on the first side changed their activity patterns after the initial training series on the second side. The activity of one of these cells is shown in Fig. 4 (see Sect. 2.3). Notably, these changes remained significant in all subsequent series, unlike temporary changes of activity of specialized neurons in the first trials of specific acts after alternation or rest

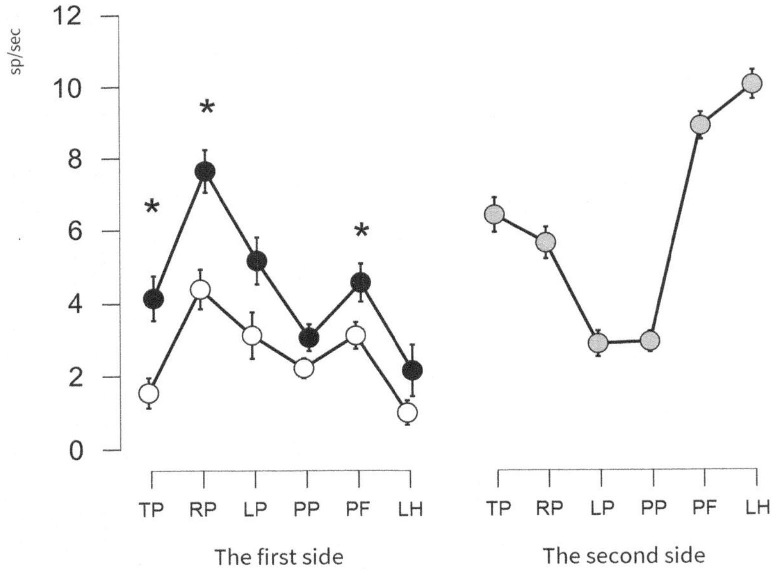

Fig. 4 Firing frequency of neuron # 40103-1 in series of behavioral acts on the first side before (white circles) and after (black circles) training on the second side. After the formation of behavior on the second side, the frequency increased in all acts on the first side (left). Significant differences revealed with the Wilcoxon criterion are marked with an asterisk ($p < 0.05$). Activity during the performance on the second side (right) includes the specific act LH. Ordinate: mean ± SEM spike frequency (spikes per second). Abscissa: acts of food-acquisition behavior (see Fig. 2 for the definition of behavioral acts; *LP* locating in the pedal corner before pedal pressing)

periods. The modifications of this kind were termed by us as "accommodative" reconsolidation (Alexandrov et al., 2001).

The mentioned reconsolidation effect associated with new learning was also detected in our studies at the level of analysis of slow electrical activity of the brain. We compared EEG potentials in rats performing an instrumental lever-pressing task before and after additional training in a slightly different environment (Gavrilov, 2018). EEG was recorded over the motor, retrosplenial posterior, and visual areas of the cortex. A similar configuration of behavior-related EEG potentials was observed in all studied areas during performing the instrumental task before and after additional learning, which suggests that, in general, common brain processes underlie the behavior in the compared conditions. However, differences in the amplitudes and latencies of components of behavior-related potentials support the hypothesis that the composition of the elements of individual experience underlying these behavior changes after additional learning.

Reorganization of a previously formed system after the acquisition of a new one was also observed in our work using immediate early gene (IEG) expression analysis (Svarnik et al., 2013). This study was designed to control for learning prior to operant food acquisition by pedal-pressing to reveal activation of the first-skill-specific neurons during the acquisition of the second one. In the experimental group of rats, the first skill was a "whisking task"—that of using left or right whiskers to receive a water drop. These animals had been overtrained for 5 days before the second skill of pedal pressing for food was introduced. The control group acquired the same food-acquisition behavior, but the first task was non-instrumental drinking instead of the "whisking." Albeit the second skill did not involve the whiskers, we have found c-Fos expression in a significantly greater number of barrel-field neurons in animals of the experimental group compared to the control. These data may suggest that c-Fos induction during the second training took place in neurons specialized in relation to the first, "whisking" task, which is a sign of accommodative reconsolidation. Therefore, along with the specialization of neurons in relation to new systems, some morphological and functional modifications occur in previously specialized neurons. These modifications do not change existing neuronal specializations but provide the inclusion of neurons comprising the new system into the existing systemic structure of individual experience.

Cognitive theories had previously proposed that the reorganization of memory may be either routine (reordering interactions of existing schemas) or heuristic (emergence of new components along with modifications of the prior ones) (Piaget, 1951). The processes of specialization and accommodative reconsolidation would belong to the second type of reorganization. The first type would involve neuronal modifications within existing systems, without the emergence of a new system, and may be referred to as "reorganizational" consolidation. Presumably, "reorganizational" consolidation results in a gradual increase of efficiency, whereas specialization and accommodative reconsolidation result in a transition to good performance, e.g., what we see during our food-acquisition training.

The difference between specialization and accommodative reconsolidation processes is essential for the investigation of the underpinnings of learning. These processes may be indistinguishable from the data on molecular and cellular learning-related processes, which is ignored in most of the studies. The similarities and differences between (re)consolidation, extinction, and forgetting (de Oliveira Alvares & Do-Monte, 2021) are also explained in terms of these two processes (Sozinov et al., 2023). Differentiation of the processes that manifest the emergence of new experience from prior experience modification is necessary for contemporary research of memory principles.

In the frame of the systems perspective that we develop, learning is the key process we study, because it covers the most essential changes of individual experience—the emergence of new systems and modifications within existing systems. Memory is viewed as active and dynamic. Therefore, further we address some of the fundamental processes that underlie or accompany learning: the mismatch between an organism's "needs" and what is available in the environment, neurons and "altruistic suicide", and long-term potentiation.

3 Fundamentals of Learning Within the Systems Perspective

3.1 Memory Formation Starts with a Mismatch

As mentioned earlier, the initial step of the cascade of subcellular molecular events that determine the morphological modifications of neurons, both in the process of morphogenesis (early ontogenesis) and in memory consolidation in adults, is the expression of immediate early genes, followed by the expression of "late" genes that might be directly related to the structural modifications of a neuron. The relations between IEG expression and learning were noted a while ago (Maleeva et al., 1989; Tischmeyer et al., 1990; Anokhin & Rose, 1991) and became widely accepted (e.g., Horn, 2004; Kubik et al., 2007; Barry & Commins, 2011; Minatohara et al., 2016).

In the framework of systems neuroscience, it seems logical to assume that the expression of IEGs and the formation of neuronal specializations are related. Indeed, we showed earlier that those structures that contained a lot of neurons specialized in relation to operant behavior also demonstrated a higher number of Fos-positive neurons after learning (Svarnik et al., 2005).

Induction of IEG expression in adults takes place not only during learning but also during stress, intoxication, lesions of the nervous system, brain ischemia, and other conditions (Herrera & Robertson, 1996; Meyer, 2015). It was also shown that an artificial change in the microenvironment of neurons causes the appearance of activity in previously silent cells and the expression of IEGs (Stone et al., 1993). Given that the activity of a neuron is determined by a mismatch between the neuron's needs and the current influx of metabolites (as discussed above), IEG expression—a specific manifestation of cellular activity (Clayton, 2000) arising in a situation of novelty (Anokhin & Sudakov, 2003; Aggleton et al., 2012)—is suggested to be evident in a general bottom line of all these situations—during the mismatch.

The mismatch arises when the previous possibilities of meeting neuronal metabolic "needs" within the existing memory structure are ineffective in the condition of a stable change in the microenvironment of neurons. The latter occurs upon change in the circumstances of corresponding behavior. Neurons, as noted earlier, may provide for their "needs" by interacting with other elements of the organism and forming a functional system. Achieving the result of the system simultaneously eliminates the mismatch between "needs" and the state of the microenvironment of neurons, which provides the desired result for the organism at the behavioral level. This may happen only when the corresponding behavior has already been formed. However, learning in normal conditions and recovery in pathology (for example, after a stroke, or traumatic brain injury) occur when the "needs" cannot be conformed with the existing matching methods (i.e., within the available individual experience). The mismatch in this situation is different from that in the definitive behavior: it is eliminated not by reactivating existing memory, but by learning a new one via systemogenesis, i.e., selection and fixation of new elements and variants of their combinations (see Sect. 2.1).

3.2 *From Mismatch Through Match to Consolidation*

The emergence of a new system (systemogenesis) may lead to the achievement of the desired result for the organism and the satisfaction of the metabolic "needs" of neurons. However, memory is dynamic. It was shown that the activity of the human brain changes not only in the process of learning but also during hours (and days) after learning criteria have been achieved (e.g., Karni et al., 1995). It was also shown in animal experiments that the parameters of neuronal activations as well as the number of activated cells change within hours and days from the first successful behavioral trial (Erickson & Desimone, 1999; Kuzina et al., 2016; Horn, 2004; McKenzie & Eichenbaum, 2011; Smith et al., 2012; etc).

Our results (Svarnik et al., 2005) showed that the number of neurons in which the IEG expression is detected exceeds by many times the number of neurons specialized in relation to the system of the newly formed behavior. We believe that, in part, this is explained by the fact that some of these genetically activated cells are neurons specialized in relation to earlier formed systems, and the IEG expression in these cells reflects the beginning of the process of accommodative reconsolidation. Others could be pre-specialized neurons, and their gene expression induction is a prerequisite for the transition of cells into a state of readiness for selection during the trials. In the case of developing a more or less complex instrumental behavior, as we hypothesized, it is in the trial-and-error process that certain neurons are selected from activated ones (activated both genetically and, presumably, electrically) and become specialized in relation to the functional system. A decrease in the number of activations as well as the heterogeneity of activity of specialized neurons in the course of memory consolidation, demonstrated in our previous studies (Kuzina et al., 2016), reflects this selection process and changes in neuronal subserving of new behavior. We compared neuronal activity patterns in rat RSC recorded during the pedal-pressing (see Sect. 2.3) within either the first 5 days after its acquisition (group 1), or on days 7–15 (group 2), which corresponds to the "early" and "later" stages of consolidation in rodents (e.g., Buitrago et al., 2004; Dudai et al., 2015). We did not find significant differences in the general proportion of neurons specialized in relation to acts in the "feeder" or "pedal" phases of the behavioral cycle. However, most "pedal" neurons in group 2 were specialized in relation to either approaching or pressing the pedal. In contrast, most of the "pedal" neurons in group 1 were specifically activated both when approaching and pressing the pedal. As noted earlier, neurons that are specifically active in more than one behavioral act are considered as specialized in relation to functional systems that arose in the process of learning at earlier stages of training (see Sect. 2.3). Also, within the first 5 days, there were significantly more pedal-specific neurons with highly differentiated nonspecific activity in other acts of the behavior, compared to group 2, especially in acts associated with the "feeder" part of the behavioral cycle. The activity of "pedal" neurons in group 2 was consistently reduced in "feeder" acts. The reduction in the variability of activity may be associated with the completion of the selection process and stabilization of the neuronal population involved in the newly formed behavior. It is possible that such stabilization requires not only time but also a certain number of repetitions of experience reactivation (Cho et al., 2022; McKenzie et al., 2013; Weible et al., 2009, 2012; McKenzie et al., 2013; Reinert et al., 2021; Weible et al., 2009, 2012).

It was shown that some cells were activated only during the initial stages of learning, and when behavior is stabilized, their activations deceased and did not occur anymore (Hwang et al., 2021; Shima et al., 1996; Wirth et al., 2003). Some of these cells probably belong to pre-specialized neurons that are activated during the first trials of new behavior. At the start of training for similar behavioral acts (i.e., pedal-pressing on the first and the second side in our setup) activity of neurons during the first-trained act may look like variable nonspecific activity, presumably reflecting the process of specialization (ref. to Fig. 13 in Alexandrov, 2008). In addition, as our data show, changes in the pattern of neuronal activity can occur due to the activation of neurons specialized in relation to previously formed behavior, i.e., various acts on the first side. For example, the activity of a neuron in Fig. 5 was recorded during the following periods of the experiment: pressing pedal 1 before pedal 2 training (Fig. 5a, above arrow, black histogram), acquisition of pedal 2 pressing (Fig. 5b), and pedal 1 pressing after acquisition of both skills (Fig. 5a, below arrow, grey histogram). Spike frequencies in behavioral acts with no specific activations (TP, PF, and LH, see Fig. 2) significantly decreased after the acquisition of the pedal-pressing on the second side. The activity of this neuron during the specific acts (RP and PP) on the first side also changed: activations started and ended earlier after the acquisition of the pedal-pressing on the second side. Additionally, the activity of neurons during the last two periods might be considered as a neuronal basis of skill transfer: it is accompanied by an increase in the speed of learning after

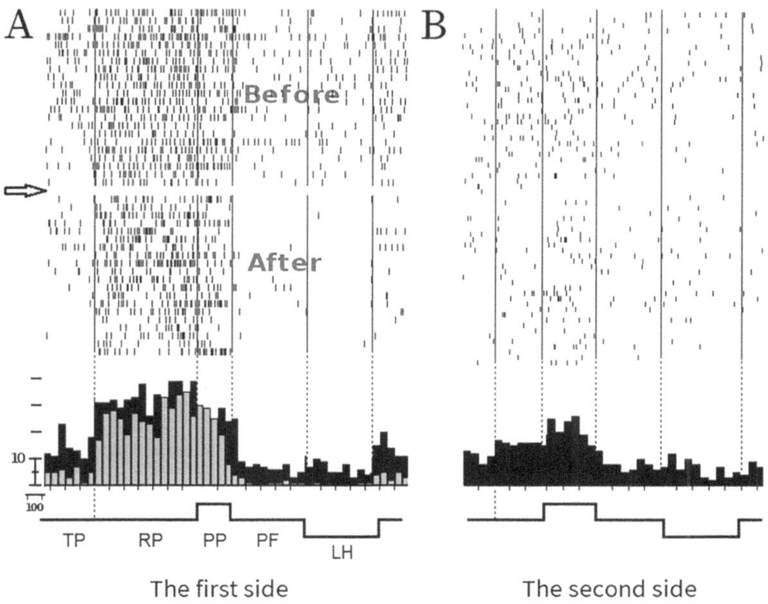

Fig. 5 The activity of a rat RSC single neuron specialized in relation to acts RP and PP of the first pedal was recorded before and after learning to press the second pedal. Recording time: 27 min. See Fig. 2 for the designation of behavioral acts. For results of statistical comparisons see text. (**a**) Combined spike raster plots (top) and histograms (bottom) aligned to the ends of 5 sequential acts on the first side. The actogram (line pattern below the histogram) and vertical lines on the raster plots delimit the behavioral acts on the basis of their mean duration. Neuronal activity during successive trials before (22 trials, above the arrow, black histogram) and after (below the arrow, gray histogram) starts learning to press the second pedal (20 trials). Ticks of the histogram correspond to tens of spikes in 50-ms bins (ordinate); and hundreds of milliseconds (abscissa). (**b**) Spike raster plot and histogram aligned to ends of the 5 acts on the second side (67 trials). All markers as in panel (**a**)

a similar experience has been acquired previously. Thus, mean spike frequencies were significantly higher along the whole period of learning in PP than in any other acts, including RP (Fig. 5b). Also, there was a significant decrease in activity in all acts, except PP, from the beginning to the end of learning to press pedal on the second side.

Finally, it might be suggested that the first trials of learning in organisms with highly developed nervous systems are subserved by co-activation of not only changing sets of specialized and pre-specialized neurons but also so-called "novelty" neurons (which are possibly specialized in relation to orienting (for further details, see Ranganath & Rainer, 2003; Aleksandrov, 2006). This co-activation may provide trial performance as well as achievements of the first positive results during learning. After stabilization of the behavior, "novelty" neurons as well as a number of other previously specialized neurons cease their activity.

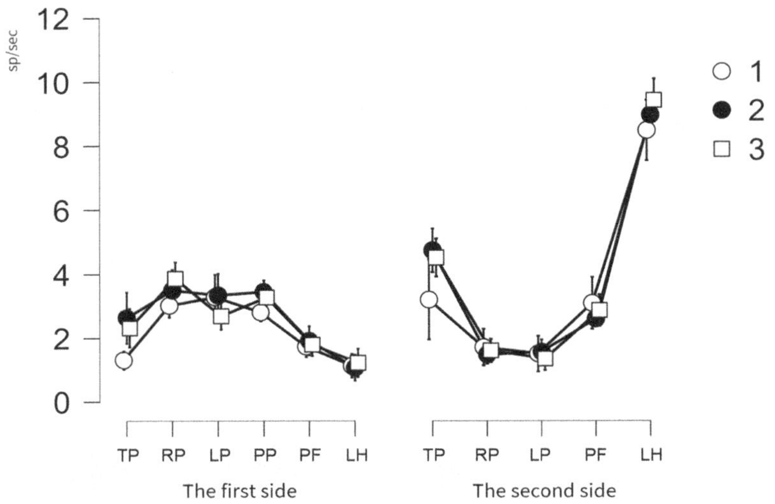

Fig. 6 The activity of neuron 32402-2 in series 1-3 of trials after the onset of training on the second side of the experimental chamber. See Figs. 2 and 4 for designations. The lines 1, 2, and 3 represent successive series of behavioral acts on the two alternated sides. There is no act PP on the second side due to the absence of these acts at the start of training

Therefore, the decline of the activity of past-specialized and "novelty" neurons corresponds to the consolidation process and signifies serious reorganization of neuronal supply of memory reactivation. It occurs despite the activation of neurons with new specializations in the first implementation of the new behavior.

The chronic tetrode recordings described earlier (Sect. 2.5) have also revealed neurons specialized in relation to the behavior on the second side. Activation of these neurons that satisfied the criterion of specific activity (see Sect. 2.3) was evident since the first relevant behavioral act was acquired during learning. An example of such neuronal activity is shown in Fig. 6. Smith et al. (2012) also found the phenomenon of the emergence of supposedly specific activity in neurons during the first trials of new behavior. However, in this study, animals were not pre-trained (except for familiarization with the new environment), as in our experiments where animals were pre-trained to press the first pedal before they learned to press the second one. Perhaps this difference is one of the reasons why the activation (in the place field of a neuron) during the first implementation of behavior was evident in the hippocampus, but not in the cingulate (retrosplenial) cortex.

Earlier (Alexandrov et al., 1991), we showed that the percentages of neurons of different specializations were not changed in the motor cortex in animals under acute ethanol treatment as compared to control. Meanwhile, the neuronal population was different under ethanol: there were significantly fewer specialized neurons in the upper cortical layers, and neurons in the lower layers were more involved into the population that subserved the behavior. Thus, at the early stages of learning, the processes of neuronal specialization may proceed differently in different brain structures.

The studies of brain activity reorganization during learning showed that the role of activity in different brain structures changes at various stages of training (Rose, 1993; Kelly & Garavan, 2005). It is known that learning scores, memory, and "cognitive control" depend on the intact cingulate cortical regions of the human brain (Hayden et al., 2010). Furthermore, outwardly the same behavior is accompanied by the activation of different areas or layers of the cingulate cortex as learning progresses (Trask et al., 2021). On the one hand, the relation between indices of activity and task execution declines in the process of learning in anterior regions of the cingulate cortex and increases in posterior regions (Fournier et al., 2021; Freeman & Gabriel Freeman Jr & Gabriel, 1999; Tracy et al., 2003). On the other hand, in other tasks, activity in the posterior cingulate (retrosplenial) cortex is evident both at early and late stages of learning (Gabriel et al., 1991; Katche et al., 2013). It was also shown that the anterior cingulate cortex is involved in the "context-freezing" task at both early and late stages of learning, and it is necessary for the reconsolidation of this memory (Vetere et al., 2011; Einarsson & Nader, 2012). If, as argued earlier, neuronal specializations are constant (i.e., neuronal differentiation is irreversible—Sect. 2.1), and evident from the first implementations of the newly formed acts, it can be assumed that the most significant contribution to the described reorganization is made not by the dynamics of the activity of the specialized neurons but by other ("unidentified") neurons.

In our experiment, we recorded the activity of neurons in the anterior and posterior areas of rabbit cingulate cortices at "early" (the first week of learning) and "late" (the second week of learning) stages of training of pedal-pressing behavior. We analyzed the average frequency of spikes, percentages of specialized and unidentified neurons, and the number of behavioral acts with nonspecific activations (i.e., acts with a probability of activation between 40% and 100% in unidentified neurons, see Sozinov et al., 2012). We found that all these variables for specialized neurons did not differ between the first and the second week of training, whereas the average frequencies and the number of acts with nonspecific activations were different for neurons with unidentified specializations. Therefore, neuronal activity appears to differ between the first and the second week of training of food-acquisition skills. These differences were primarily indicators of activity of unidentified neurons, rather than specialized neurons.

Based on these results, it could be possible that, in unidentified neurons of the anterior cingulate cortex, the frequency of activations decreases from the first to the second week of training, but in such neurons of the posterior cingulate cortex, the number of activations in new behavioral acts increases during the same period. In other words, the dynamics of brain activations are less accounted for by newly specialized neurons but are more likely to be due to the activity of neurons, which specializations are not identified. As we argued earlier (Alexandrov et al., 1993), unidentified neurons are probably specialized in relation to systems of other behavioral acts and not those formed in our training. Therefore, the established differences in characteristics of neuronal activity at successive learning stages may be connected to the processes of reorganization of experience, which served as a basis for the newly formed behavior, rather than to changes in the cohort of specialized neurons.

These data support the aforementioned idea that it is necessary to differentiate the characteristics of the new experience formation and the old experience reorganization (Alexandrov et al., 2001; Grosmark & Buzsáki, 2016; McKenzie & Eichenbaum, 2011). However, the manifestation of reconsolidative changes also maintains the stability of differently specialized neurons as well as changes in neurons of different specializations.

Treating a neuron as an active living organism has additional consequences for several well-known phenomena. Among them are "altruistic suicide" and long-term potentiation discussed in the remainder of this paragraph.

3.3 *"Altruistic Suicide"*

As it was aforementioned, the IEG expression is induced when the organism does not have experience that would work to satisfy the metabolic "needs" of its cells in a certain situation, or when repetitive impulses of co-activated neurons do not lead to the achievement of the required result (goal achievement). The IEG expression can be considered not only as the first step of the consolidation process but also as the induction of other transcriptional factors underlying the cell's "decision to live or die" (Lee et al., 1998, p. 2736). If the mismatch between the "needs" of neurons and their microenvironment is prolonged, neurons become hyperactive, and waves of IEG expression follow one after another. In such cases, "death" gene expression may be induced, resulting in neuronal death—apoptosis (see Fig. 7). Thus, when a mismatch between the "needs" of the neuron and its microenvironment cannot be eliminated in the conditions of existing experience, the neuron has two alternatives: to change during systemogenesis (new system formation) or die (Fig. 7). These two alternatives exist in both normal conditions (during early ontogenesis and adulthood) and pathological conditions. The involvement into systemogenesis may reside in the system specialization process or the process of accommodative (reorganizational) reconsolidation. Cell death is often observed during early development and in pathology when the existing experience of the organism is not equipped to agree metabolisms of different cells of the organism. However, there are data showing that apoptosis is also evident in the brains of healthy adults and is essential for their normal functioning (e.g., Leist & Jäättelä, 2001). As systemogenesis is a general principle of early development and learning at any age, including adaptation and recovery, these data suggest that the options to "change or die" exist in normal conditions. It was shown (Sherstnev et al., 2013) that the elimination of neurons observed as neuronal selection in early ontogenesis, which is important for behavioral repertoire formation at that stage, also contributes to the process of systemogenesis during adulthood (Fig. 7). Thus, instead of looking at systemogenesis or death as two alternatives, they

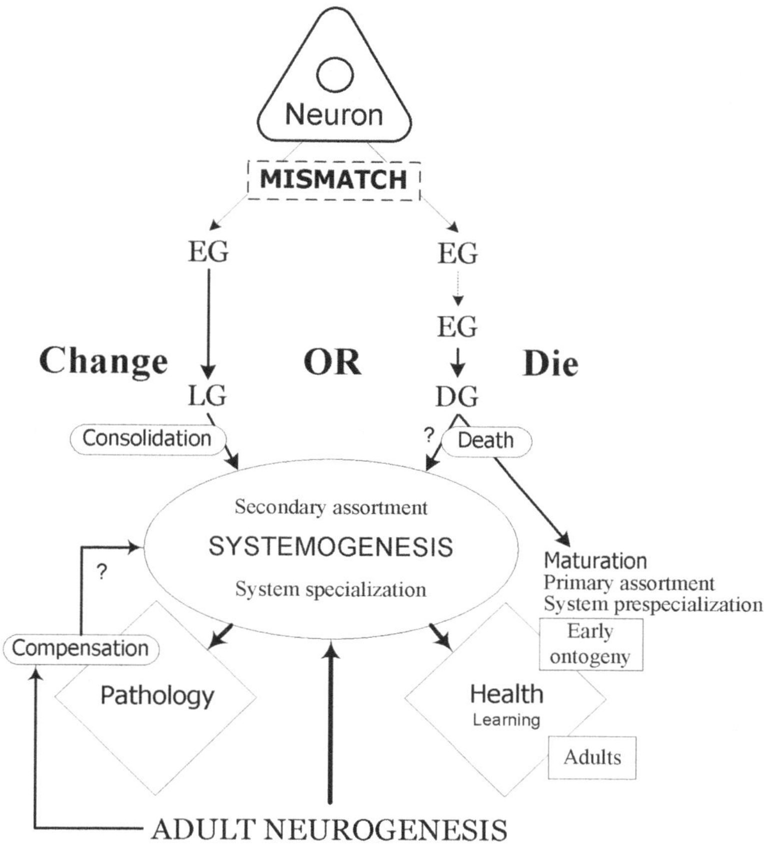

Fig. 7 The general framework of the systemic organization of behavior: Theoretical schema of the "Change or Die" principle. *EG* "early" genes, *LG* "late" genes, *DG* "death" genes. See Sects. 2.1 and 3.3 for explanations

can be considered as interconnected paths to systemogenesis: modification of a neuron or its death. It may be assumed that the death of some cells is a necessary process required for successful systemogenesis. Importantly, the activity principle is applicable for all periods and aspects of the existence of a neuron, including the processes underlying the "change or die" alternative. Each stage of cellular elimination is an active process (Raoul et al., 2000); and thus, neuronal elimination can be viewed as a suicide (Leist & Jäättelä, 2001). This suicide is altruistic in a sense that the neuron turns on the program of self-elimination in order to solve a metabolic conflict and support the survival of other neurons that belong to the same cellular clone. Some other authors also suggested arguments for the existence of "altruistic cellular suicide" in the nervous system (Allsopp & Fazakerley, 2000) and unicellular organisms (Strassmann et al., 2000).

3.4 Long-Term Potentiation: Traditional and Systems Approach

We have described the system approach to learning and memory processes. If someone were to argue for the alternative traditional framework of learning mechanisms, they would probably refer to the phenomenon of long-term potentiation (LTP). LTP is considered to be a physiological mechanism of long-term memory and is regarded as an experimental model of activity-dependent plasticity. For many years, the studies of LTP have been seen as the most important and urgent approach, mainly because this phenomenon was well demonstrated in the framework describing the formation of memory as an increase in synaptic efficiency of impulse conductance in neuronal networks. Within the systems approach, LTP is regarded as an electrophysiological description of the mismatch (see Sect. 3.1). If we consider the neuronal activity as determined by the mismatch, we may conclude that an artificial electrical (or chemical) stimulation used to elicit an influx, which is not accordant with the neuron's preceding activity and not caused by it, serves as a powerful mismatch factor. And the increased cell excitability persistently found during testing is a reflection of this mismatch. Not only theoretical framework but also experimental data point to the link between LTP and the mismatch process. For instance, LTP was shown to be similar to the processes that take place during pathological conditions, when the metabolic cellular environment is dramatically changed (McEachern & Shaw, 1996; Vikman et al., 2003). Thus, although using tetanization is not intended to induce the mismatch, it does. The mismatch is, as argued earlier, the initial stage of learning and the formation of a new memory. Then, we may consider LTP as a phenomenon related to the mechanisms of learning and memory, but for different reasons, because it models the initial stages of learning—the mismatch condition. However, it is not known whether the mismatch obtained during the experimental induction of LTP has the properties of a natural mismatch during learning. Note, that the discrepancy between the traditional concept of LTP and data accumulated from studies of this phenomenon requires an alternative explanation even for those authors who have no doubt that the increase in synaptic efficiency between neurons provides the basis for the formation of memories. McEachern and Shaw (1996) believe that the mechanisms of receptor regulation provide neurons with an attempt to prevent long-term changes in their synaptic excitability, which is harmful to neurons. LTP (like depression), acting against this regulation, is not the basis of learning but is an initial manifestation of the cascade of processes leading to the reorganization of activity of a neuronal group, which "strives" to maintain homeostasis. Shors and Matzel (1997) also concluded that there is a non-correspondence between the properties of LTP, particularly its duration, and those required if LTP was to support the retention of long-term memory. They believe that LTP is a mechanism related not to the maintenance of long-term memory but to the initial stage of its formation. As we can see, the presented conclusions support the suggestion that the mismatch is the initial stage of systemogenesis and that LTP is an electrophysiological description of the mismatch process. If we consider this suggestion about LTP, we may conclude that although the duration of LTP is insufficient for it to be regarded as the basis of long-term memory, it may be adequate for it to be considered as an electrophysiological manifestation of prolonged mismatch leading to cell death. Put more simply, the logic of the ideas proposed here suggests a link between LTP and neuronal death. There are data demonstrating this connection (McEachern & Shaw, 1996; Manahan-Vaughan et al., 1999; Ambrogini et al., 2004). Thus, within the systems approach, the phenomenon of LTP might be related to the mechanisms of learning and memory, not because it models increased effectiveness of impulse transmission in neuronal networks, but because it models the mismatch process, which is a characteristic of the initial stages of learning.

4 The History of Memory Formation and the Memory Structure Are Interrelated

Our previous studies had supported the hypothesis that any behavior is subserved by activation of not only new systems formed during learning, but also older systems formed at earlier stages of individual development (see Alexandrov, 2008; Alexandrov et al., 2000). Thus, behavior reflects the history of its formation (phylogenetic history as well as ontogenetic), i.e., the actualization of multiple systems, each of which had fixated its own stage of acquisition. This suggests that the system organization of overt similar actions is different if the history of their formation is different. For example, our experiments with complex operant behavior in rabbits demonstrated that neuronal activity in the cingulate cortex varied significantly when rabbits learned the same behavior but in a different order (Gorkin & Shevchenko, 1996).

In other experiments, we tested the hypothesis about a connection between the activity of specialized neurons and the number of stages the learning involved. This hypothesis arises from the following logic: neurons are activated during different acts of the acquired (on a daily basis) instrumental behavior (approach to the feeder, turn from the feeder to the pedal, approach to the pedal, and the pedal pressing), and these acts constitute a sequence during the performance; therefore, this behavior is based on the reactivation of all systems of these acts; thus, learning entails the transformation of its stages into systems of the learned behavioral acts, and if the number of learning stages differs, then organization of neuronal activity during overtly identical behavior would also be different. To check this assumption, we compared neuronal activity in the RSC in two groups of rats that either learned the instrumental behavior in one stage (only pedal pressing was reinforced) or in several stages (each stage of instrumental behavior described earlier was reinforced). We found that the number of all newly specialized neurons did not differ between the two groups. However, aggregated activations of all new system-specialized neurons two sequential acts of moving to a pedal and pressing it (RP and PP, see Fig. 2 for identification of acts) were higher in the group with multiple stages of learning. In the group trained in one stage, we found more neurons with specific activations during the acts of turning from the feeder (TP) and higher spike frequency of specialized neurons in all acts. Interestingly, the same significant difference in the proportion of neurons with specialized activations in RP and PP between the two groups of rats persisted after they learned to press the second pedal in the same experimental chamber. These data suggest that there is a connection between the activity of neurons specialized in relation to newly acquired behavior and the number of stages used to acquire this behavior (Kuzina & Aleksandrov, 2020). Thus, in different species we find a consistent general principle—the organization of neuronal activity during behavior depends on the history of its acquisition. Moreover, the history of behavior formation is also reflected in neuronal replays during decision-making in animals: neurons specialized in relation to certain behavioral acts formed recently are activated earlier than neurons related to movements implemented across various categories of behavior, presumably formed earlier in life (Alexandrov et al., 2022).

The complexity of previous experience can be assessed not only via the fraction of neurons involved in new behavior but also through the selectivity of their activations in behavioral acts (Arlt et al., 2022; Gorkin et al., 2018). However, the occurrence of different types of selective activity in neurons of a particular area of the brain, certainly, depends on the features of the task (Dang et al., 2021).

This principle is also manifested at the molecular level. We showed that the number of neurons that changed their gene expression (detected by Fos expression) during new behavior learning depended on the number of acquisition stages of the previous training (Svarnik et al., 2013). In these experiments, we trained animals to press a pedal on one side of the experimental cage in one or several stages and then compared the number of Fos-positive neurons after the acquisition of the same skill on the other side of the cage. Despite the fact that the overt behavior during the second acquisition was similar in these two cases, the number of neurons with changes in gene expression was significantly different. It might be suggested that such changes depended on the processes of accommodative reconsolidation described earlier. Thus, learning involves not only the acquisition of new information but also the assimilation of this information into earlier established experience structures or schemas (e.g., Tse et al., 2011).

5 Conclusion

To conclude, we suggest the following sequence of memory formation and functioning, summarized in Fig. 8. Learning starts with a mismatch between individual needs and available possibilities to meet these needs that exist in memory. At the cellular level, it is manifested as a mismatch between metabolic cellular "needs" and recent metabolic input. In a familiar situation, such mismatch may be cleared by performing previously formed behavior: Fig. 8a shows memory reactivation during

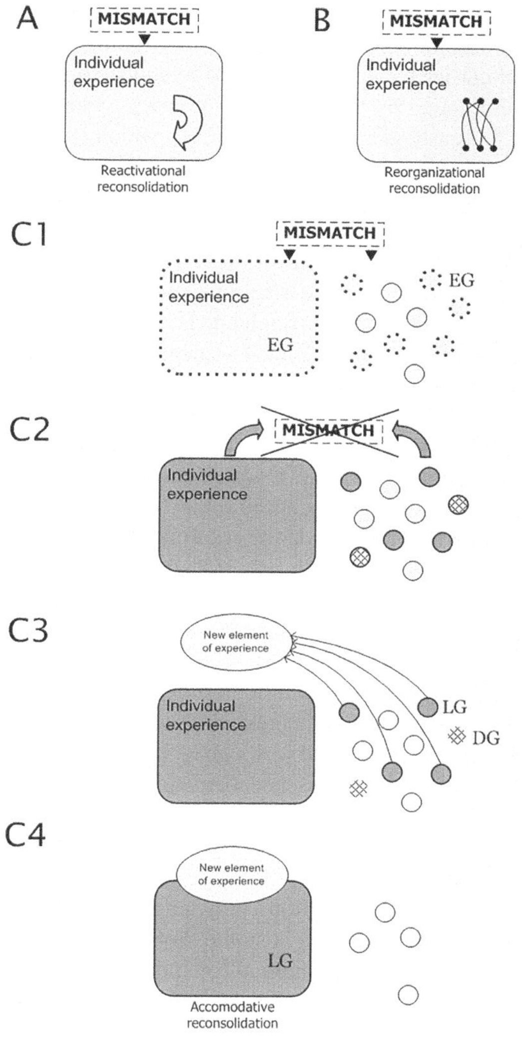

Fig. 8 The general framework of the systemic organization of behavior: Types and stages of modification of the individual experience structure. See Fig. 7 for abbreviations and Conclusion for explanations

previously learned behavior. Memory reactivation may be connected to changes in the structure of experience due to "reactivation reconsolidation". This type of reconsolidation likely appears as continuous memory formation.

In many experimental models of reconsolidation, the acquisition is usually followed by the presentation of a reminder—namely, by additional training (see, e.g., Davis et al., 2010). Therefore, Fig. 8b shows a modification of the structure of individual experience in a new situation, limited by the reorganization of previously formed systems. In this situation, a new element of experience (a new system) is not formed. This type of modification occurs due to "reorganizational reconsolidation."

The type of modification shown in Fig. 8c is one where the mismatch can neither be eliminated by reactivation of existing memory (as in A) nor by reorganization of earlier formed systems and intersystem relations (as in B). Then the mismatch is eliminated by the formation of a new system. This process involves several stages: C1—early gene expression (EG), which is manifested at early stages of acquisition in pre-specialized neurons (circles), as well as in neurons that belong to the systems of prior individual experience; C2—selection during trial behavior: a necessary combination (darker circles) is selected among activated pre-specialized neurons (that may also appear during adult neurogenesis); C3—during the process of selection, a neuron has two options, either to change and get involved into a new system consolidated later through late gene expression (LG), or die (crossed circles; "death" gene expression, DG); C4—accommodative modification of neurons specialized in relation to earlier formed systems (the rectangle), determined by inclusion of a new system into the structure of individual experience.

The formation of a new integration, which is proceeded by "internal" testing and hypothesis selection, is also manifested in the trial behavior. At the cellular level, this trial behavior is based on testing combinations of activated neurons; successful combinations provide the result achievement and elimination of the mismatch (Fig. 8, C2). The success is accomplished by modifications of some cells and elimination of others (Fig. 8, C3). After the first results are achieved, the cells presumably pre-specialized in relation to searching behavior, as well as other cells that belong to earlier formed systems, gradually decrease and eliminate their activity. It is probably manifested in temporal changes in overt behavior (that appears as already formed) and in changes of the set of activated cells. Gradual stabilization of the set of activated cells may be manifested in more stable relations between neuronal activations and behavior. Late gene expression provides reorganization of selected neurons and their transition to constantly specialized in relation to a newly formed system. This system modifies earlier specialized neurons during the process of accommodative reconsolidation (Fig. 8, C4). Thus, the stability of neuronal specializations in a sense of belonging to certain systems does not mean that formed memory is stable. Some of the processes proposed by the illustration in Fig. 8 remain hypothetical (including the modification of intersystem relations), but we consider this as a consistent framework that provides testable propositions.

Acknowledgments The text updating and part of the research was performed in the framework of the State assignment of Ministry of Science and Higher Education of Russia # 0138-2023-0002 for the Institute of Psychology, Russian Academy of Sciences.

Conflict of Interest The authors declare that they have no conflict of interest.

References

Abdou, K., Shehata, M., Choko, K., Nishizono, H., Matsuo, M., Muramatsu, S. I., & Inokuchi, K. (2018). Synapse-specific representation of the identity of overlapping memory engrams. *Science, 360*(6394), 1227–1231. https://doi.org/10.1126/science.aat3810

Adams, N. E., Sherfey, J. S., Kopell, N. J., Whittington, M. A., & LeBeau, F. E. (2017). Heterogeneity in neuronal intrinsic properties: A possible mechanism for hub-like properties of the rat anterior cingulate cortex during network activity. *eNeuro, 4*(1), ENEURO-0313.

Aggleton, J. P., Brown, M. W., & Albasser, M. M. (2012). Contrasting brain activity patterns for item recognition memory and associative recognition memory: Insights from immediate-early gene functional imaging. *Neuropsychologia, 50*(13), 3141–3155. https://doi.org/10.1016/j.neuropsychologia.2012.05.018

Aleksandrov, Y. I. (2006). Learning and memory: Traditional and systems approaches. *Neuroscience and Behavioral Physiology, 36*(9), 969–985.

Alexandrov, Y. I. (2008). How we fragment the world: The view from inside versus the view from outside. *Social Science Information, 47*, 419–457.

Alexandrov, Y. I. (2015). Cognition as Systemogenesis. In M. Nadin (Ed.), *Anticipation: Learning from the past. Cognitive systems monographs* (Vol. 25, pp. 193–220). Springer.

Alexandrov, Y. I., & Alexandrov, I. O. (1982). Specificity of visual and motor cortex neurons activity in behavior. *Acta Neurobiologiae Experimentalis., 42*, 457–468.

Alexandrov, Y. I., & Pletnikov, M. V. (2022). Neuronal metabolism in learning and memory: The anticipatory activity perspective. *Neuroscience & Biobehavioral Reviews, 137*, 104664. https://doi.org/10.1016/j.neubiorev.2022.104664

Alexandrov, Y. I., & Sams, M. (2005). Emotion and consciousness: Ends of a continuity. *Cognitive Brain Research, 25*(2), 387–405.

Alexandrov, Y. I., Grinchenko, Y. V., Laukka, S., Jarvilehto, T., Maz, V. N., & Svetlaev, I. A. (1990). Acute effect of ethanol on the pattern of behavioral specialization of neurons in the limbic cortex of the freely moving rabbit. *Acta Physiologica Scandinavica., 140*, 257–268.

Alexandrov, Y. I., Grinchenko, Y. V., Laukka, S., Jarvilehto, T., & Matz, V. N. (1991). Acute effects of alcohol on unit activity in the motor cortex of freely moving rabbits: Comparison with the limbic cortex. *Acta Physiologica Scandinavica., 1991*(142), 429–435.

Alexandrov, Y. I., Grinchenko, Y. V., Laukka, S., Jarvilehto, T., Maz, V. N., & Korpusova, A. V. (1993). Effect of ethanol on hippocampal neurons depends on their behavioral specialization. *Acta Physiologica Scandinavica., 149*, 429–435.

Alexandrov, Y. I., Grechenko, T. N., Gavrilov, V. V., et al. (2000). Formation and realization of individual experience: a psychophysiological approach. In R. Miller, A. M. Ivanitsky, & P. M. Balaban (Eds.), *Conceptual advances in brain research* (Conceptual advances in Russian neuroscience: Complex brain functions) (Vol. 2, pp. 181–200). Harwood Academic Publishers.

Alexandrov, Y. I., Grinchenko, Y. V., Shevchenko, D. G., Averkin, R. G., Matz, V. N., Laukka, S., & Korpusova, A. V. (2001). A subset of cingulate cortical neurons is specifically activated during alcohol-acquisition behavior. *Acta Physioljgica Scandinavica, 171*, 87–97.

Alexandrov, Y. I., Grinchenko, Y. V., Shevchenko, D. G., Averkin, R. G., Matz, V. N., Laukka, S., & Sams, M. (2013). The effect of ethanol on the neuronal subserving of behavior in the hippocampus. *Journal of Behavioral and Brain Science, 3*, 107–130.

Alexandrov, Y. I., Svarnik, O. E., Rozhdestvin, A. V., & Grinchenko, Y. V. (2022). Brain mechanisms of embodied decision-making. *International Journal of Cognitive Research in Science, Engineering and Education, 10*(2), 163–171. https://doi.org/10.23947/2334-8496-2022-10-2-163-171

Allsopp, T. E., & Fazakerley, J. K. (2000). Altruistic cell suicide and the specialized case of the virus-infected nervous system. *Trends in Neuroscience., 23*, 284–290.

Ambrogini, P., Orsini, L., Mancini, C., Ferri, P., Ciaroni, S., & Cuppini, R. (2004). Learning may reduce neurogenesis in adult rat dentate gyrus. *Neuroscience Letters, 359*, 13–16.

Anacker, C., & Hen, R. (2017). Adult hippocampal neurogenesis and cognitive flexibility—Linking memory and mood. *Nature Reviews. Neuroscience, 18*, 335–346. https://doi.org/10.1038/nrn.2017.45

Anokhin, P. K. (1974). *Biology and neurophysiology of the conditioned reflex and its role in adaptive behavior*. Pergamon Press.

Anokhin, K. V., & Rose, S. P. (1991). Learning-induced increase of immediate early gene messenger RNA in the Chick forebrain. *The European Journal of Neuroscience, 3*(2), 162–167.

Anokhin, K. V., & Sudakov, K. V. (2003). Genome of brain neurons in organization of systemic mechanisms of behavior. *Bulletin of Experimental Biology and Medicine, 135*(2), 107–113.

Anokhin, K. V., Tiunova, A. A., & Rose, S. P. R. (2002). Reminder effects -reconsolidation or retrieval deficit? Pharmacological dissection with protein synthesis inhibitors following reminder for a passive-avoidance task in young chicks. *European Journal of Neuroscience., 15*, 1759–1765.

Arlt, S., Barroso-Luque, R., Kira, S., Bruno, C. A., Xia, N., Chettih, S. N., Soares, S., Pettit, N. L., & Harvey, C. D. (2022). Cognitive experience alters cortical involvement in goal-directed navigation. *eLife, 11*, e76051. https://doi.org/10.7554/eLife.76051

Barry, D. N., & Commins, S. (2011). Imaging spatial learning in the brain using immediate early genes: Insights, opportunities and limitations. *Reviews in the Neurosciences, 22*, 131–142. https://doi.org/10.1515/RNS.2011.019

Bartlett, F. C. (1995). *Remembering: A study in experimental and social psychology*. Cambridge University Press.

Bonner, J. T. (1988). *The evolution of complexity by means of natural selection*. Princeton University Press.

Brecht, M., Sceneider, M., & Manns, I. D. (2005). Silent neurons in sensorimotor cortices: Implication for cortical plasticity. In F. F. Ebner (Ed.), *Neural plasticity in adult somatic sensory-motor systems* (pp. 1–19). Taylor & Francis Group, LLC.

Buitrago, M. M., Ringer, T., Schulz, J. B., Dichgans, J., & Luft, A. R. (2004). Characterization of motor skill and instrumental learning time scales in a skilled reaching task in rat. *Behavioural Brain Research, 155*, 249–256.

Bunge, M. A. (1963). *Causality: The place of the causal principle in modern science*. Harvard University Press.

Burgess, N., & O'Keefe, J. (2011). Models of place and grid cell firing and theta rhythmicity. *Current Opinion in Neurobiology, 21*(5), 734–744. https://doi.org/10.1016/j.conb.2011.07.002

Buzsaki, G., McKenzie, S., & Davachi, L. (2022). Neurophysiology of remembering. *Annual Review of Psychology, 4*, 187–215. https://doi.org/10.1146/annurev-psych-021721-110002

Cacioppo, J. T., & Gardner, W. L. (1999). Emotion. *Annual Review of Psychology., 50*, 191–214.

Carleton, A., Petreanu, L. T., Lansford, L., & Lledo, P.-M. (2003). Becoming a new neuron in the adult olfactory bulb. *Nature Neuroscience, 6*, 507–518.

Changeux, J. P., & Connes, A. (1999). *Conversations on mind, matter, and mathematics*. Princeton University Press.

Chestek, C. A., Batista, A. P., Santhanam, G., Yu, B. M., Afshar, A., Cunningham, J. P., Gilja, V., Ryu, S. I., Churchland, M. M., & Shenoy, K. V. (2007). Single-neuron stability during repeated reaching in macaque premotor cortex. *The Journal of Neuroscience., 27*, 10742–10750.

Cho, H.-Y., Lee, H.-S., Jeong, Y., Han, J., Yoo, M., & Han, J.-H. (2022). Excitability-independent memory allocation for repeated event. *Frontiers in Behavioral Neuroscience, 16*, 860027. https://doi.org/10.3389/fnbeh.2022.860027

Cisek, P. I., & Kalaska, J. F. (2010). Neural mechanisms for interacting with a world full of action choices. *Annual Review of Neuroscience., 33*, 269–298. https://doi.org/10.1146/annurev.neuro.051508.135409

Clayton, D. F. (2000). The genomic action potential. *Neurobiology of Learning and Memory., 74*, 185–216.

Collins, A. G. E., & Shenhav, A. (2022). Advances in modeling learning and decision-making in neuroscience. *Neuropsychopharmacology, 47*(1), 104–118. https://doi.org/10.1038/s41386-021-01126-y

Dang, W., Jaffe, R. J., Qi, X. L., & Constantinidis, C. (2021). Emergence of nonlinear mixed selectivity in prefrontal cortex after training. *The Journal of Neuroscience, 41*(35), 7420–7434. https://doi.org/10.1523/JNEUROSCI.2814-20.2021

Davis, S., Renaudineau, S., Poirier, R., Poucet, B., Save, E., & Laroche, S. (2010). The formation and stability of recognition memory: What happens upon recall? *Frontiers in Behavioral Neuroscience., 4*, 177. https://doi.org/10.3389/fnbeh.2010.00177

DeFelipe, J. (2011). The evolution of the brain, the human nature of cortical circuits, and intellectual creativity. *Frontiers in Neuroanatomy, 5*, 29. https://doi.org/10.3389/fnana.2011.00029

Dudai, Y. (2002). *Memory from A to Z. Keywords, concepts and beyond*. Oxford Univ. Press.

Dudai, Y. (2012). The restless engram consolidations never end. *The Annual Review of Neuroscience., 35*, 227–247. https://doi.org/10.1146/annurev-neuro-062111-150500

Dudai, Y., & Eisenberg, M. (2004). Rites of passage of the engram: Reconsolidation and the lingering consolidation hypothesis. *Neuron, 44*, 93–100.

Dudai, Y., Karni, A., & Born, J. (2015). The consolidation and transformation of memory. *Neuron, 88*(1), 20–32. https://doi.org/10.1016/j.neuron.2015.09.004

Edelman, G. M. (1987). *Neural Darwinism: The theory of neural group selection*. Basic Books.

Einarsson, E. O., & Nader, K. (2012). Involvement of the anterior cingulate cortex in formation, consolidation, and reconsolidation of recent and remote contextual fear memory. *Learning & Memory., 19*, 449–452. https://doi.org/10.1101/lm.027227.112

Engel, A. K., Fries, P., & Singer, W. (2001). Dynamic predictions: Oscillations and synchrony in top–down processing. *Nature Reviews Neuroscience., 2*(10), 704–716. https://doi.org/10.1038/35094565

Erickson, C. A., & Desimone, R. (1999). Responses of macaque perirhinal neurons during and after visual stimulus association learning. *Journal of Neuroscience, 19*, 10404–10416.

Feld, G. B., & Born, J. (2017). Sculpting memory during sleep: Concurrent consolidation and forgetting. *Current Opinion in Neurobiology, 44*, 20–27. https://doi.org/10.1016/j.conb.2017.02.012

Fodor, J. (2007). Against Darwinism. In S. Vosniadou, D. Kayser, & A. Protopapas (Eds.), *Proceedings of EuroCogSci07* (pp. 23–28). Lawrence Erlbaum Associates.

Fournier, D. I., Cheng, H. Y., Tavakkoli, A., Gulledge, A. T., Bucci, D. J., & Todd, T. P. (2021). Retrosplenial cortex inactivation during retrieval, but not encoding, impairs remotely acquired auditory fear conditioning in male rats. *Neurobiology of Learning and Memory, 185*, 107517. https://doi.org/10.1016/j.nlm.2021.107517

Frankland, P. W., Kohler, S., & Josselyn, S. A. (2013). Hippocampal neurogenesis and forgetting. *Trends in Neuroscience., 36*, 497–503. https://doi.org/10.1016/j.tins.2013.05.002

Fraser, G. W., & Schwartz, A. B. (2012). Recording from the same neurons chronically in motor cortex. *Journal of Neurophysiology, 107*, 1970–1978. https://doi.org/10.1152/jn 01012.2010

Freeman, J. H., Jr., & Gabriel, M. (1999). Changes of cingulothalamic topographic excitation patterns and avoidance response incubation over time following initial discriminative conditioning in rabbits. *Neurobiology of Learning and Memory, 72*, 259–272.

Gabriel, M., Vogt, B. A., Kubota, Y., Poremba, A., & Kang, E. (1991). Training-stage related neuronal plasticity in limbic thalamus and cingulate cortex during learning: A possible key to mnemonic retrieval. *Behavioral Brain Research, 46*, 175–185.

Gavrilov, V. (2018). Changes in brain activity during instrumental behavior after additional learning in rats. In J. M. Delgado-García, X. Pan, R. Sánchez-Campusano, & R. Wang (Eds.), *Advances in cognitive neurodynamics (VI)* (pp. 55–61). Springer.

Gavrilov, V., Grinchenko, Y. V., & Alexandrov, Y. I. (2002). Do neurons in homologous cortical areas of rabbits and rats have similar behavioral specialization? *FENS Abstracts, 1*, A040.8.

Gisquet-Verrier, P., & Riccio, D. C. (2022). Revisiting systems consolidation and the concept of consolidation. *Neuroscience and Biobehavioral Reviews, 132*, 420–432. https://doi.org/10.1016/j.neubiorev.2021.12.003

Gorkin, A. G., & Shevchenko, D. G. (1996). Distinctions in the neuronal activity of the rabbit limbic cortex under different training strategies. *Neuroscience and Behavioral Physiology, 26*(2), 103–112.

Gorkin, A. G., Kuzina, E. A., Ivlieva, N. P., & Solov'eva OA, Aleksandrov YuI. (2018). Activity patterns in neurons in the retrosplenial area of the cortex in operant food-procuring behavior in

rats of different ages. *Neuroscience and Behavioral Physiology, 48*, 1014–1018. https://doi.org/10.1007/s11055-018-0663-8
Greenberg, P. A., & Wilson, F. A. W. (2004). Functional stability of dorsolateral prefrontal neurons. *Journal of Neurophysiology, 92*, 1042–1055.
Gregory, T. R. (2001). Coincidence coevolution, or causation? DNA content, cell size, and the C-value enigma. *Biological Reviews, 76*, 65–101.
Grosmark, A. D., & Buzsáki, G. (2016). Diversity in neural firing dynamics supports both rigid and learned hippocampal sequences. *Science, 351*(6280), 1440–1443. https://doi.org/10.1126/science.aad1935
Hartley, T., Lever, C., Burgess, N., & O'Keefe, J. (2013). Space in the brain: How the hippocampal formation supports spatial cognition. *Philosophical Transactions of the Royal Society of London. Series B, Biological Sciences, 369*(1635), 20120510. https://doi.org/10.1098/rstb.2012.0510
Hayden, B. Y., Smith, D. V., & Platt, M. L. (2010). Cognitive control signals in posterior cingulate cortex. *Frontiers in Human Neuroscience, 4*, 1–8. https://doi.org/10.3389/fnhum.2010.00223
Hennies, N., Ralph, M. A., Kempkes, M., Cousins, J. N., & Lewis, P. A. (2016). Sleep spindle density predicts the effect of prior knowledge on memory consolidation. *The Journal of Neuroscience., 36*(13), 3799–3810. https://doi.org/10.1523/JNEUROSCI.3162-15.2016
Herrera, D. G., & Robertson, H. A. (1996). Activation of c-fos in the brain. *Progress in Neurobiology, 50*(2–3), 83–107.
Horn, G. (2004). Pathways of the past: The imprint of memory. *Nature Review of Neuroscience, 5*, 108–121.
Hoshiba, Y., Wada, T., & Hayashi-Takagi, A. (2017). Synaptic ensemble underlying the selection and consolidation of neuronal circuits during learning. *Frontiers in Neural Circuits, 11*, 12. https://doi.org/10.3389/fncir.2017.00012
Hupbach, A., Gomez, R., Hardt, O., & Nadel, L. (2008). The dynamics of memory: Context-dependent updating. *Learning & Memory, 15*, 574579. https://doi.org/10.1101/lm.1022308
Hwang, E. J., Dahlen, J. E., Mukundan, M., & Komiyama, T. (2021). Disengagement of motor cortex during long-term learning tracks the performance level of learned movements. *The Journal of Neuroscience, 41*(33), 7029–7047. https://doi.org/10.1523/JNEUROSCI.3049-20.2021
Jackson, A., Mavoori, J., & Fetz, E. E. (2007). Correlations between the same motor cortex cells and arm muscles during a trained task, free behavior, and natural sleep in the macaque monkey. *Journal of Neurophysiology, 97*, 360–374.
Kandel, E. R. (2006). *In search of memory: The emergence of a new science of mind*. WW Norton & Company.
Karni, A., Meyer, G., Jezzard, P., Adams, M. M., Turner, R., & Ungerleider, L. G. (1995). Functional MRI evidences for adult motor cortex plasticity during motor skill learning. *Nature, 377*, 155–158.
Katche, C., Dorman, G., Slipczuk, L., Cammarota, M., & Medina, J. H. (2013). Functional integrity of the retrosplenial cortex is essential for rapid consolidation and recall of fear memory. *Learning & Memory, 20*, 170–173. https://doi.org/10.1101/lm.030080.112
Kelly, A. M. C., & Garavan, H. (2005). Human functional neuroimaging of brain changes associated with practice. *Cerebral Cortex, 15*, 1089–1102.
Kempermann, G., Kuhn, G. H., & Gage, F. H. (1998). Experience-induced neurogenesis in the senescent dentate gyrus. *Journal of Neuroscience, 18*, 3206–3212.
Kitamura, T., Ogawa, S. K., Roy, D. S., Okuyama, T., Morrissey, M. D., Smith, L. M., Redondo, R. L., & Tonegawa, S. (2017). Engrams and circuits crucial for systems consolidation of a memory. *Science, 356*(6333), 73–78. https://doi.org/10.1126/science.aam6808
Korhonen, O., Saarimäki, H., Glerean, E., Sams, M., & Saramäki, J. (2017). Consistency of regions of interest as nodes of fMRI functional brain networks. *Network Neuroscience, 1*(3), 254–274. https://doi.org/10.1162/NETN_a_00013
Kubik, S., Miyashita, T., & Guzowski, J. F. (2007). Using immediate-early genes to map hippocampal subregional functions. *Learning & Memory, 14*, 758–770.

Kuzina, E., & Aleksandrov, Y. (2020). Characteristics of the neuronal support for operative behavior formed by mono- and multistep methods. *Neuroscience and Behavioral Physiology, 50*, 710–722. https://doi.org/10.1007/s11055-020-00959-2

Kuzina, E. A., Gorkin, A. G., & Aleksandrov, Y. I. (2016). Neuron activity in the retrosplenial cortex of the rat at the early and late stages of memory consolidation. *Neuroscience and Behavioral Physiology, 46*(7), 789–793. https://doi.org/10.1007/s11055-016-0312-z

Lee, Y., Park, K. H., Baik, S. H., & Cha, C. I. (1998). Attenuation of c-Fos basal expression in the cerebral cortex of aged rat. *Neuroreport, 9*, 2733–2736.

Leist, M., & Jäättelä, M. (2001). Four deaths and a funeral: From caspases to alternative mechanisms. *Nature Reviews, 2*, 1–10.

Lohmann, G., Stelzer, J., Zuber, V., Buschmann, T., Margulies, D., Bartels, A., & Scheffler, K. (2016). Task-related edge density (TED)—A new method for revealing dynamic network formation in fMRI data of the human brain. *PLoS One, 11*(6), e0158185. https://doi.org/10.1371/journal.pone.0158185

Luczak, A., McNaughton, B. L., & Kubo, Y. (2022). Neurons learn by predicting future activity. *Nature Machine Intelligence, 4*, 62–72. https://doi.org/10.1038/s42256-021-00430-y

Maleeva, N. E., Ivolgina, G. L., Anokhin, K. V., & Limborskaia, S. A. (1989). Analysis of the expression of the c-fos proto-oncogene in the rat cerebral cortex during learning. *Genetika, 25*(6), 1119–1121. (in Russian).

Manahan-Vaughan, D., Behnish, T., & Reymann, K. G. (1999). ACPD-mediated slow-onset potentiation is associated with cell death in the rat CA1 region in vivo. *Neuropharmacology, 38*, 487–494.

Mann, K., Deny, S., Ganguli, S., & Clandinin, T. R. (2021). Coupling of activity, metabolism and behaviour across the Drosophila brain. *Nature, 593*(7858), 244–248. https://doi.org/10.1038/s41586-021-03497-0

McEachern, J. C., & Shaw, C. A. (1996). An alternative to the LTP orthodoxy: A plasticity-pathology continuum model. *Brain Research Reviews, 22*, 51–92.

McKenzie, S., & Eichenbaum, H. (2011). Consolidation and reconsolidation: Two lives of memories? *Neuron, 71*, 224–233. https://doi.org/10.1016/j.neuron.2011.06.037

McKenzie, S., Robinson, N. T., Herrera, L., Churchill, J. C., & Eichenbaum, H. (2013). Learning causes reorganization of neuronal firing patterns to represent related experiences within a hippocampal schema. *The Journal of Neuroscience, 33*(25), 10243–10256. https://doi.org/10.1523/JNEUROSCI.0879-13.2013

McMahon, D. B., Jones, A. P., Bondar, I. V., & Leopold, D. A. (2014). Face-selective neurons maintain consistent visual responses across months. *Proceedings of the National Academy of Sciences of the United States of America, 111*(22), 8251–8256. https://doi.org/10.1073/pnas.1318331111

Meyer, R. E. (2015). Physiologic measures of animal stress during transitional states of consciousness. *Animals (Basel), 5*(3), 702–716. https://doi.org/10.3390/ani5030380

Minatohara, K., Akiyoshi, M., & Okuno, H. (2016). Role of immediate-early genes in synaptic plasticity and neuronal ensembles underlying the memory trace. *Frontiers in Molecular Neuroscience, 8*, 78. https://doi.org/10.3389/fnmol.2015.00078

Moscovitch, M., Cabeza, R., Winocur, G., & Nadel, L. (2016). Episodic memory and beyond: The hippocampus and neocortex in transformation. *Annual Review of Psychology, 67*, 105–134. https://doi.org/10.1146/annurev-psych-113011-143733

Nader, K. (2003). Response to Arshavsky: Challenging the old views. *Trends in Neuroscience, 26*, 466–468.

Nader, K. (2015). Reconsolidation and the dynamic nature of memory. *Cold Spring Harbor Perspectives in Biology, 7*, 1–16. https://doi.org/10.1101/cshperspect.a021782

Nader, K., Schafe, G. E., & Le Doux, J. E. (2000). Fear memories require protein synthesis in the amygdala for reconsolidation after retrieval. *Nature, 406*, 722–726.

Neisser, U. (1976). *Cognition and reality: Principles and implications of cognitive psychology*. Freeman.

O'Keefe, J. (1976). Place units in the hippocampus of the freely moving rat. *Experimental Neurology, 51*, 78–109.
de Oliveira Alvares, L., & Do-Monte, F. H. (2021). Understanding the dynamic and destiny of memories. *Neuroscience & Biobehavioral Reviews, 125*, 592–607. https://doi.org/10.1016/j.neubiorev.2021.03.009
Paton, J. A., & Nottebohm, F. N. (1984). Neurons generated in the adult brain are recruited into functional circuits. *Science, 225*, 1046–1048.
Paxinos, G., & Watson, C. (1997). *The Rat Brain in stereotaxic co-ordinates*. Academic.
Piaget, J. (1951). *Play, dreams, and imitation in childhood*. Norton.
Pinotsis, D. A., & Miller, E. K. (2023). In vivo ephaptic coupling allows memory network formation. *Cerebral Cortex, 33*(17), 9877–9895. https://doi.org/10.1093/cercor/bhad251.
Prickaerts, J., Koopmans, G., Blokland, A., & Scheepens, A. (2004). Learning and adult neurogenesis: Survival with or without proliferation? *Neurobiology of Learning and Memory, 81*, 1–11.
Ranganath, C., & Rainer, G. (2003). Neural mechanisms for detecting and remembering novel events. *Nature Review Neuroscience, 4*, 193–202.
Raoul, C., Pettmann, B., & Henderson, C. E. (2000). Active killing of neurons during development and following stress: A role for p75NTR and Fas? *Current Opinion in Neurobiology, 10*, 111–117.
Reinert, S., Hübener, M., Bonhoeffer, T., & Goldstein, P. M. (2021). Mouse prefrontal cortex represents learned rules for categorization. *Nature, 593*, 411–417. https://doi.org/10.1038/s41586-021-03452-z
Rose, S. (1993). *The making of memory: From molecules to mind*. Bantam Books.
Sara, S. J. (2000). Retrieval and reconsolidation: Toward a neurobiology of remembering. *Learning & Memory, 7*, 73–84.
Schmidt, E. M., Bak, M. J., & McIntosh, J. S. (1976). Long-term chronic recordings from cortical neurons. *Experimental Neurology, 52*, 496–506.
Sherstnev, V. V., Gruden, M. A., Alexandrov, Y. I., Storozheva, Z. I., Golubeva, O. N., & Proshin, A. T. (2013). Different populations of neurons in relevant brain structures are selectively engaged in the functioning of long-term spatial memory. *Neurochemical Journal, 7*(4), 278–283.
Shima, K., Mushiake, H., Saito, N., & Tanji, J. (1996). Role for cells in the presupplementary motor area in updating motor plans. *Proceedings of the National Academy of Sciences of the United States of America, 93*, 8694–8698.
Shors, T. J., & Matzel, L. D. (1997). Long-term potentiation [LTP]: what's learning got to do with it? *Behavioral and Brain Sciences, 20*, 597–655.
Shors, T. J., Miesegaes, G., Beylin, A., Zhao, M., Rydel, T., & Gould, E. (2001). Neurogenesis in the adult is involved in the formation of trace memories. *Nature, 410*, 372–376.
Shvyrkov, V. B. (1986). Behavioral specialization of neurons and the system-selection hypothesis of learning. In F. Klix & H. Hagendorf (Eds.), *Human memory and cognitive capabilities* (pp. 599–611). Elsevier.
Smith, D. M., Barredo, J., & Mizumori, S. J. Y. (2012). Complimentary roles of the hippocampus and retrosplenial cortex in behavioral context discrimination. *Hippocampus, 22*, 1121–1133. https://doi.org/10.1002/hipo.20958
Sozinov, A. A., Kazymaev, S. A., Grinchenko, Y. V., & Alexandrov, Y. I. (2012). Percent of task-specialized cingulate cortex neurons does not change during training. *FENS Abstracts, 6*, 114.08.
Sozinov, A. A., Alexandrov, I. O., Gorkin, A. G., Grechenko, T. N., & Alexandrov, Y. I. (2023). Consolidation, reconsolidation of memory, extinction and forgetting: A review. *Russian Psychological Journal, 20*(3), 131–156. https://rpj.ru.com/index.php/rpj/article/view/1488/1729
Stone, E. A., Zhang, Y., John, S., Filer, D., & Bing, G. (1993). Effect of locus coeruleus lesion on c-fos expression in the cerebral cortex caused by yohimbine injection or stress. *Brain Research, 603*, 181–185.
Strassmann, J. E., Zhu, Y., & Queller, D. C. (2000). Altruism and social cheating in the social amoeba Dictyostelium discoideum. *Nature, 408*, 965–967.

Svarnik, O. E., Alexandrov, Y. I., Gavrilov, V. V., Grinchenko, Y. V., & Anokhin, K. V. (2005). Fos expression and task-related neuronal activity in rat cerebral cortex after instrumental learning. *Neuroscience, 136*, 33–42.

Svarnik, O. E., Bulava, A. I., & Alexandrov, Y. I. (2013). Expression of c-Fos in the rat retrosplenial cortex during instrumental re-learning of appetitive bar-pressing depends on the number of stages of previous training. *Frontiers in Behavioral Neuroscience, 7*, 78. https://doi.org/10.3389/fnbeh.2013.00078

Swadlow, H. A., & Hicks, T. P. (1997). Subthreshold receptive fields and baseline excitability of «silent» S1 callosal neurons in awake rabbits: Contributions of AMPA/kainate and NMDA receptors. *Experimental Brain Research, 115*, 403–409.

Thompson, L. T., & Best, P. J. (1990). Long-term stability of the place-field activity of single units recorded from the dorsal hippocampus of freely behaving rats. *Brain Research, 509*, 299–308.

Tischmeyer, W., Kaczmarek, L., Strauss, M., Jork, R., & Matthies, H. (1990). Accumulation of c-fos mRNA in rat hippocampus during acquisition of a brightness discrimination. *Behavioral and Neural Biology, 54*(2), 165–171.

Tolman, E. C. (1948). Cognitive maps in rats and men. *The Psychological Review, 55*(4), 189–208.

Tracy, J., Flanders, A., Madi, S., Laskas, J., Stoddard, E., Pyrros, A., Natale, P., & DelVecchio, N. (2003). Regional brain activation associated with different performance patterns during learning of a complex motor skill. *Cerebral Cortex, 13*, 904–910.

Trask, S., Ferrara, N. C., Jasnow, A. M., & Kwapis, J. L. (2021). Contributions of the rodent cingulate-retrosplenial cortical axis to associative learning and memory: A proposed circuit for persistent memory maintenance. *Neuroscience and Biobehavioral Reviews, 130*, 178–184. https://doi.org/10.1016/j.neubiorev.2021.08.023

Tse, D., Langston, R. F., Kakeyama, M., Bethus, I., Spooner, P. A., Wood, E. R., Witter, M. P., & Morris, R. G. M. (2007). Schemas and memory consolidation. *Science, 316*, 76–82.

Tse, D., Takeuchi, T., Kakeyama, M., Kajii, Y., Okuno, H., Tohyama, C., Bito, H., & Morris, R. G. M. (2011). Schema-dependent gene activation and memory encoding in neocortex. *Science, 333*, 891–895. https://doi.org/10.1126/science.1205274

Venerin, A. A., Venerina, Y. A., & Alexandrov, Y. I. (2020). Cell functioning in norm and pathology in terms of the activity paradigm: Oncogenesis. *Medical Hypotheses, 144*, 110240. https://doi.org/10.1016/j.mehy.2020.110240

Vetere, G., Restivo, L., Novembre, G., Aceti, M., Lumaca, M., & Ammassari-Teule, M. (2011). Extinction partially reverts structural changes associated with remote fear memory. *Learning & Memory, 18*, 554–557. https://doi.org/10.1101/lm.2246711

Vikman, K. S., Duggan, A. W., & Siddall, P. J. (2003). Increased ability to induce long-term potentiation of spinal dorsal horn neurons in monoarthritic rats. *Brain Research, 990*, 51–57.

Vogt, B. A., Sikers, R. W., Swaldow, H. A., & Weyand, T. G. (1986). Rabbit cingulate cortex: Cytoarchitecture, physiological border with visual cortex, and different cortical connections of visual, motor, postsubicular and intracingulate origin. *Journal of Comparative Neurology, 248*, 74–94.

von Stein, A., Chiang, C., & König, P. (2000). Top-down processing mediated by interareal synchronization. *National Academy of Sciences of the United States of America, 97*(26), 14748–14753. https://doi.org/10.1073/pnas.97.26.14748

Weber, A., Prokazov, Y., Zuschratter, W., & Hauser, M. J. B. (2012). Desynchronisation of glycolytic oscillations in yeast cell populations. *PLoS One, 7*(9), e43276. https://doi.org/10.1371/journal.pone.0043276

Weible, A. P., Rowland, D. C., Pang, R., & Kentros, C. (2009). Neural correlates of novel object and novel location recognition behavior in the mouse anterior cingulate cortex. *Journal of Neurophysiology, 102*, 2055–2068. https://doi.org/10.1152/jn.00214.2009

Weible, A. P., Rowland, D. C., Monaghan, C. K., Wolfgang, N. T., & Kentros, C. G. (2012). Neural correlates of long-term object memory in the mouse anterior cingulate cortex. *Journal of Neuroscience, 32*, 5598–5608. https://doi.org/10.1523/JNEUROSCI.5265-11.2012

Westlin, C., Theriault, J. E., Katsumi, Y., Nieto-Castanon, A., Kucyi, A., Ruf, S. F., et al. (2023). Improving the study of brain-behavior relationships by revisiting basic assumptions. *Trends in Cognitive Sciences, 27*(3), 246–257. https://doi.org/10.1016/j.tics.2022.12.015

Whishaw, I. Q., Sarna, J. R., & Pellis, S. M. (1998). Evidence for rodent-common and species-typical limb and digit use in eating, derived from a comparative analysis of ten rodent species. *Behavioural Brain Research, 96*, 79–91.

Williams, J. C., Rennaker, R. L., & Kipke, D. R. (1999). Stability of chronic multichannel neural recordings: Implications for a long-term neural interface. *Neurocomputing, 26*, 1069–1076.

Wilson, M. A., & McNaughton, B. L. (1993). Dynamics of the hippocampal ensemble code for space. *Science, 261*, 1055–1058.

Wirth, S., Yanike, M., Frank, L. M., Smith, A. C., Brown, E. N., & Wendy, A. S. (2003). Single neurons in the monkey hippocampus and learning new associations. *Science, 300*, 1578–1581.

Wright, R. (1995). *The moral animal*. Vintage Books.

Xue, Z. M. (1998). The studies on neurogenesis induced by brain injury in adult ring dove. *Cell Research, 8*, 151–162.

Zhang, Y., Denman, A. J., Liang, B., Werner, C. T., Beacher, N. J., Chen, R., Lin, Y., Shaham, Y., Barbera, G., & Lin, D.-T. (2022). Detailed mapping of behavior reveals the formation of prelimbic neural ensembles across operant learning. *Neuron, 110*(4), 674–685. https://doi.org/10.1016/j.neuron.2021.11.022

Emergence of the Hippocampus as a Vector for Goal-Directed Spatial Navigation

Susumu Takahashi, Fumiya Sawatani, and Kaoru Ide

Abstract The hippocampus, which is deeply involved in episodic memory, plays a pivotal role in spatial navigation, an essential animal behavior. Spatial navigation requires the calculation of the distance and direction from a current to the final position, i.e., a vector to a goal. Place cells in the mammalian hippocampus maximally increase their firing rates when the animal passes a particular location and then encode the animal's current location. The entorhinal cortex, one synapse upstream of the hippocampus, contains both grid and head direction cells that encode distance and direction information, respectively. However, the question of whether the hippocampus generates a vector for goal-directed navigation during the integration of distance and direction to the destination remains unclear. Mounting evidence of the cell types involved in spatial navigation has been obtained mainly in mammalian model animals such as rats and mice. Recent advances in wireless and miniaturized neural activity monitoring devices have begun to yield results not only in model organisms but also in wild mammals, birds, fish, and insects. A scrutiny of the literature examining neural correlates of spatial navigation across multiple animal species reveals that few place cells or grid cells have been found, but that head direction cells are commonly present in multiple animal species. Exceptionally, rodent-like place cells were only found in the medial pallium of tufted titmice, a food-caching bird. The medial pallium is an avian brain region homologous to the mammalian hippocampus. By contrast, rodent-like head direction cells are found in the medial pallium of quails. Head direction cells are also found in the medial pallium of streaked shearwaters, a migratory bird. The avian hippocampus contains information about the animal's current location or direction, but the neural encoding may differ depending on the ecological characteristics of the bird species. The place cells of bats, which are mammals, fly in three-dimensional space and encode vectorial information toward the goal. Training rats with an ingenious task that required them to choose a direction for each run in a maze suggested that place cells encode a vector for goal-directed spatial navigation. Thus, the scrutiny of

S. Takahashi (✉) · F. Sawatani · K. Ide
Graduate School of Brain Science, Doshisha University, Kyoto, Japan
e-mail: stakahas@mail.doshisha.ac.jp

A. C.-H. Yu et al. (eds.), *Systems Neuroscience*, Advances in Neurobiology 41,
https://doi.org/10.1007/978-3-031-69188-1_2

the literature on spatial navigation-related neuronal activity across multiple animal species suggests that depending on a combination of external conditions such as the context in which the animal is situated (e.g., the context or the framework composed of landmarks in the environment) and internal conditions such as the ecological and behavioral characteristics of the animal, hippocampal neurons can be identified as place cells or head direction cells. We thus propose a conjecture that primitively, the hippocampus, or its homolog, contains information about the travel direction and that the emergence of the hippocampus during evolution has enabled the generation of vector information to the goal for advanced spatial navigation such as the search for the shortcut path and episodic memory capacity.

Keywords Hippocampus · Spatial navigation · Place cells · Goal-directed navigation · Head direction cells

1 Introduction

Numerous experimental pieces of evidence suggest that the hippocampus is profoundly involved in learning and memory. Damage to the hippocampus and its surrounding region called hippocampal formation causes dysfunction in the formation and recall of episodic memory and concomitant anterograde amnesia (Scoville & Milner, 1957), whereas damage to other brain regions does not cause these symptoms. Based on these cases, the hippocampal formation is thought to be critically involved in memory for facts and events that occur within a specific time and space framework of what, where, when, and how (Tulving, 1983). The mounting evidence supporting these findings is derived from human case studies and experimental studies on animal models, including rats and mice. Can the view that the hippocampus supports episodic memory be shared across other animal species?

Spatial navigation from the current position to the goal forms episodic memories about events that occurred during movement. Recalling these memories allows for detours and the search for the shortest path (Tolman, 1948). It is well known that in cases of severe Alzheimer's disease, neurons in the hippocampus are degenerated and depleted. As patients are known to get lost and wander, these cases suggest that the hippocampus is deeply involved in spatial navigation as well as episodic memory. However, it is difficult to distinguish between episodic memory and spatial navigation. Spatial navigation is an essential ability of life because it is an indispensable behavior for life activities including foraging and escape behaviors. Even primitive organisms without episodic memory capacity exhibit spatial navigational behaviors. Therefore, we can consider that the hippocampus was originally utilized for spatial navigation, and then episodic memory capacity emerged to achieve spatial navigation with memory. In addition, an overview of evolutionary phylogenetic differences in hippocampal function allows us to explore this hypothesis in more depth. In this chapter, we will compare differences among animal species regarding hippocampal function to elaborate on the hippocampus as the locus of spatial navigation.

Consider spatial navigation in a space with no obstacles. If the goal location is memorized or visible, then the hippocampus would be able to navigate to the end of the journey by knowing the direction and distance from the current location to the final goal. In other words, the most important information in spatial navigation is the recognition of the current location and the goal location and the direction and distance to the goal connecting them, i.e., the vector. This consideration shows that the recognition of the current location is essential for spatial navigation and episodic memory. In the rodent hippocampus, place cells are highly active when the animal passes one or several specific points, and cells with the same function as these place cells are found only in the hippocampal proper, which includes the CA1, CA2, and CA3 areas, and dentate gyrus (DG). Therefore, when the hippocampal proper and DG are damaged, place cells are lost, leading to severe impairment of spatial navigation and episodic memory capacity.

How are goal locations for spatial navigation recognized? In the rodent entorhinal cortex and subiculum, cells with multiple place fields are found in a two-dimensional space, their locations are geometrically located at the vertices of an equilateral triangular lattice that fills the environment (Hafting et al., 2005). These cells are referred to as grid cells because of their geometric arrangement. Because each grid cell is arranged on a unique equilateral triangular grid, the activities of multiple grid cells can be combined to identify a particular location. Given the input of a group of grid cells, information on the current location and goal position can be obtained (McNaughton et al., 2006). Furthermore, head direction cells have also been found in the entorhinal cortex and subiculum, which are highly active when the animal's head is pointing in a certain direction. It is expected that the vector to the goal can be calculated if the head direction at the current position can be associated with the goal position. The hippocampus, one synapse downstream of the entorhinal cortex, is a prime candidate for the locus that receives input from the entorhinal cortex and computes the vector to the goal.

It is known that the hippocampal proper and DG contain place cells but no head direction cells. When animals forage for randomly scattered food pellets in an open field, place cells are not direction-dependent. However, if the animal traverses between both ends of a linear track to forage foods, the locations indicated by the place receptive fields on the outward and return paths are different. This suggests that depending on the contextual framework in which the animal is situated, where the food is located, or which path it takes, the place cells may be equipped with direction dependence.

Can the direction dependence of place cells be interpreted as a hippocampal vectorial encoding to a goal? All of these details of hippocampal function have been clarified by experimental evidence in mammalian model animals such as rats and mice. By utilizing model animals, robust evidence has been accumulated through systematic experiments including the neural correlate of behaviors (O'Keefe & Dostrovsky, 1971; Sakurai, 1990; Takahashi, 2013, 2015). Selective lesions of neurons, and more recently, the causal relationship between neuronal activity and behaviors by activity manipulation including optogenetics have also been researched (Deisseroth & Schnitzer, 2013). On the other hand, morphological or developmental studies have reported that brain regions homologous to the mammalian hippocampus exist in other animal species. Therefore, although systematic and robust demonstration in model animals is not possible, hippocampal homologous regions in other animal species must provide clues to elucidate the functional significance of the hippocampus. Recent technological developments have led to the development of monitoring devices called neurologgers that can record neural activity wirelessly. These devices will make it possible to study wild animals and other species outside the laboratory (Ide & Takahashi, 2022). Currently, a series of studies on bats, which are flying mammals, is taking place. Place cells have been found in the hippocampus of bats, similar to rodents (Ulanovsky & Moss, 2011), and by taking advantage of the ecological traits of bats, the characteristics of place cells in three-dimensional space are being elucidated (Yartsev & Ulanovsky, 2013). As these findings demonstrate, experiments across animal species enable us to clarify the information possessed by hippocampal neurons that cannot be revealed if we focus only on rodents.

In this chapter, we review the literature on functional cell types involved in spatial navigation that is shared by the brains of mammals, fish, birds, and insects and discuss hippocampal function from the perspective of cross-species comparisons.

2 Head Direction Cells Are Shared across the Brains of Mammals, Fish, Birds, and Insects

Spatial navigation is an essential behavior shared among mammals, fish, birds, and insects. Place cells are unique in that are deeply involved in spatial navigation but are found in a limited number of animal species including some mammals and birds; in contrast, head direction cells are commonplace. Head direction cells are thought of as cells that indicate the direction in which an animal is moving and are referred to as an internal compass in the brain. Rodent head direction cells have been found in various brain regions, including the thalamus (Goodridge & Taube, 1997; Stackman & Taube, 1997), entorhinal cortex (Solstad et al., 2008), subiculum (Boccara et al., 2010; Taube et al., 1990), lateral mammillary nucleus (Stackman & Taube, 1998), retrosplenial cortex (Smith et al., 2012), dorsal tegmental nucleus (Sharp et al., 2001), and striatum (Mizumori et al., 2000). The directional selectivity and the response to the external environment differ among these regions. What is shared across brain regions is that a single head direction cell prefers a specific direction, but the preference of a group of cells as a whole is not biased toward a particular direction, such as north, as in a geomagnetic compass. Thus, the collection of directional receptive fields to which multiple head direction cells respond uniformly fills the entire 360 degrees. Leaving the details of head direction cells to the numerous review articles (Yoder & Taube, 2014), here we discuss head direction cells found in various animal species and infer the function of head direction cells from an evolutionary perspective. Head direction cells have been found in the brain of primates (Robertson et al., 1999), suggesting that they are widely shared among mammalian species. Because head direction cells were first discovered in the brain of rodents (Taube et al., 1990), one may define head direction cells in a narrow sense as those having the complex characteristics of rodent head direction cells influenced by vestibular inputs and environmental cues including visual and other sensory inputs, but in this chapter, we define head direction cells in a broad and simple sense as neurons that maximally increase their firing rates when the animal faces in a certain direction.

Head direction cells are not only found in the brain of mammals but also in birds; the medial pallium of quail contains a large number of head direction cells with a broad directional receptive field (Ben-Yishay et al., 2021). Head direction cells found in the anterior thalamus and other areas of the rodent brain have narrow directional receptive fields for specific orientations (Stackman & Taube, 1997), whereas those in the entorhinal cortex and subiculum have broader directional receptive fields (Boccara et al., 2010; Solstad et al., 2008). However, since the avian pallium does not have a layered structure like the mammalian cortex (Jarvis et al., 2005), it may contain a mixture of homologous regions such as the mammalian entorhinal cortex and subiculum. Recently, it was reported that head direction cells are also present in the fish telencephalon (Vinepinsky et al., 2020). In addition to head direction cells, the goldfish telencephalon contains cells that respond along the walls of the water tank, and the frequency of their activity is closely related to the swimming speed (Vinepinsky et al., 2020). These findings were followed by our discovery of head direction cells in the telencephalon of trout, a salmonid fish (Takahashi et al., 2021). It seems that head direction information plays navigational roles in fish as well.

Thus, head direction cells are spatial navigation-related cells shared among mammals, birds, and fish of different divergence ages (Fig. 1). The question arises whether cells with head direction information have been evolutionarily conserved or whether they have converged to achieve their spatial navigation ability. Recently, head direction cells have been found in the brains of insects (Seelig & Jayaraman, 2015). Since chordates and arthropods diverged more than 500 million years ago, it can be inferred that head direction information is an essential function inherited from primitive animals, but whether it is evolutionarily conserved or due to convergent evolution remains an open question.

In the next section and beyond, we will discuss in more depth the neurons involved in spatial navigation that have been discovered in the brains of birds, fish, and insects.

3 Enigmatic Presence of Place Cells and Head Direction Cells of Avian Species

The avian medial pallium is located on a surface easily accessible from the top of the head, whereas the mammalian hippocampus is folded and stored deep in the brain as

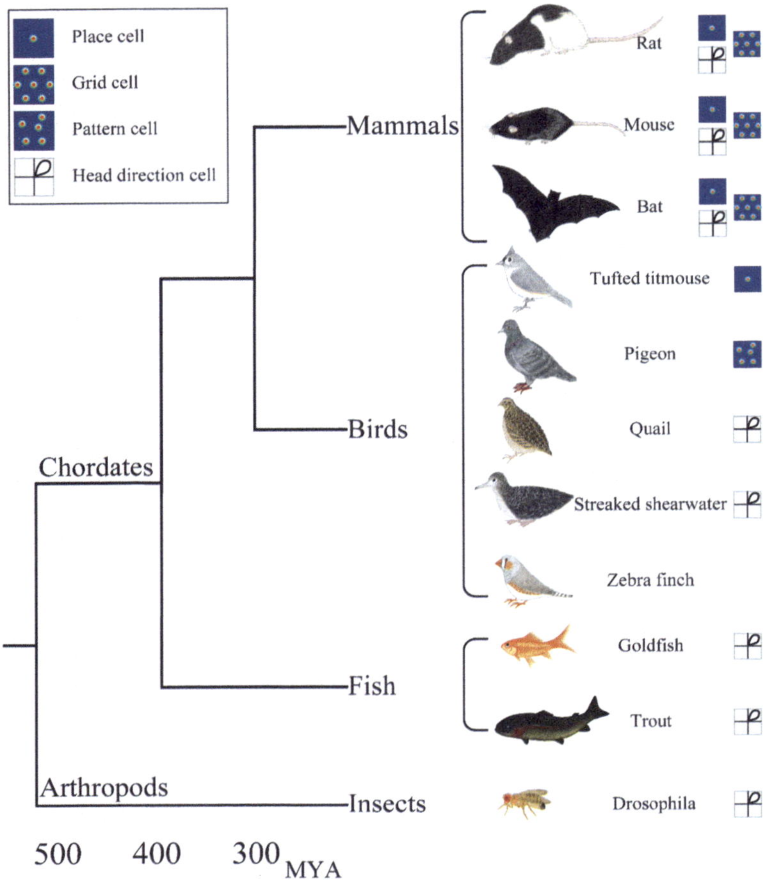

Fig. 1 Phylogeny of animal species discussed in this chapter and spatial cell types found in their brain. (Left panel) Simplified phylogenetic tree of mammals, birds, fish, and insects with the timeline in MYA (million years ago). (Right panel) Drawings of animal species. If place cells, grid cells, pattern cells, and/or head direction cells were found in their brain, icons illustrated in the left top were drawn on the right side. (All illustrations were created using Adobe illustrator and Fresco)

the deep layer of the CA1 area of the rodent hippocampus is located dorsally and the superficial layer is located ventrally. However, it is thought to be embryologically or morphologically homologous to the mammalian hippocampus and surrounding regions (Jarvis et al., 2005).

For birds, a pioneering study in pigeons led by Bingman et al. electrophysiologically recorded neuronal activity from the medial pallium of pigeons and analyzed the correlation of the firing rates with pigeon movement. An area called the place receptive field was examined, and it was discovered that the firing rates of place cells were elevated (Hough & Bingman, 2004; Kahn et al., 2008; Siegel et al., 2005). This implies that cells with specific place receptive fields are found in the ventral part of the medial pallium. They increase their firing rates when pigeons are present in particular locations. However, they do not have one single receptive field like the place cells in the CA1 area of the rodent hippocampus, but rather multiple place receptive fields. They are referred to as pattern cells because they form a geometric pattern in the environment. Because of the view that the ventral part of the medial pallium is morphologically homologous to the mammalian hippocampal formation, including the hippocampal CA1 and CA3 areas and the DG, the authors argue that avian place cells have different functional characteristics from those of rodents. In addition, because the pattern cells and grid cells in the medial entorhinal cortex of rodents (Hafting et al., 2005) were discovered almost simultaneously, the arrangement of place receptive fields shown by pattern cells was not compared with the equilateral triangular lattice of grid cells. The pattern of grid cells is geometric, but it is controversial whether this geometric property is essential for spatial navigation. Thus, the ventral part of the medial pallium of pigeons can be considered a functionally homologous region of the rodent hippocampal formation including the entorhinal cortex.

It is difficult to show experimentally what functions grid cells require for spatial navigation. Even with optogenetics, it is not possible with existing technology to manipulate the activity of multiple grid cells at will because there are no markers to genetically identify grid cells. Real-time identification of grid cells by two-photon microscopy is also difficult without invasion of the prism (Gu et al., 2018) because of their location in the deep lateral areas of the brain. In addition, granule cells in the entorhinal cortex and subiculum have the characteristics of head direction cells and border cells as well as grid cells (Boccara et al., 2010; Solstad et al., 2008), making it impossible to selectively manipulate grid cells alone. Therefore, a study using a deep learning model was conducted to investigate the causal relationship between grid cells and spatial navigation (Banino et al., 2018). In the study, grid cells emerged in a deep learning model that network-like mimics rodent hippocampal formation. By artificially feeding the model with the current location supplied by place cells and the direction supplied by head direction cells, and then performing a manipulation called dropout, in which half of the synapses are randomly deleted at each learning session, the researchers were able to generate cells in the model that looked exactly like rodent grid cells. The important point is that the grid cells, conjunctive cells with characteristics of both grid cells and head direction cells, and geometric border cells, which had been found in the rodent entorhinal cortex, were also found in similar proportions. When this model with grid cells was mounted on an artificial agent exploring a maze in virtual space, it was able to find the shortcut path as if it were a rodent. In the model, the grid cells can be selectively destroyed, allowing the functional significance of the grid cells to be investigated. When grid cells were excluded from the model, the shortcut path could no longer be found, suggesting that grid cells are an essential element for finding the shortcut path and for calculating vectors to a goal. If it is true, pigeons may not be able to find the shortcut path.

Considering the multiple place receptive fields in the pigeon pattern cells and their non-geometric arrangement, the ventral part of the medial pallium is considered the DG (Nitz & McNaughton, 2004) rather than the brain region homologous to the mammalian entorhinal cortex.

Payne et al. conducted a spatial recognition cell study focusing on the dorsal medial pallium of the tufted titmouse, a food-caching bird (Payne et al., 2021). Place cells are expected to be necessary because hiding food requires episodic-like memory to remember when and where the food was hidden. They found place cells along with a small number of place receptive fields, like the hippocampal CA1 area in rodents. They also found that the number of place fields with which a single place cell varies along the axis from rostral to caudal, and in particular, place cells with a single confined place field were found on the rostral side. They conducted a simultaneous comparison with zebra finches, a songbird, and found few such place cells. Therefore, it appears that not all bird species have place cells. Payne et al. argue that ecological characteristics, i.e., rodent-like foraging behavior based on episodic-like memory to find food, contribute significantly to the formation of place receptive fields. Considering these results together with the pioneering work on pigeons, the dorsomedial pallium is considered to be the mammalian homolog of the hippocampal proper and DG.

Almost simultaneously, Ben-Yishay et al. electrophysiologically recorded neuronal activity from the dorsomedial pallium of spontaneously roaming quails (Ben-Yishay et al., 2021). As a result, about 10% of the recorded cells were identified as head direction cells. Their directional receptive fields are broad and resemble head direction cells found in the entorhinal cortex of rodents. Combined with the aforementioned findings from pigeons, tufted titmice, and zebra finches, it appeared that the presumption that the dorsomedial pallium was the homolog of the hippocampal proper and DG, may be inappropriate because head direction cells have not been found in there. Under the hypothesis that ecological characteristics form place cells, it is also possible that the ecological characteristics of the quail caused the formation of head direction cells, rather than place cells, in the medial pallium. If this is correct, then rodent place cells are not produced by the morphological position of the hippocampus, but by the ecological characteristics of rodents.

It is noteworthy that the brain structures of mammals and birds have stark differences: the mammalian cortex is layered, while the avian pallium is not. Therefore, these lines of evidence suggest that it is difficult to simply separate the dorsal and ventral sides and determine functional homology. In other words, the abovementioned argument in this chapter is that the medial pallium is functionally homologous to the rodent hippocampal proper and DG is inappropriate. No functional subdivision of the medial pallium may have occurred in bird species (Fig. 2).

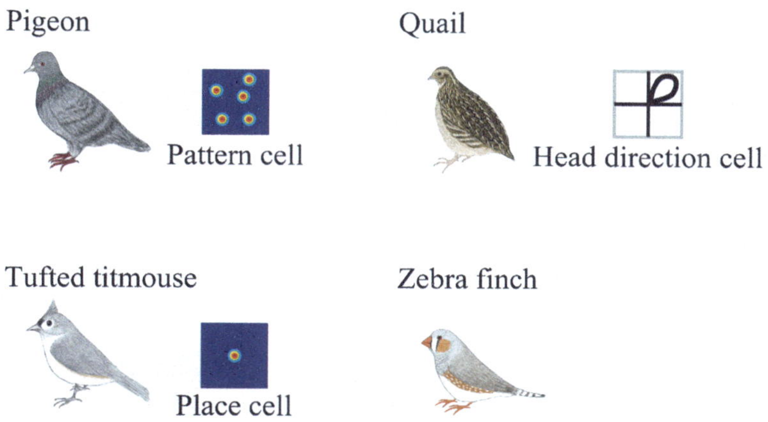

Fig. 2 Variety of spatial cell types in the avian medial pallium. (Left top) Historically, pattern cells with multiple place fields were found in the medial pallium of pigeons. Following this seminal study, two research groups simultaneously examined neural correlates of spatial navigation in the medial pallium of tufted titmice, zebra finches, and quails. (Left Bottom) A group reported the presence of place cells but no head direction cells in tufted titmice. (Right Bottom) Absence of place cells in zebra finches. (Right Top) Presence of head direction cells but no place cells in quails. (All illustrations were created using Adobe illustrator and Fresco)

4 Head Direction Cells in the Fish Telencephalon

The relationship between swimming movements and neuronal activity in the brain of fish has been largely unknown. Conventionally, a tethered cable to the head is a prerequisite for monitoring neuronal activity from the brain of a walking animal. Since in water the torque of the tethered cable hampers the natural swimming behavior, few experiments have been conducted, except for those on electric fish. Neuronal activity recordings have been made in studies of electric fish, such as their ability to sense electric fields, but these results have little relevance to spatial navigation. Canfield and Mizumori have developed a technique to explore spatial recognition cells in the telencephalon of goldfish and cichlid fish (Canfield & Mizumori, 2004). They found a few neurons in the telencephalon of goldfish and cichlid fish that, like place cells, have place receptive fields. As the tethered recordings inhibited free-swimming behavior, only a very few cells were detected. Thus, quantitative identification of place receptive fields has not been achieved.

Vinepinsky et al. were the first to successfully record neural activity from the telencephalon of swimming goldfish without cables by utilizing a neurologger that can store neuronal activity directly to a memory card, thereby making it wireless (Vinepinsky et al., 2020). By utilizing this innovative neurologger, they succeeded in identifying the head direction cells in the goldfish's telencephalon. They simultaneously identified edge cells and speed cells in the telencephalon that responded with the boundaries of the water tank and the swimming speed of

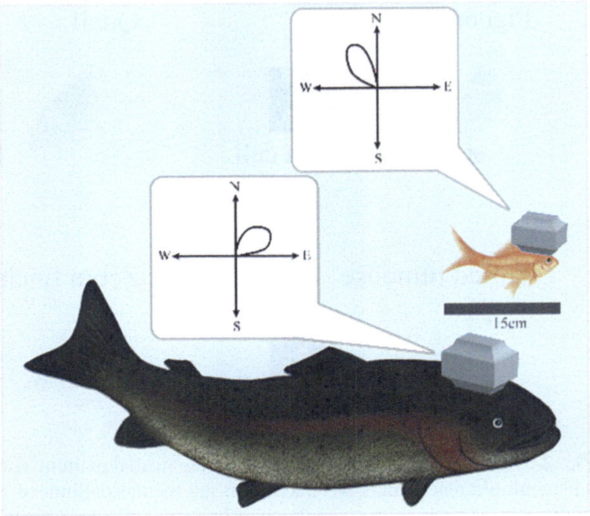

Fig. 3 Neurologger recording neuronal activity from the brain of swimming bony fish. The neurologger is a wireless lightweight logging device enclosed in a waterproof case that stores neuronal activity including action potentials and local field potentials extracellularly recorded from the brain into an SD memory card. Thanks to the neurologger, head direction cells were found in the telencephalon of both goldfish and trout. (All illustrations were created using Adobe illustrator and Fresco)

the goldfish, respectively. However, they have not found cells with a single place receptive field, such as rodent place cells.

As discussed in the findings of the bird species in the previous section, ecological characteristics could have a profound effect on neuronal activity. Therefore, we hypothesized that migratory animals that navigate long distances, such as mother-river homing, have different types of neurons than animals that primarily navigate short distances, such as rodents. Therefore, we attempted an experiment on trout, a salmonid fish, using the underwater wireless neurologger technology developed for goldfish as a reference (Takahashi et al., 2021). Even though they are classified as bony fish, trout, which weigh about 2 kg, and goldfish, which weigh about 100 g, differ greatly in swimming ability and the associated torque applied to their heads. The most significant difference in monitoring the neuronal activity of the bony fish is the location of the brain in the head. The brain of trout is located about 3–4 cm deep below the surface of the head. Brain position in fish is generally deeper in proportion to body size. The brain position of the goldfish is less than 1 cm deep from the head surface, making it easily accessible. Therefore, it has been the subject of numerous studies on the neuronal activity of the brain. This suggests that the techniques developed in the goldfish cannot be simply transferred to salmonids in their original form due to increased torque and other factors caused by differences in brain location and body weight. Therefore, we have established a technique to record neuronal activity in the telencephalon of trout, which is located deep within the head surface, by combining the brain fixation technique developed for rodent multi-unit recording (Takahashi, 2013, 2015) and making the waterproof case more rigid. Similar to the goldfish, when neuronal activity was recorded from the telencephalon of swimming trout, approximately 30% of the recorded cells were identified as head direction cells (Fig. 3).

These studies of goldfish and trout demonstrate that the telencephalon of bony fish contains head direction information. However, it remains unclear whether bony fish have self-location information like place cells. Based on the aforementioned pieces of literature, it appears that the cells involved in spatial navigation that are shared across the brains of mammals, birds, and fish are the head direction cells (Fig. 1).

5 Insect Head Direction Cells Connected in a Ring Shape

Head direction cells are also present in the brains of insects. The most popular theoretical model for the formation of head direction cells is called the ring attractor (Zhang, 1996). In this model, head direction cells are arranged in a ring-like formation, each of which inhibits the other cells (Fig. 4). For example, suppose that four head direction cells are responsible for either east, west, north, or south. When the animal turns to the east, the head direction cell responsible for the east maximally increases its firing rate. As the range of direction to which the head direction cells respond, i.e., the directional receptive field has a certain width, not only the cells in charge of the east but also those in charge of the north and south slightly increase their firing rate. Considering the reading of direction information

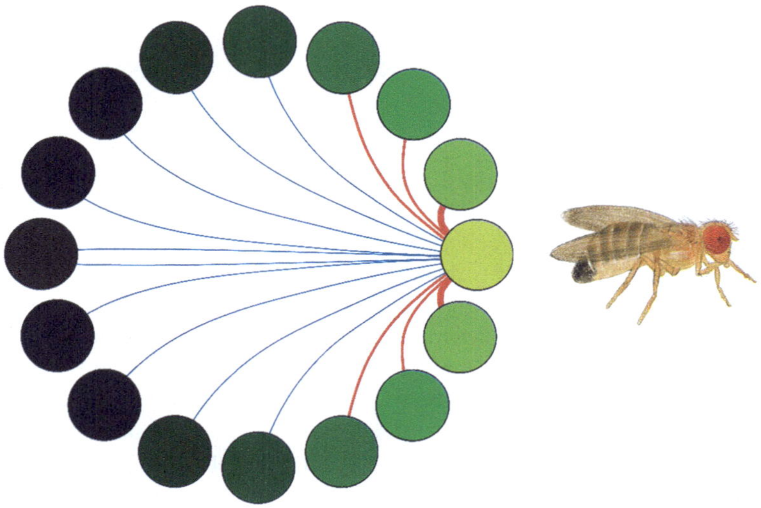

Fig. 4 Schematic of the ring-like connection of head direction cells found in the brain of Drosophila. When the Drosophila faces the right orientation, a head direction cell with the right orientation preference (Left, light green circle) is activated and simultaneously inhibits cells connected in a ring with other orientation preferences. The inhibition is stronger for cells with similar orientation preferences. As the ring attractor model predicted, the ring-connection shapes the sharp directional preference of the head direction cell. (All illustrations were created using Adobe illustrator and Fresco)

from the head direction cells, if the directional receptive fields of each cell are wide enough to affect cells responsible for adjacent orientations, the direction information read out is ambiguous and poorly resolved. This is contrary to the evidence that the directional receptive fields of head orientation cells found in the anterodorsal thalamus of rodents, for example, have sharp directional selectivity (Goodridge & Taube, 1997; Stackman & Taube, 1997). In the ring attractor model, the cells responsible for the east maximally inhibit others connected on the ring: the cells responsible for the west, north, and south. Only the cells responsible for the east are selectively active, which fits well with the experimental evidence. Such morphological ring connections have not been identified in the mammalian brain. The identification of ring connections requires the identification of the firing properties of the head direction cells as well as the simultaneous identification of the morphological relationships in which they are synaptically connected. The identification of the functional units to effectively drive downstream neurons is also a prerequisite since an excitatory single neuron cannot drive a postsynaptic cell via a single synapse. In the brain of Drosophila, which has neurons with relatively simpler anatomical connections than mammals, synaptically connected head direction cells on a ring have been found (Fig. 4) (Kim et al., 2017). This provides strong evidence that ring attractors are a strong candidate for supporting the presence of head direction cells. This is one example of the importance of identifying a specific type of neurons across animal species for a deeper understanding of their function, as it was difficult to identify them only by studying mammals.

6 Hippocampal Place Cells in Bats and Rats Have Vector Information for Goals

To summarize the aforementioned findings, the cell type involved in spatial navigation and shared in the brain of mammals, birds, fish, and insects is the head direction cell. However, place cells and their relatives have not been found in fish and insects, leaving the possibility that recording methodologies or brain regions are problematic. Therefore, we revisit the study of birds, in which head direction cells and place cells were found in the hippocampal homolog of mammals. Although there is a major conflict in that only head direction cells were found in quails, only place cells were found in tufted titmice. We are hopeful that incorporating the recently reported evidence held by rodent place cells may resolve this conflict.

Sarel et al. reported neuronal activity from the bat hippocampus that was modulated by direction and distance to the goal. Bats flew freely within an experimental room with a central landing site that allowed for reward and rest, and neural activity was recorded during navigation behavior to reach a goal (Fig. 5a) (Sarel et al., 2017). A subset of neurons in the hippocampal CA1 area of bats exhibited firing activity modulated by the goal-direction angle. This is defined as the angle between the head direction and the goal direction during a flight. The

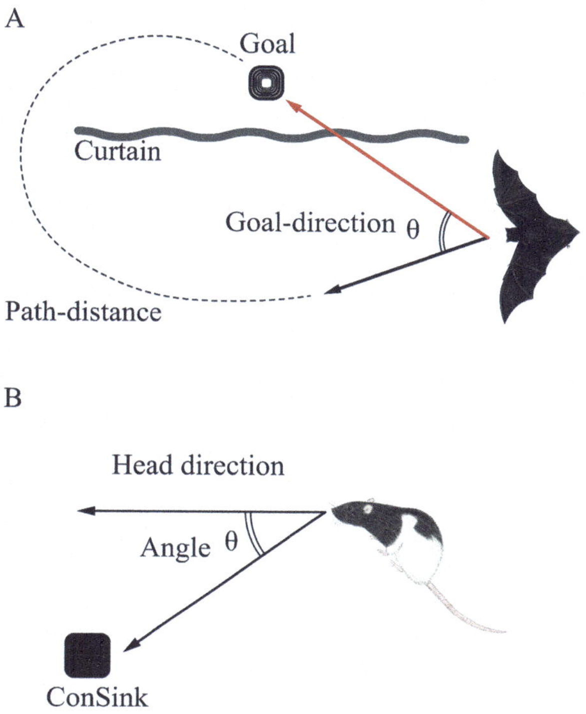

Fig. 5 Hippocampal place cells of bats and rats encode a vector for goal-directed spatial navigation. (**a**) The firing rate of hippocampal place cells of bats was biased toward cells tuned to a goal-direction angle even when the goal is hidden by a curtain. The cells were modulated by the distance to the goal as well as the direction of the goal. These goal-distance cells showed a peak firing rate when bats approached not a path distance (dotted line) but a Euclidean distance (red line) from the goal landing. (**b**) The firing rate of hippocampal place cells of rats encodes a vector field centered at a location called the convergence sink (ConSink). The ConSink can be explained as the modulation of the firing rate of place cells based on the distance toward the assumed ConSink and the angle (q) between the rat's current head direction and the assumed ConSink. The Consinks are concentrated near the goal. (All illustrations were created using Adobe illustrator and Fresco)

firing modulation of these groups of cells spanned the entire 360° range but was biased toward cells tuned to a goal-direction angle of 0°. In other words, many goal-direction cells exhibited a peak firing frequency relative to the goal direction. When these goal-direction cells were examined after the goal position was changed, more than half of them no longer showed firing modulation to the previous goal direction when the goal was lost. The majority of the cells adjusting to the new goal position were different from the previous goal-direction cells, suggesting that the goal direction is reorganized when the goal position changes. Most goal-direction cells lost their modulation to the previous goal position immediately after the goal position changed to a new position. Some goal-direction cells did show a gradual attenuation and disappearance of goal-direction modulation during 10–15 min. This may reflect memory traces of previous goal positions. Spatial navigation performed in natural environments may require memory of the goal position for navigation, as the goal may not be visible (Eichenbaum, 2017). Such activity, which may be a memory trace of previous goal location, suggests that hippocampal goal-direction cells contribute to memory-based rather than sensory-based navigation.

Furthermore, the cells were shown to be modulated by the distance to the goal as well as the direction of the goal. These goal-distance cells showed a peak firing frequency when bats approached a certain distance within 0–2 m from the goal landing. In this experiment, the bats were flying in a nonlinear complex flight, and most of the goal-distance cells were modulated with the path distance to the goal with respect to the landing at the goal, but some cells showed a strong modulation in the Euclidean distance to the goal. Path distance-dependent activity may be useful for spatial navigation during nonlinear flight. On the other hand, bats flying in a straight line during long-distance flights would be similar to the pathway distance and Euclidean distance during natural long-distance flights. In this case, neural activity in the hippocampus may form a complete vector representation that encodes information about the Euclidean distance to the goal and the direction of the goal. Cells that are strongly modulated by goal distance are also strongly modulated by goal direction, and there is a significant correlation between the two. These cells suggest that they form a bound representation of distance and direction to the goal and that a single neuron may contribute to navigation as a vector representation containing information about direction and distance to the goal.

These lines of evidence indicate that the hippocampus not only has signals for self-position (O'Keefe & Dostrovsky, 1971) and head direction (Taube et al., 1990) but also for goal direction and distance. Spatial navigation has been considered to require vectorial encoding of the current and goal positions and the direction and distance to the goal. This study in bats demonstrated a previously unknown vector representation of single neurons for spatial goals and their potential to contribute to vector-based navigation strategies. A major ecological difference between rodents and bats is that rodents move through two-dimensional space, whereas bats move through three-dimensional space. Although this difference is simple, the comparison is instructive in interpreting the neuronal activity involved in spatial navigation. Previous studies on rodent place cells have not shown that place cells have vector information to a goal. Thus, the outstanding contribution by Ulanovsky et al. on navigation-related cells utilizing bats has synergistic effects that go beyond simple comparisons with previously accumulated findings in model animals.

Following the findings of Ulanovsky et al., Ormond and O'Keefe reported place cells with information not only about the animal's self-location but also about its direction (Fig. 5b) (Ormond & O'Keefe, 2022). These unique place cells have features that change dynamically as learning progresses, unlike the firing features fixed to the direction and distance of the goal of Sarel et al. (2017).

The honeycomb maze developed by O'Keefe et al. (Wood et al., 2018) is a maze consisting of 37 platforms that can be elevated and lowered, allowing rats to be confined to the raised platform. In the task, two platforms near a platform where the rat was stationed were elevated, and the rat had to successively select the platform that could reach the goal in the shortest time. In this honeycomb maze task, the rats also sampled directions other than the goal, and the direction and distance of the goal gradually changed with selection. This behavior is useful for examining hippocampal neural activity in terms of the heading direction.

About 30% of the recorded rat hippocampal CA1 place cells exhibited a firing pattern represented by a vector field centered at a featureless location called the convergence sink (ConSink). The ConSink was able to explain the modulation of the firing rate of place cells based on the distance toward the assumed ConSink and the angle between the rat's current head direction and the assumed ConSink. The ConSink locations were scattered in all directions in the environment but were somewhat concentrated near the goal, and averaged over the goal area, suggesting that it was a signal related to spatial navigation.

The ConSink of each cell became more concentrated near the goal position as the task was learned. When the goal position was changed, ConSinks that had been organized around the original goal gradually moved to the new goal position. In addition, together with place cells the new ConSinks were organized around the new goal. This suggests that ConSinks may be pointing in the direction of the goal.

When the neuronal activity from the honeycomb maze task was recorded in the foraging task with all platforms elevated, it showed significantly less ConSink entrainment of place cells than in the honeycomb maze task. Furthermore, those ConSinks were not localized around the goal position. Thus, unlike the task with a single goal direction, the hippocampus did not show strong modulation to a fixed direction in the task without a specific direction. Firing toward the goal and respective ConSink was also reduced during error platform selection in the honeycomb maze task. These lines of evidence suggest that directional ConSink place cells may guide correct platform selection and support goal-directed navigation.

In light of these findings obtained from mammalian bats and rats, we can interpret the directional dependence of place cells as described. When an animal is foraging on randomly scattered foods on an open field, the place cells are directionless (Fig. 6a). The reason for this can be interpreted as the absence of a goal because the foods are randomly scattered in the extreme vicinity of the animal. Thus, only the property of being active at a specific location remains. On the other hand, in the case of traversing between the ends of a linear track to seek food, the goal is only at both ends of the run (Fig. 6b). This interpretation explains why the place cells have both self-position and distance properties.

Next, we similarly interpret the experimental data for the quail (Ben-Yishay et al., 2021). Since the quail was roaming spontaneously in an open field, we assume that the goals for each cell are randomly or uniformly scattered in all directions at an infinitely distant point (Fig. 6c). If each neuron in the medial pallium pointed to a single goal, it would be identified as a head direction cell. The goal was located at the infinity point, and so it points in the same direction no matter where the animal is situated. On the other hand, in experiments with tufted titmice, goals

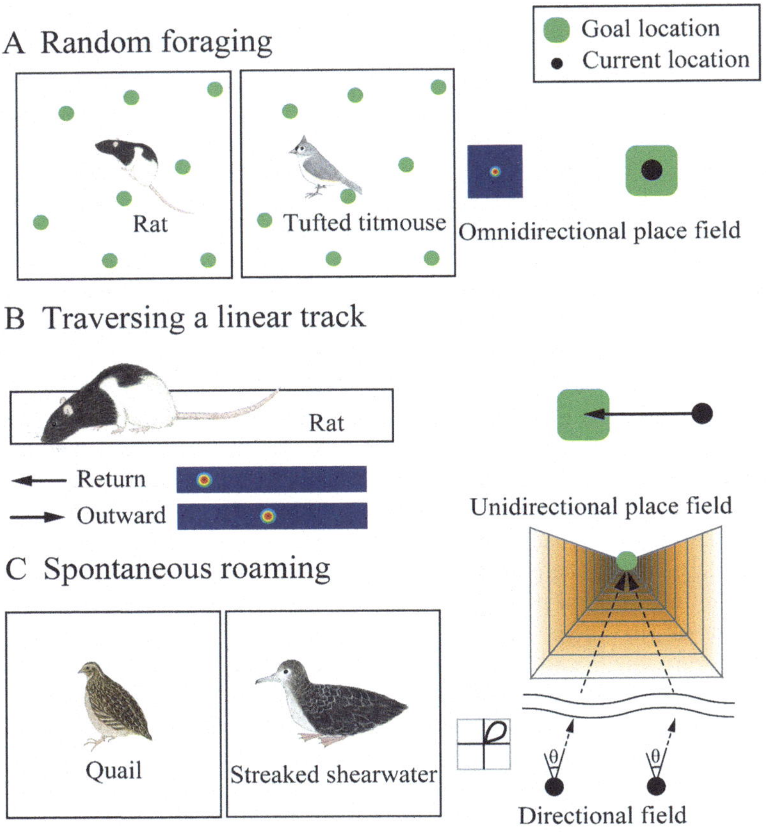

Fig. 6 Schematics for our conjecture of the hippocampal vectorial representation. (**a**) Left: When animals are foraging randomly scattered rewards such as food pellets in an open field, place cells with omnidirectional place fields emerge. Rodent-like place cells were found in the medial pallium, a hippocampal homolog, in the avian brain of tufted titmice. Right: During the random foraging, we propose that the goal and the animal's current location are always the same. Consequently, hippocampal cells encode a zero vector to goals: an omnidirectional place field. (**b**) Left: When rats are traversing between both ends of a linear track, a single hippocampal place cell fires at different locations on the outward and return trajectories. Right: During traversing a linear track, we conjecture that each cell has a place preference with a different direction to a goal on the outward and return trajectories. Consequently, hippocampal cells encode a trajectory-dependent vector: a unidirectional place field. (**c**) When animals are spontaneously roaming without rewards in an open field, head direction cells emerge. Head direction cells were found in the medial pallium in the avian brain of quails and streaked shearwaters. Right: During spontaneous roaming, the direction to the goal is always constant if animals roam because the distance between the goal and the animal's current location is distant enough to neglect the roaming distance. Consequently, cells in the medial pallium encode a place-invariant vector to a goal: a directional receptive field. (All illustrations were created using Adobe illustrator and Fresco)

were uniformly distributed in the extreme vicinity of the animal, similar to the experiments with rodents. These two species forage for randomly scattered foods. Therefore, the directional information to the goal was masked (Fig. 6a), leaving only self-location information, which was identified as place cells. Thus, as the experiments of Ormond and O'Keefe demonstrated, if we consider that conventional place cells represent not only self-locations but also vectors to goals, we can interpret the seemingly contradictory cross-species hippocampal neuronal activity in a unified manner.

7 Cells in the Medial Pallium of Juvenile Migratory Seabirds Prefer the Opposite Direction of the Wintering Grounds

Wild animals exhibit astonishing navigational abilities, as exemplified by migration and mother-river homing. Migratory birds seem to know their long-distant goal on a global scale, as they migrate without hesitation to the wintering grounds hundreds of kilometers away from the breeding grounds. In the previous sections, to explain why the representation of information in the medial pallium is different between quails and tufted titmice, we proposed a conjecture that the goal is so far away that the angle and distance to that destination cannot be measured when a reward such as food is not nearby. Is this assumption appropriate? We recently performed experiments on seabirds that annually migrate from breeding grounds to wintering grounds to assess neuronal activity from their dorsal medial pallium (Takahashi et al., 2022). Migratory birds perform a unique long-distance navigational behavior in which their destination is set at a long distance so that the angle and distance to that destination cannot be measured, which provides conditions to test this assumption.

The streaked shearwaters are a migratory seabird that breeds on the islands of Japan, Korea, and China, with the adult birds migrating to the Philippines, Indonesia, or northern Australia when they have finished raising their chicks (Fig. 7a). This migration requires fledgling chicks, but strangely enough, they do not follow their parents. Once the parents have finished raising their chicks, they migrate before the chicks can fly. How do the chicks perform long-distance migration? The streaked shearwater has a skeleton optimized to maintain its wings spread for what is called dynamic soaring, in which it glides along the water's surface without frequent flapping of wings. This means that they are not adept at flapping their wings, and in situations where it is necessary for them to continuously flap, their energy is severely depleted. They are likely to lose their lives if they flap their wings over long distances. More than half of the individuals are known to die within a month of leaving the nest. Yoda et al. captured the migratory routes of adult birds and fledgling chicks using the GPS (global positioning system) data loggers (Yoda et al. 2017, 2021). Japan is an island nation, surrounded by the sea, and its land is a series of mountain ranges. To go straight from Niigata to Indonesia, which is located in the south, it is necessary to cross the mountain range. It was determined that the

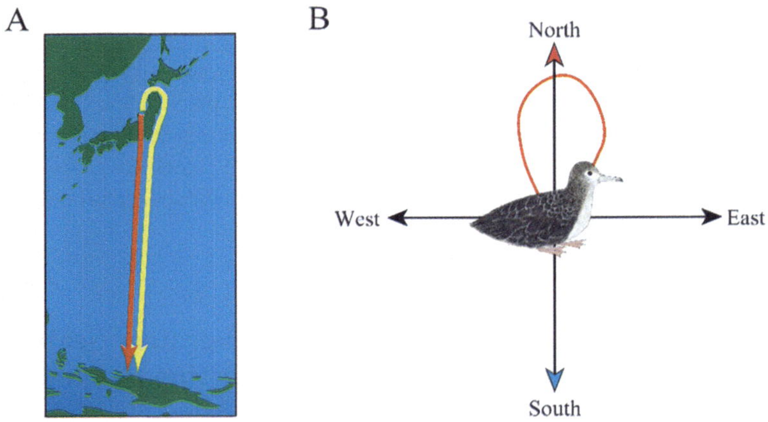

Fig. 7 Northward overrepresentation of head direction cells of migratory birds. (**a**) Streaked shearwaters, migratory seabirds, annually migrate over a thousand kilometers away from the breeding grounds (Awashima island, Niigata, Japan) to the wintering grounds (Indonesia). Whereas adults avoid crossing mountain ranges (yellow arrow), juveniles fly direct routes to the south (orange arrow) as if they only know the geomagnetic compass orientation. The illustration was created using Adobe illustrator and Fresco. (**b**) Head direction cells were found in the medial pallium of streaked shearwater juveniles. Unlike head direction cells found in the brains of other animal species, their orientation preference was confined to the geomagnetic north like a geomagnetic compass. (The illustration was created using Adobe illustrator and Fresco)

adult birds avoid the route over the mountains by flying over the sea, where they can travel by dynamic soaring. Since they have migrated many times, they fly as if they understand the terrain and choose a detour that minimizes energy expenditure. Fledgling chicks fly straight over the mountains when they believe their wintering grounds are to the south. Streaked shearwaters do not seem to have an innate global map. The study by Yoda et al. suggested that the chicks cross long distances relying only on geomagnetic cues to the south, which is considered the primary candidate.

Building on this study, a collaboration between ourselves and the Yoda ecology team, conducted an experiment to monitor neuronal activity on the hypothesis that there are cells in the medial pallium of the streaked shearwater chicks that are involved in migration and that have some relationship to the geomagnetic field. In the experiment, as in the quail experiment, we did not scatter food in an open field and allowed the chicks to roam spontaneously (Fig. 7b). We recorded neuronal activity in the medial pallium of the chicks while they were roaming and analyzed the relationship between head direction and firing rates. As in quails, about 20% of all recorded cells were identified as head-direction cells. However, the head direction cells in the medial pallium of the shearwater chicks had characteristics that were decisively different than those found in other animal species. When the direction was roughly delimited as east, west, north, and south, they were biased toward the northern direction as if they were a compass (azimuthal magnetic needle). This experiment was first conducted indoors, but because wireless neural loggers were employed as a method of logging neural activity, the experiment could also be conducted outdoors. The head direction cells maintained their preference for the north direction even in the outdoors, several kilometers away from the indoor experiment. In the case of mammalian head direction cells, their preference changes depending on cues in the vicinity. Based on this idea, we conducted an experiment in which the outer frame of the open field was transformed from circular to square. The results showed that the cells identified as head direction cells changed with each shape modification in the open field. Those originally identified as head direction cells lost their directional selectivity, while the newly identified cells still preferred the north direction. This evidence can be interpreted as the head direction cells of the shearwater chicks prefer the opposite direction to the goal because of the southern location of their wintering grounds.

To explain the presence of quail head direction cells, we introduced the conjecture that if no food is scattered in the open field and the birds are roaming spontaneously, then the goal is evenly distributed in all directions at the far point. However, in the case of streaked shearwater chicks, the goal is a single point in the far south, their wintering site. In this case, the head direction cells do not exhibit a variety of orientations, and we would expect any of the cells to face south, but why would they face north?

Based on our findings, we propose a hypothesis that the head direction cells in the brain of the chicks signal the moving direction and guide migration paths like a compass. If contrary to our finding, the head direction cells preferred the south, they would have to be continuously active during the thousands of kilometers of southbound migration, which would be energy-inefficient. Therefore, the reason why the head direction cells of the streaked shearwater chicks prefer north may be a "false alarm", letting them know that they are flying in the wrong direction if the head direction cells are active and correctly heading south. This is their destination even if they are not active.

Recently, head direction cells were found in the brain of monarch butterflies, a migratory insect (Beetz et al., 2022). Like rodent head direction cells, the directional preference of the cells changed in response to visual cues, whereas about half of the spatially tuned cells were unchanged. Interestingly, like the head direction cells of streaked shearwater chicks, the directional preference of the cue-invariant head direction cells of monarch butterflies was confined to the opposite of the animal's goal. This might suggest that the anti-goal direction signal shared between streaked shearwaters and monarch butterflies plays an important role in long-distance migration.

8 Summary

Here we focused on spatial navigation and scrutinized the literature on the correlation between spatial navigation and neuronal activity to make a cross-species comparative study of mammals, birds, fish, and insects. Cells involved in spatial navigation such as place cells, head direction cells, and grid cells have been identified in studies of hippocampal formation primarily in rodent model animals such as rats and mice. These cell types involved in spatial navigation exist across a wide variety of mammalian species including primates and bats. However, place cells and grid cells have not yet been found in the brains of fish or insects. This may be because fish and insects do not have brain regions homologous to the mammalian hippocampus in their brain.

The avian medial pallium is thought to be homologous to rodent hippocampal formation. Pattern cells with multiple place receptive fields have been found in the medial pallium of pigeons. Place cells have been identified in the medial pallium of tufted titmice, similar to those in the CA1 area of the rodent hippocampus with a single place receptive field. In quails, few cells with place receptive fields were found. Instead, head direction cells were identified. Taken together, these results suggest that only head direction cells exist across mammals, birds, fish, and insects.

Recent studies of mammalian rats and bats suggest that place cells represent not only the current position but also information about the direction of the goal. In light of these reports, head direction cells and place cells may be two sides of the same cell type. In the case of foraging behavior where the goal is set in the vicinity of the animal and the goal is randomly scattered in two-dimensional space, the goal becomes indefinite, and the distance from the current position to the goal is always zero, and there is no specific direction, so it becomes a zero vector. This is considered to be the spatial representation of the place cell. If there is only one goal, the goal responds when it arrives at a certain distance and direction from the current position to the goal. The reason why place cells are detected in the hippocampus of mammals including rats and mice may be because the experiments are conducted under such conditions. On the other hand, if the goal is set sufficiently far away relative to the distance traveled by the animal, it behaves like a head direction cell because the direction and distance from the current position to the goals are constant. Even in birds, in the case of experiments without random foraging, such as studies of quails and streaked shearwaters, this condition is compatible, and thus, head direction cells are detected.

Developmental studies in rodents have reported that head direction cells have adult-like activity characteristics shortly after birth, while place cells gradually reach adult levels, requiring about 4 weeks of experience to reach grid cells (Langston et al., 2010; Wills et al., 2010). This can be interpreted as direction and place information going through different developmental stages. Even rodents do not engage in foraging behavior in the early stages of development. This is a similar situation to the reports of quails and streaked shearwaters. Provided that the hippocampal neurons of juvenile rats encode a vector toward the goal and conjunctively encode self-location and head direction, our conjecture may account for these phenomena.

In this chapter, we propose a conjecture that the essential function of hippocampal cells is not the representation of the current position of the animal but the vector for goal-directed navigation. This conjecture can account for research findings on the hippocampus and its homologous regions in the brains of mammals, birds, fish, and insects. Following this work, we hope that comparative studies involving not only model animals but also wild animals will advance our understanding of episodic memory ability as well as spatial navigation in the hippocampus.

Acknowledgments This work was supported by JPSP KAKENHI (JP23H00502, JP21H05296).

References

Banino, A., Barry, C., Uria, B. et al. (2018). Vector-based navigation using grid-like representations in artificial agents. *Nature, 557*, 429–433. https://doi.org/10.1038/s41586-018-0102-6

Beetz, M. J., Kraus, C., Jundi, B., et al. (2022). Neural representation of goal direction in the monarch butterfly brain. *Nature Communications, 14*(1), 5859. https://doi.org/10.1101/2022.10.15.512348

Ben-Yishay, E., Krivoruchko, K., Ron, S., Ulanovsky, N., Derdikman, D., & Gutfreund, Y. (2021). Directional tuning in the hippocampal formation of birds. *Current Biology, 31*, 2592–2602. https://doi.org/10.1016/j.cub.2021.04.029

Boccara, C. N., Sargolini, F., Thoresen, V. H., Solstad, T., Witter, M. P., Moser, E. I., & Moser, M. B. (2010). Grid cells in pre-and parasubiculum. *Nature Neuroscience, 13*, 987–994. https://doi.org/10.1038/nn.2602

Canfield, J. G., & Mizumori, S. J. Y. (2004). Methods for chronic neural recording in the telencephalon of freely behaving fish. *Journal of Neuroscience Methods, 133*, 127–134. https://doi.org/10.1016/J.JNEUMETH.2003.10.011

Deisseroth, K., & Schnitzer, M. J. (2013). Engineering approaches to illuminating brain structure and dynamics. *Neuron, 80*, 568–577. https://doi.org/10.1016/j.neuron.2013.10.032

Eichenbaum, H. (2017). The role of the hippocampus in navigation is memory. *Journal of Neurophysiology, 117*, 1785–1796. https://doi.org/10.1152/jn.00005.2017

Goodridge, J. P., & Taube, J. S. (1997). Interaction between the postsubiculum and anterior thalamus in the generation of head direction cell activity. *Journal of Neuroscience: The Official Journal of the Society for Neuroscience, 17*, 9315–9330.

Gu, Y., Lewallen, S., Kinkhabwala, A. A., Domnisoru, C., Yoon, K., Gauthier, J. L., Fiete, I. R., & Tank, D. W. (2018). A map-like micro-organization of grid cells in the medial entorhinal cortex. *Cell, 175*, 736–750.e30. https://doi.org/10.1016/j.cell.2018.08.066

Hafting, T., Fyhn, M., Molden, S., Moser, M., & Moser, E. I. (2005). Microstructure of a spatial map in the entorhinal cortex. *Nature, 436*, 801–806. https://doi.org/10.1038/nature03721

Hough, G. E., & Bingman, V. P. (2004). Spatial response properties of homing pigeon hippocampal neurons: Correlations with goal locations, movement between goals, and environmental context in a radial-arm arena. *Journal of Comparative Physiology. A, Neuroethology, Sensory, Neural, and Behavioral Physiology, 190*, 1047–1062. https://doi.org/10.1007/s00359-004-0562-z

Ide, K., & Takahashi, S. (2022). A review of neuloggers for extracellular recording of neuronal activity in the brain of freely behaving wild animals. *Micromachines, 13*, 1529. https://doi.org/10.3390/mi13091529

Jarvis, E., Güntürkün, O., Bruce, L., Csillag, A., Karten, H., Kuenzel, W., Medina, L., Paxinos, G., Perkel, D. J., Shimizu, T., Striedter, G., Martin Wild, J., Ball, G. F., Dugas-Ford, J., Durand, S. E., Hough, G. E., Husband, S., Kubikova, L., Lee, D. W., Mello, C. V., Powers, A., Siang, C., Smulders, T. V., Wada, K., White, S. A., Yamamoto, K., Yu, J., Reiner, A., & Butler, A. B. (2005). Avian brains and a new understanding of vertebrate brain evolution. *Nature Reviews. Neuroscience*. https://doi.org/10.1038/nrn1606

Kahn, M. C., Siegel, J. J., Jechura, T. J., & Bingman, V. P. (2008). Response properties of avian hippocampal formation cells in an environment with unstable goal locations. *Behavioural Brain Research, 191*, 153–163.

Kim, S. S., Rouault, H., Druckmann, S., & Jayaraman, V. (2017). Ring attractor dynamics in the Drosophila central brain. *Science, 356*, 849–853. https://doi.org/10.1126/science.aal4835

Langston, R. F., Ainge, J. A., Couey, J. J., Canto, C. B., Bjerknes, T. L., Witter, M. P., Moser, E. I., & Moser, M.-B. B. (2010). Development of the spatial representation system in the rat. *Science, 328*, 1576–1580. https://doi.org/10.1126/science.1188210

McNaughton, B. L., Battaglia, F. P., Jensen, O., Moser, E. I., & Moser, M. (2006). Path integration and the neural basis of the "cognitive map". *Nature Reviews. Neuroscience, 7*, 663–678. https://doi.org/10.1038/nrn1932

Mizumori, S. J. Y., Ragozzino, K. E., & Cooper, B. G. (2000). Location and head direction representation in the dorsal striatum of rats. *Psychobiology, 28*, 441–462. https://doi.org/10.3758/BF03332003

Nitz, D., & McNaughton, B. (2004). Differential modulation of CA1 and dentate gyrus interneurons during exploration of novel environments. *Journal of Neurophysiology, 91*, 863–872. https://doi.org/10.1152/jn.00614.2003

O'Keefe, J., & Dostrovsky, J. (1971). The hippocampus as a spatial map. Preliminary evidence from unit activity in the freely-moving rat. *Brain Research, 34*, 171–175.

Ormond, J., & O'Keefe, J. (2022). Hippocampal place cells have goal-oriented vector fields during navigation. *Nature, 607*, 741–746. https://doi.org/10.1038/s41586-022-04913-9

Payne, H. L., Lynch, G. F., & Aronov, D. (2021). Neural representations of space in the hippocampus of a food-caching bird. *Science, 373*, 343–348. https://doi.org/10.1126/SCIENCE.ABG2009/SUPPL_FILE/ABG2009-PAYNE-SM.PDF

Robertson, R. G., Rolls, E. T., Georges-François, P., & Panzeri, S. (1999). Head direction cells in the primate pre-subiculum. *Hippocampus, 9*, 206–219. https://doi.org/10.1002/(SICI)1098-1063(1999)9:3<206::AID-HIPO2>3.0.CO;2-H

Sakurai, Y. (1990). Hippocampal cells have behavioral correlates during the performance of an auditory working memory task in the rat. *Behavioral Neuroscience, 104*, 253–263.

Sarel, A., Finkelstein, A., Las, L., & Ulanovsky, N. (2017). Vectorial representation of spatial goals in the hippocampus of bats. *Science, 355*, 176–180. https://doi.org/10.1126/science.aak9589

Scoville, W. B., & Milner, B. (1957). Loss of recent memory after bilateral hippocampal lesions. *Journal of Neurology, Neurosurgery, and Psychiatry, 20*, 11–21.

Seelig, J. D., & Jayaraman, V. (2015). Neural dynamics for landmark orientation and angular path integration. *Nature, 521*, 186–191. https://doi.org/10.1038/nature14446

Sharp, P. E., Tinkelman, A., & Cho, J. (2001). Angular velocity and head direction signals recorded from the dorsal tegmental nucleus of gudden in the rat: Implications for path integration in the head direction cell circuit. *Behavioral Neuroscience, 115*, 571–588.

Siegel, J. J., Nitz, D., & Bingman, V. P. (2005). Spatial-specificity of single-units in the hippocampal formation of freely moving homing pigeons. *Hippocampus, 15*, 26–40. https://doi.org/10.1002/hipo.20025

Smith, D. M., Barredo, J., & Mizumori, S. J. Y. (2012). Complimentary roles of the hippocampus and retrosplenial cortex in behavioral context discrimination. *Hippocampus, 22*, 1121–1133. https://doi.org/10.1002/hipo.20958

Solstad, T., Boccara, C. N., Kropff, E., Moser, M. B., & Moser, E. I. (2008). Representation of geometric borders in the entorhinal cortex. *Science, 322*, 1865–1868. https://doi.org/10.1126/science.1166466

Stackman, R. W., & Taube, J. S. (1997). Firing properties of head direction cells in the rat anterior thalamic nucleus: Dependence on vestibular input. *The Journal of Neuroscience, 17*, 4349–4358.

Stackman, R. W., & Taube, J. S. (1998). Firing properties of rat lateral mammillary single units: Head direction, head pitch, and angular head velocity. *Journal of Neuroscience: The Official Journal of the Society for Neuroscience, 18*, 9020–9037. https://doi.org/10.3389/fnbeh.2016.00040

Takahashi, S. (2013). Hierarchical organization of context in the hippocampal episodic code. *eLife, 2*, e00321. https://doi.org/10.7554/eLife.00321

Takahashi, S. (2015). Episodic-like memory trace in awake replay of hippocampal place cell activity sequences. *eLife, 4*, e08105.

Takahashi, S., Hombe, T., Takahashi, R., Ide, K., Okamoto, S., Yoda, K., Kitagawa, T., & Makiguchi, Y. (2021). Wireless logging of extracellular neuronal activity in the telencephalon of free-swimming salmonids. *Animal Biotelemetry, 9*, 1–12. https://doi.org/10.1186/S40317-021-00232-4/TABLES/1

Takahashi, S., Hombe, T., Matsumoto, S., Ide, K., & Yoda, K. (2022). Head direction cells in a migratory bird prefer north. *Science Advances, 8*, 6848. https://doi.org/10.1126/SCIADV.ABL6848/SUPPL_FILE/SCIADV.ABL6848_SM.PDF

Taube, J. S., Muller, R. U., & Ranck, J. B. (1990). Head-direction cells recorded from the postsubiculum in freely moving rats. I. Description and quantitative analysis. *The Journal of Neuroscience: The Official Journal of the Society for Neuroscience, 10*, 420–435. https://doi.org/10.1212/01.wnl.0000299117.48935.2e

Tolman, E. C. (1948). Cognitive maps in rats and men. *Psychological Review, 55*, 189–208. https://doi.org/10.1037/h0061626

Tulving, E. (1983). *Elements of episodic memory*. Oxford University Press.

Ulanovsky, N., & Moss, C. F. (2011). Dynamics of hippocampal spatial representation in echolocating bats. *Hippocampus, 21*, 150–161. https://doi.org/10.1002/HIPO.20731

Vinepinsky, E., Cohen, L., Perchik, S., Ben-Shahar, O., Donchin, O., & Segev, R. (2020). Representation of edges, head direction, and swimming kinematics in the brain of freely-navigating fish. *Scientific Reports, 10*, 1–16. https://doi.org/10.1038/s41598-020-71217-1

Wills, T. J., Cacucci, F., Burgess, N., & O'Keefe, J. (2010). Development of the hippocampal cognitive map in preweanling rats. *Science, 328*, 1573–1576. https://doi.org/10.1126/science.1188224

Wood, R. A., Bauza, M., Krupic, J., Burton, S., Delekate, A., Chan, D., & O'Keefe, J. (2018). The honeycomb maze provides a novel test to study hippocampal-dependent spatial navigation. *Nature, 554*, 102–105. https://doi.org/10.1038/nature25433

Yartsev, M. M., & Ulanovsky, N. (2013). Representation of three-dimensional space in the hippocampus of flying bats. *Science, 340*, 367–372. https://doi.org/10.1126/science.1235338

Yoda, K., Yamamoto, T., Suzuki, H., Matsumoto, S., Müller, M., & Yamamoto, M. (2017). Compass orientation drives naïve pelagic seabirds to cross mountain ranges. *Current Biology, 27*, R1152–R1153. https://doi.org/10.1016/J.CUB.2017.09.009

Yoda, K., Okumura, M., Suzuki, H., Matsumoto, S., Koyama, S., & Yamamoto, M. (2021). Annual variations in the migration routes and survival of pelagic seabirds over mountain ranges. *Ecology, 102*, e03297. https://doi.org/10.1002/ecy.3297

Yoder, R. M., & Taube, J. S. (2014). The vestibular contribution to the head direction signal and navigation. *Frontiers in Integrative Neuroscience, 8*, 32. https://doi.org/10.3389/fnint.2014.00032

Zhang, K. (1996). Representation of spatial orientation by the intrinsic dynamics of the head-direction cell ensemble: A theory. *The Journal of Neuroscience, 16*, 2112–2126. https://doi.org/10.1523/jneurosci.16-06-02112.1996

Modelling and Controlling System Dynamics of the Brain: An Intersection of Machine Learning and Control Theory

Quanying Liu, Chen Wei, Youzhi Qu, and Zhichao Liang

Abstract The human brain, as a complex system, has long captivated multidisciplinary researchers aiming to decode its intricate structure and function. This intricate network has driven scientific pursuits to advance our understanding of cognition, behavior, and neurological disorders by delving into the complex mechanisms underlying brain function and dysfunction. Modelling brain dynamics using machine learning techniques deepens our comprehension of brain dynamics from a computational perspective. These computational models allow researchers to simulate and analyze neural interactions, facilitating the identification of dysfunctions in connectivity or activity patterns. Additionally, the trained dynamical system, serving as a surrogate model, optimizes neurostimulation strategies under the guidelines of control theory. In this chapter, we discuss the recent studies on modelling and controlling brain dynamics at the intersection of machine learning and control theory, providing a framework to understand and improve cognitive function, and treat neurological and psychiatric disorders.

Keywords Modelling brain dynamics · Controlling brain dynamics · Machine learning · Control theory

1 Introduction

The human brain, as an intricate network of billions of neurons, has fascinated scientists and researchers in their pursuit of understanding its structure, function, and dynamics. Advancing our knowledge of cognition, behavior, and neurological disorders hinges upon delving into the complex mechanisms that underlie brain function and dysfunction (Bassett & Gazzaniga, 2011; Park & Friston, 2013). Research on neural dynamics is driven by a multifaceted motivation to decode

Q. Liu (✉) · C. Wei · Y. Qu · Z. Liang
Department of Biomedical Engineering, Southern University of Science and Technology, Shenzhen, GD, P.R. China
e-mail: liuqy@sustech.edu.cn

A. C.-H. Yu et al. (eds.), *Systems Neuroscience*, Advances in Neurobiology 41,
https://doi.org/10.1007/978-3-031-69188-1_3

the intricacies of brain function, to address neurological and psychiatric disorders, and to enhance human cognition (Muldoon et al., 2016; Gu et al., 2015). By integrating the insights gleaned from modelling and controlling neural dynamics, we can not only deepen our comprehension of the brain but also develop innovative therapeutic strategies to tackle the challenges posed by neurological and psychiatric disorders (Cocchi et al., 2017; Lynn & Bassett, 2019; Singh et al., 2022).

Modelling neural dynamics is a fundamental approach to understanding the complex relationships and interactions among neurons, neural circuits, and their emergent properties (Vyas et al., 2020). Motivated by the need to decipher the mechanisms governing neural function and behavior, computational models enable researchers to simulate and analyze these interactions, ultimately advancing our knowledge of neurological and psychiatric disorders (Byrne et al., 2020; Deco et al., 2011). By developing accurate models of neural circuits, researchers can identify dysfunction in connectivity or activity patterns and facilitate the creation of targeted therapies and interventions. Moreover, these models offer the predictive capacity to inform experimental design and probe brain function while providing a framework to understand the impact of environmental factors on cognition and behavior (Singh et al., 2022). Through these endeavors, modelling neural dynamics serves as a cornerstone in the quest to unravel the brain's inner workings and address the challenges posed by neurological and psychiatric conditions.

Controlling neural dynamics, driven by the desire to intervene and manipulate neural networks' activity and connectivity, offers immense potential for both scientific inquiry and practical applications. Utilizing techniques such as transcranial magnetic stimulation (TMS) and transcranial electrical stimulation (TES), researchers can develop innovative therapeutic strategies targeting neurological and psychiatric disorders, such as Alzheimer's disease, Parkinson's disease, depression, and schizophrenia (McNamara et al., 2001; Bikson et al., 2016; Coffman et al., 2014). Additionally, controlling neural dynamics presents the opportunity for cognitive enhancement in healthy individuals by modulating neural circuits involved in memory, attention, and decision-making (Muldoon et al., 2016; Coffman et al., 2014). Furthermore, this ability serves as a powerful tool for testing hypotheses about brain function and validating computational models, as experimentally manipulating neural activity can provide valuable insights into the causal relationships between neural dynamics and cognitive processes, informing the development of more accurate models of brain function.

The general framework of modelling and controlling system dynamics of the brain is shown in Fig. 1. Studying neural dynamics is crucial for understanding, investigating, and improving cognitive function, as well as addressing neurological and psychiatric disorders. In the following subsections, we will delve into modelling and controlling neural dynamics, enriching our understanding of the brain and its intricate functions from these perspectives.

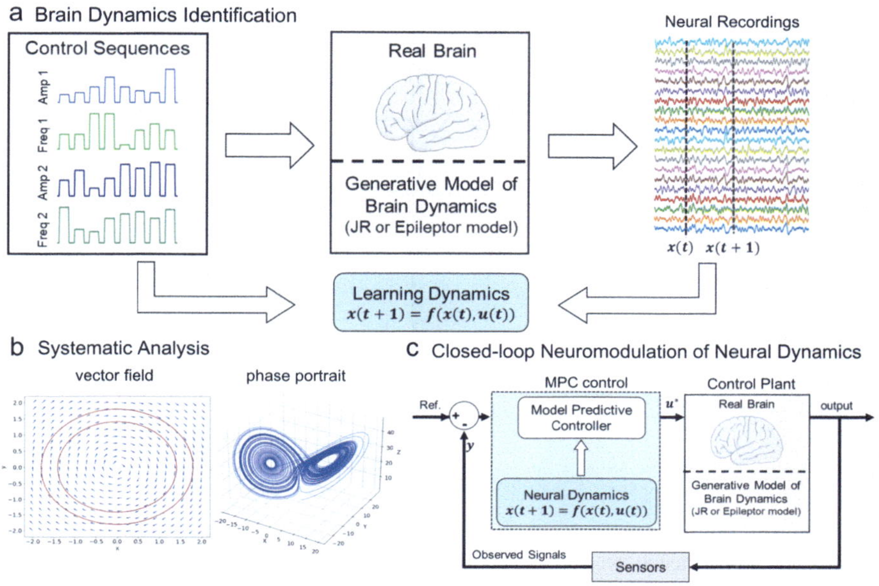

Fig. 1 The framework of modelling and controlling system dynamics of the brain. (**a**) Brain dynamics identification. (**b**) Example of systematic analysis: the vector field and phase portrait. (**c**) Closed-loop neuromodulation framework of neural dynamics (Liang et al., 2022)

2 Modelling the Neural Dynamics

Neural dynamics modelling characterizes the evolution of neural activity across different spatial and temporal scales. For instance, neural population-level dynamics study the collective activity patterns and responses of multiple neurons in a neural circuit (Vyas et al., 2020; Shenoy & Kao, 2021). It helps understand how neuron interactions and inputs influence neural activity and behavior. Mathematical models, like dynamical systems, describe these complex patterns using a lower-dimensional neural population state (Breakspear, 2017a). In a large spatial scale, brain network-level dynamics modelling depicts the neural dynamics of how information flows or information integrates across spatially densely interconnected and segregated regions (Breakspear, 2017b; Luo et al., 2022).

In a general dynamical systems framework, neural dynamics is expressed as

$$\mathbf{dx/dt} = f(\mathbf{x}(t), \mathbf{u}(t)),$$

where $\mathbf{x}$ is an N-dimensional vector representing regional-level functional dynamics or neuronal-level firing rates (neural population state) evaluated around time t. The function f captures the system's inherent dynamics based on the state x and external influences or inputs. The term $\mathbf{u}(t)$ represents these external influences or control

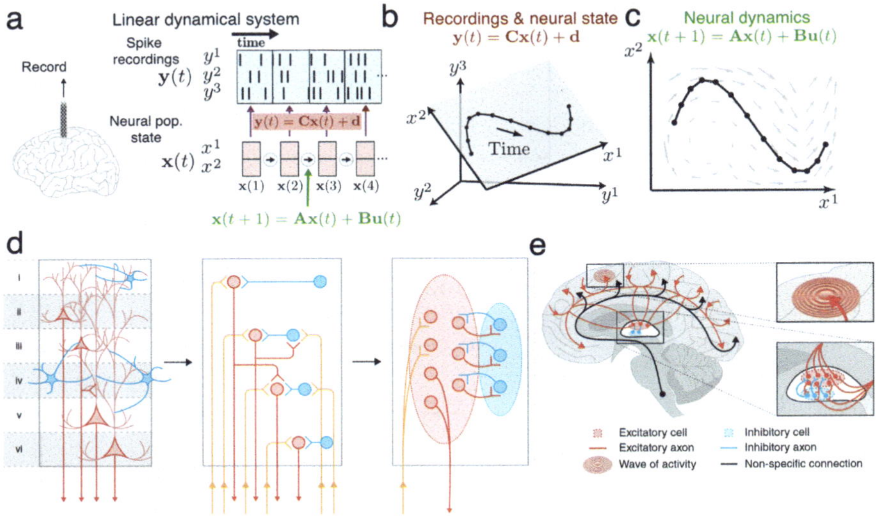

Fig. 2 Overview of neural dynamics and large-scale brain models. (**a**) Linear dynamical system (LDS). Binned spiking activity, $y(t)$, relates to a latent neural population state, $x(t)$, evolving according to linear dynamics, A, with inputs from other cortical areas, $Bu(t)$. The neural state is typically a low-dimensional trajectory in a flow field. (**b**) The neural state is represented as a low-dimensional trajectory within a subspace of higher-dimensional recordings. (**c**) In this low-dimensional subspace, the neural state evolves following neural dynamics that define a flow field. In an LDS, the dynamics of A can be contractive, expansive, rotational, or exhibit a fixed point. External inputs may result in the flow field displaying more complex motifs. Generally, both system dynamics and dimensionality reduction can be modeled using nonlinear models. Panels (**a**)–(**c**) are adopted from Shenoy and Kao (2021), and Pandarinath et al. (2018). (**d**) Neural mass models (NMMs). By averaging the states of all neurons of each class (e.g., pyramidal, inhibitory) in a local population, NMMs reduce population dynamics to a low-dimensional differential equation. Variability in the cell threshold smooths the action potential of each neuron into a sigmoidal map between average membrane potential and firing rate. (**e**) Neural field models (NFMs). Treating the cortex as a smooth sheet, NFMs support waves of propagating activity. Panels d and e adopted from Breakspear (2017a)

inputs that act on the system at time t. This captures the influence of regional dynamics or intracellular dynamics and inputs on the neural response.

Neural dynamics can be simplified to a linear dynamical system (LDS) (Fig. 2a), with a dynamic equation (Fig. 2c) and observation equation (Fig. 2b).

$$\begin{aligned} &\text{A linear dynamic model:} && \mathbf{x}(t+1) = \mathbf{Ax}(t) + \mathbf{Bu}(t), \\ &\text{An observation model:} && \mathbf{y}(t) = \mathbf{Cx}(t) + \mathbf{Du}(t), \end{aligned} \tag{1}$$

where $\mathbf{y}(t) \in \mathcal{R}^n$ represents observed n-dimensional neural activity recordings, such as action potential, local field potential, EEG signal, or fMRI BOLD signal. $\mathbf{x}(t) \in \mathcal{R}^m$ represents the m-dimensional latent neural state, for instance, neural population state or source signal at time t. The latent neural state $\mathbf{x}(t)$ evolves as a neural trajectory. The linear dynamic model captures how the state $\mathbf{x}(t)$ evolves over time based on a dynamics state-transition matrix $\mathbf{A}$, an input matrix $\mathbf{B}$, and inputs $\mathbf{u}(t)$ from other brain areas and sensory pathways. The state-transition matrix $\mathbf{A}$ characterizes how information flows or how the brain region interacts among brain regions or neuron populations. The observation model connects the latent neural state $\mathbf{x}(t)$ to the observed signal $\mathbf{y}(t)$ through an observation matrix $\mathbf{C}$. The observation matrix $\mathbf{C}$ can be defined as the feed-forward leadfield matrix in modelling EEG dynamics (Tu et al., 2019). The direct transition matrix $\mathbf{D}$ maps the input $u(t)$ to the observed signal.

There are several ways to obtain the matrices $\mathbf{A}, \mathbf{B}, \mathbf{C}, \mathbf{D}$ in LDS. On the one hand, based on the neuroscience prior, the state-transition matrix $\mathbf{A}$ in large-scale brain network models can be defined as the structural connection among the spatially interconnected regions (Gu et al., 2015). On the other hand, the matrices $\mathbf{A}, \mathbf{B}, \mathbf{C}, \mathbf{D}$ can be estimated by system identification techniques in a data-driven fashion, such as linear regression-based inference, dynamic mode decomposition (Brunton & Kutz, 2022), and Bayesian inference of state-space model (Tu et al., 2019).

2.1 *Modelling the Large-Scale Brain Dynamics and Its Parameter Inference*

Modelling large-scale brain dynamics helps understand how neural populations interact at different scales to produce complex brain functions (Byrne et al., 2020). These models provide a bridge between the microscopic level of individual neurons and the macroscopic level of brain-wide activity, while also incorporating parameter inference techniques to identify the best-fitting model parameters for observed neural data. They include neural mass models (NMMs), brain network models (BNMs), and neural field models (NFMs), offering complementary perspectives on brain dynamics. Being more computationally tractable than detailed simulations of every neuron, these models facilitate the study of large-scale brain dynamics and their role in various brain processes.

Neural Mass Models are computational approaches used to study large-scale brain dynamics by simplifying the complexity of neural populations (Fig. 2d). Instead of modelling each individual neuron within a population, NMMs focus on the average activity of distinct neuronal classes, such as excitatory and inhibitory neurons (Deco et al., 2011). This reduction in complexity allows researchers to explore the interactions between local neural populations and gain insights into the underlying mechanisms of various brain processes. NMMs can be derived from single-neuron models or based on empirical observations of neural population activity. In these models, the all-or-nothing firing of individual neurons is replaced with a sigmoid-shaped activation function that maps the average membrane potential to the mean firing rate. This approach implicitly incorporates the variance of individual neural thresholds and the dispersion of their states. The primary advantage of using NMMs is the reduction in computational complexity, enabling the study of brain dynamics on a larger scale. There are several types of NMMs, including Wilson–Cowan models (Wilson & Cowan, 1972) and Jansen–Rit models (Jansen & Rit, 1995). NMMs can exhibit different types of behavior, such as steady-state, periodic, or chaotic oscillations (Larter et al., 1999; Breakspear et al., 2003). They have been instrumental in understanding local population dynamics and their role in various brain processes.

Brain Network Models are a computational approach used to investigate large-scale brain dynamics by connecting multiple NMMs into a network. BNMs allow researchers to study the interactions between local neural populations at a mesoscopic (Wong & Wang, 2006) or macroscopic (Breakspear et al., 2004) level, providing insights into the underlying mechanisms of various brain processes. The coupling of NMMs in BNMs is informed by anatomical connectivity data, known as the connectome. This connectivity data can be obtained from invasive tracing studies or inferred from noninvasive techniques such as diffusion MRI-based tractography. BNMs represent the cortex as a discrete network of dynamic nodes, with each node reflecting local population activity and influences from distal regions.

Neural Field Models are a class of computational models used to study large-scale brain dynamics by treating the cortex as a continuous sheet rather than a discrete network of nodes, as in BNMs (Amari, 1977) (Fig. 2e). NFMs focus on the macroscopic scale, taking into account the dense short-range connections within the cortex and their rapid decrease in number with increasing distance between areas (Coombes, 2010). NFMs are formulated using a combination of differential equations for time dynamics and integral equations for spatial coupling and time delays. Under biologically realistic assumptions, these equations can often be expressed as wave equations, which are partial differential equations containing both temporal and spatial derivatives (Jirsa & Haken, 1996; Robinson et al., 1997; Coombes et al., 2007). NFMs have been employed to investigate various aspects of brain function, such as the contribution of large-scale cortical systems to cortical rhythms (Robinson et al., 2002), transitions between different states of consciousness like sleep (Phillips & Robinson, 2007) and wakefulness, epileptic seizures (Breakspear et al., 2006; Robinson et al., 2002), and evoked potentials (Rennie et al., 2002; Pinotsis and Friston, 2011). NFMs are intrinsically nonlinear wave models, which allows them to account for a wide spectrum of wave-like empirical data, such as the propagating patterns of oscillatory activity observed in sensory and motor cortices (Muller et al., 2014; Rubino et al., 2006; Heitmann et al., 2013).

Parameter Inference plays a vital role in modelling large-scale brain dynamics. The Virtual Brain project is a comprehensive framework that integrates various NMMs to simulate large-scale brain networks with biologically realistic connectivity, offering insights into individualized brain dynamics. Jirsa et al. (2017) took a significant step in personalized healthcare by developing a Virtual Epileptic Patient (VEP) brain model, demonstrating how to create patient-specific brain network models and develop personalized strategies for therapy and intervention. Hashemi et al. (2023) introduced the simulation-based inference for the VEP model (SBI-VEP), leveraging state-of-the-art deep learning algorithms for conditional density estimation to efficiently estimate the parameters related to the epileptogenic and propagation zones from sparse intracranial electroencephalography recordings. Gonçalves et al. (2020) employed deep neural density estimators to carry out Bayesian inference on mechanistic models of neural dynamics, allowing rapid analysis of new data after initial training and helping close the gap between data-driven and theory-driven models of neural dynamics. These approaches contribute to understanding the underlying mechanisms of neural interactions and brain functions, with potential applications in diagnosing and treating brain disorders.

By accurately estimating parameters in large-scale brain dynamics models, researchers can better understand the underlying mechanisms of neural interactions and brain functions. Furthermore, the intersection of machine learning and control theory can provide new insights and tools for the analysis and control of these complex systems, opening new avenues for understanding and treating brain disorders.

2.2 Using Neural Networks for Modelling Neural Dynamics

Neural networks, especially recurrent neural networks (RNNs), have emerged as powerful tools for modelling complex input–output relationships and nonlinear dynamical systems in neuroscience (Hazan et al., 2017; Lanzetti et al., 2019; Bieker et al., 2020; Yang et al., 2019; Mante et al., 2013; Song et al., 2016; Remington et al., 2018; Banino et al., 2018; Michaels et al., 2016). These models help neuroscientists nail down hypotheses about how behavior is shaped through distributed computation across multiple cortical areas (Pinto et al., 2019; Yang et al., 2019; Kleinman et al., 2019) and provide insights into the functioning of neural systems under multiple task conditions (Yang et al., 2019).

RNNs, such as universal approximators, are capable of accurately approximating any dynamical system. A basic RNN can be expressed as follows:

$$\mathbf{dx/dt} = f_\theta(\mathbf{x}(t), \mathbf{u}(t)), \tag{2}$$

or

$$\mathbf{dx/dt} = -\frac{1}{\tau}\mathbf{x}(t) + f(W_{rec}\mathbf{x}(t) + W_{in}\mathbf{u}(t)), \tag{3}$$

where $\mathbf{x}(t)$ represents the state variables that characterize the evolution of neural dynamics; $\mathbf{u}(t)$ denotes the input; $f()$ is the nonlinear or linear activation function $\theta = \{W_{rec}$ and $W_{in}\}$ is the parameter that needs to be learned. In Eq. (3), W_{rec} is the connection matrix, W_{in} is the input matrix, and the time constant $\tau > 0$ characterizes the time of decay. RNNs often have an explicitly defined linear readout, expressed as $\mathbf{y}(t) = \mathbf{Cx}(t)$, in which $\mathbf{y}(t)$ represents the output, and $\mathbf{C}$ is the readout matrix. Usually, θ can be learned by backpropagation through time to minimize the predictive error,

$$\mathcal{L}_{\text{pred}}(\theta) = \frac{1}{T_p} \sum_{t=1}^{T_p} \| y(t) - \tilde{y}(t) \|_2^2 . \tag{4}$$

However, no biological constraints or prior knowledge of neuroscience is incorporated into these RNN models. There is a realism gap between RNNs and biological neural circuits. Future investigations can help understand the features of neural circuit-level computation that can be abstracted in RNNs and in the brain (Genkin & Engel, 2020a). For example, multi-area RNN models can incorporate anatomical constraints and represent distinct brain areas (Song et al., 2016; Kleinman et al., 2019).

Neural networks can fit complex data representation, which involves identifying underlying dynamics from recordings of a system's internal state (Genkin et al., 2021), and task-based modelling, where the focus is on transforming inputs into outputs based on input–output pairs from a behavioral task (Shine et al., 2021). In both approaches, flexible models are increasingly used for generating hypotheses, but researchers must ensure accurate interpretation to avoid drawing incorrect conclusions from the data (Genkin & Engel, 2020b).

Advances in learning techniques for computational models have embraced the temporally evolving interconnected structure of the brain, representing it as a dynamic graph (Khambhati et al., 2018). Computational models linking cellular mechanisms of neuromodulation to large-scale neural dynamics provide a framework for understanding the relationship between microscale neuronal activity and systems-level brain function (Shine et al., 2021). Nonparametric frameworks, such as the one presented by Genkin and Engel (2020b), can help uncover the transient dynamics of complex systems. As a data-driven computational framework, neural perturbational inference (NPI) maps the human effective brain connectome (EBC) (Luo et al., 2022). The EBC characterizes the causal relationship in the whole brain and the information flow at both local and global scales and provides theoretical guidance for personalized neurostimulation.

In summary, neural networks and various computational models can advance our understanding of neural dynamics and latent representations. It is crucial to fill in the gap between models and neural circuits, refine model interpretations, and explore the relationships between different scales of neural activity.

3 Analyzing the System Property of Neural Dynamics

Neural dynamics refers to the study of the temporal patterns of neuronal activity and their underlying mechanisms in the brain. These patterns emerge from complex interactions between individual neurons, local neural circuits, and large-scale brain networks (Buzsaki & Draguhn, 2004; Van Vreeswijk & Sompolinsky, 1996; Friedman et al., 2012; Muller et al., 2018). By understanding the principles governing these dynamics, we can gain valuable insights into the neural basis of cognition, perception, learning, and other brain functions. Neuroscientists have long been fascinated by the intricate dynamics of the brain, as it is believed that they hold the key to understanding how the brain processes information and generates behavior (Bassett & Gazzaniga, 2011; Vyas et al., 2020; Wang et al., 2021). The study of neural dynamics encompasses a wide range of topics, from the microscopic level of individual neurons and their synapses to the macroscopic level of brain networks and cognitive functions. The dynamical properties or computational mechanisms of the learned neural dynamics that characterize the complex cognitive functions or behavior attract computational neuroscientists' interest. The analysis of system property advances a deep understanding of the neural mechanisms of the brain. In this section, we will discuss the following system properties: stability, criticality, reachability, and controllability (Fig. 3).

3.1 Stability

Stability constitutes a core principle in the examination of neural dynamics, denoting the capacity of a neural system to revert to a stable state or fixed point following disturbances from external stimuli or intrinsic fluctuations. The topological architecture of neural networks inherently influences the stability of their respective dynamics (Watts & Strogatz, 1998; Barabási & Albert, 1999). The concepts of fixed point and attractor are the core to understand the stability of neurodynamics and the organization of neural circuits (Rabinovich et al., 2006; Litwin-Kumar & Doiron, 2014).

Fixed points are states in which the system dynamics do not change over time, whereas attractors are regions in the state space where the system's trajectories tend to converge. Attractors can take various forms, such as point attractors, limit cycles, and strange attractors, each with distinct implications for neural processing. Point attractors represent stable states that the neural system converges to over time. Point attractors can encode specific pieces of information or decision-making states, and their stability is crucial for maintaining memories or executing cognitive tasks (Wang, 2001; Buonomano & Maass, 2009; Shenoy et al., 2013). Limit cycles are closed trajectories in the state space that the system converges to over time. The Wilson–Cowan model laid the foundation for studying limit cycles in neural networks, showing that interactions between excitatory and inhibitory neurons can lead to oscillatory activity (Wilson & Cowan, 1972). Neural networks with

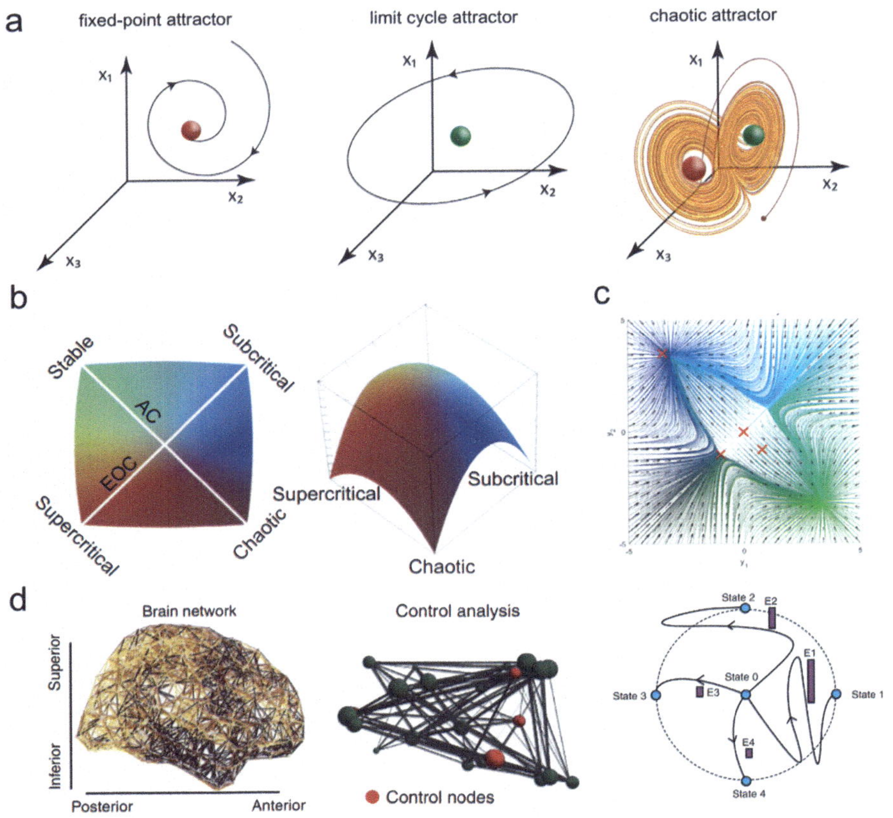

Fig. 3 Analyzing the system property of neural dynamics. (**a**) Three types of attractors in dynamical systems, including the fixed-point attractor (left), the limit cycle attractor (center), and the chaotic attractor (right), adapted from Uthamacumaran (2021). (**b**) The criticality of brain dynamics, adapted from O'Byrne and Jerbi (2022). (Left) The high-dimensional space is characterized by multiple phase transitions, highlighting the intersection of the avalanche criticality (AC) and the edge of chaos (EOC) transitions. (Right) As the system approaches these transitions, the computational features (e.g., optimal input representation) of criticality are maximized. (**c**) The reachability in a control-free dynamic, displaying the uncontrolled vector field along with its associated integral curves, adapted from Singh et al. (2022). The color indicates a distinct initial condition for each curve. (**d**) The controllability in brain networks. (Left) Brain network control points are identified. (Middle) Control nodes are identified as brain regions with significant influence over large-scale brain activity, which are capable of driving the system toward diverse cognitive functions. (Right) The state space is represented, including the original system state (state 0) and four equidistant potential target states (blue circles). The purple bar indicates the energy cost for reaching these target states via each trajectory. Adapted from Gu et al. (2015)

recurrent connections can exhibit stable dynamics through a balance of excitatory and inhibitory interactions (Van Vreeswijk & Sompolinsky, 1996). Subsequently, a series of theoretical and experimental studies on limit cycles in neural dynamics have been conducted, emphasizing their essential role in understanding oscillatory activity within the brain (Izhikevich, 2000; Ermentrout & Kopell, 1986; Latham et al., 2000; Breakspear et al., 2010). An attractor is a strange attractor if it has a fractal structure. When such a strange attractor additionally induces chaotic behavior in dynamical systems, characterized by trajectories that exhibit extreme sensitivity to initial conditions, it is additionally referred to as a chaotic attractor. Investigating the role of chaotic attractors in neural dynamics reveals that the chaotic nature of neural activity might contribute to flexible information processing in the brain (Rabinovich et al., 2006; Freeman, 1991; Lai & Tél, 2011). Emerging from complex interdependencies among neural elements, chaotic attractors promote transitions between functional states (Van Vreeswijk & Sompolinsky, 1996; Breakspear & Terry, 2002). The brain operates near a critical state, allowing for the coexistence of stable and flexible activity patterns, which optimizes information processing and adaptability (Sussillo & Abbott, 2009). The analysis of stability in neural dynamics helps us understand how neural circuits maintain stable activity patterns, enabling the brain to perform specific functions.

3.2 Criticality

Criticality is a concept originating from the study of phase transitions in physical systems, where a system exhibits a fine balance between order and disorder at a critical point (Stanley, 1971; Levina et al., 2009). This balance enables the system to respond adaptively to external stimuli and display optimal information processing capabilities. The neural system maintains a critical state near the boundary of stability and chaos, and it exhibits the phenomenon of self-organized criticality, which means having mechanisms that "self-tuning" to the critical point. Criticality is thought to play a crucial role in the brain's ability to efficiently process and transmit information, as well as adapt to changes in its environment (O'Byrne & Jerbi, 2022; Beggs & Plenz, 2003; Chialvo, 2010). At this critical state, the brain can optimally respond to inputs, store information, and generate complex, adaptive behaviors (Cocchi et al., 2017; Shew et al., 2015).

Avalanche criticality and the edge of chaos are important critical types of the brain. Avalanche criticality refers to a large class of continuous phase transitions, characterized by size distributions and durations that approximate power-law distributions. Theoretical models are essential for investigating criticality in neural networks, as they provide a framework to comprehend the underlying mechanisms. The Ising model is a simplistic physical system known for exhibiting phase transitions (Ising, 1924). It can be employed in neural dynamics to characterize the synaptic weights and interactions among neurons. Adjusting these parameters can lead the model of neural dynamics to reach avalanche criticality, suggesting that the brain may modulate these interactions to preserve a critical state (Schneidman et al., 2006). The edge of chaos refers to the phase transition between the stable phase and the chaotic phase. Through analyzing the transition from ordered to chaotic in RNN it was discovered that dynamical systems near the edge of chaos are capable of performing complex computational tasks (Bertschinger & Natschläger, 2004). An increasing number of studies have identified criticality in the brain through both in vivo and in vitro observations, as well as various imaging modalities (Beggs & Plenz, 2003; Ma et al., 2019; Linkenkaer-Hansen et al., 2001; Shriki et al., 2013; Tagliazucchi et al., 2012).

Criticality plays an essential role in neural computation and information processing (O'Byrne & Jerbi, 2022; Beggs & Plenz, 2003; Chialvo, 2010). Neural networks operating near the critical point exhibit optimal sensitivity and information transmission, resulting in heightened computational and processing capabilities (Kinouchi & Copelli, 2006; Shew et al., 2009). Neural networks also demonstrate a balance between stability and flexibility, fostering long-term information storage while preserving adaptability to new inputs (Haldeman & Beggs, 2005). Deviations from criticality may contribute to various brain disorders, such as epilepsy, schizophrenia, and Alzheimer's disease (Meisel et al., 2012, 2015; Alamian et al., 2020; Montez et al., 2009). Studying criticality not only deepens the understanding of its underlying mechanisms but also paves the way for novel therapeutic interventions aimed at restoring critical states and improving brain function.

3.3 *Reachability*

As a concept from control theory and dynamical systems, reachability aims to analyze the accessibility of different states in a system and the transformations that the system undergoes during its evolution (Lygeros, 2004). In neural dynamics, reachability refers to the ability of a neural network to transition between distinct activity patterns or states, depending on its current state and external inputs (Gu et al., 2015; Singh et al., 2022). Reachability is essential for comprehending the adaptability and flexibility of neural systems, as it reflects the ease with which states can be reached (Samsonovich & McNaughton, 1997).

Neural systems often produce attractors that generate infinite-horizon reachable sets under minor control perturbations. The computation of finite-time reachable sets is highly important for analyzing the reachability of neural dynamic (Singh et al., 2022). While the invariant set can be obtained using the Lyapunov function, which ensures the stability of model predictive control, employing the reachable set often generates better results in many cases (Rungger & Tabuada, 2017; El-Guindy et al., 2017). Control theory techniques have been widely employed in neural systems, focusing on the design and analysis of controllers that can manipulate neural system dynamics to achieve specific goals, such as desired activity patterns or trajectories. Key techniques include optimal control theory, model predictive control, linear-quadratic regulator control, and robust control theory (Todorov,

2004; Qin & Badgwell, 2003). Studies applying these techniques have demonstrated their effectiveness in understanding and controlling neural network dynamics, designing adaptive and effective neural stimulation strategies, and maintaining stability in models (Lozano & Lipsman, 2013).

3.4 Controllability

Controllability refers to the capability of a system to transition to a target state when provided with external inputs (Kalman, 1960). A system is deemed controllable if every state within the system is reachable, which implies that it is possible to navigate the system from any initial state to any desired target state. In the realm of neural dynamics, controllability analysis aims to pinpoint the crucial nodes or brain regions that have a controlling effect on the system's activity patterns (Gu et al., 2015). To investigate controllability in neural systems, researchers utilize control theory techniques, such as the controllability matrix and the minimum input energy criterion.

For the linear dynamical systems mentioned in Sect. 2, the system is represented by the equation $d\mathbf{x}/dt = \mathbf{A}\mathbf{x}(t) + \mathbf{B}\mathbf{u}(t)$. The controllability Gramian $\mathbf{P}$ can be defined as follows:

$$\mathbf{P} = \int_0^{\infty} \exp(t\mathbf{A})\mathbf{B}\mathbf{B}^T \exp(t\mathbf{A}^T)dt,$$

where $\mathbf{P}$ can be obtained by solving $\mathbf{A}\mathbf{P} + \mathbf{P}\mathbf{A}^T + \mathbf{B}\mathbf{B}^T = 0$, according to Kao and Hennequin (2019).

The controllability Gramian matrix $\mathbf{P}$ can help identify which brain regions or nodes can be controlled by external inputs, such as electrical stimulation or targeted drug delivery. The minimum amount of input to shift the system's state along direction $\mathbf{v}$ is proportional to $\mathbf{v}^T\mathbf{P}^{-1}\mathbf{v}$. The minimum input energy criterion is an optimization technique used to identify the optimal control signals required to guide a neural system toward a preferred state while minimizing the energy expended. This property is vital for comprehending how the brain produces and regulates intricate behaviors, and it is also essential for devising targeted interventions, like brain stimulation therapies, to manipulate neural activity (Muldoon et al., 2016).

Controllability is essential for cognitive performance, as it facilitates the effective modulation of neural activity patterns and adaptation to various cognitive demands (Lee et al., 2020; Bassett et al., 2013; Cui et al., 2020). The neural basis of cognitive control is attributed to a network of interconnected brain regions, including the prefrontal cortex, anterior cingulate cortex, and parietal cortex. These regions interact dynamically to enable flexible and adaptive cognitive functioning (Dosenbach et al., 2008). Higher controllability levels correlate with better cognitive performance, as they enable rapid adaptation to changing task demands and efficient neural resource allocation (Gu et al., 2015). Furthermore, individual differences in network control can significantly impact cognitive abilities, with higher controllability associated with enhanced cognitive flexibility, attentional control, and working memory capacity. Understanding these individual differences may inform targeted interventions to improve cognitive performance.

Controllability can be applied to the clinical treatment of neural disorders, particularly in the context of brain stimulation interventions, such as transcranial magnetic stimulation, which aims to activate or inhibit specific brain regions (Muldoon et al., 2016; Scheid et al., 2021). Traditional approaches have relied on empirical rules to target particular areas, but the incorporation of network control theory offers a more systematic and theoretically grounded approach for designing optimized treatment plans and exploratory outcomes. This shift allows for a more scientifically robust framework for neural feedback and interventions and providing a stronger theoretical foundation for understanding and treating neural disorders.

4 Controlling the Neural Dynamics

Besides modelling neural dynamics and analyzing the system property in detail, we are also interested in controlling neural dynamics, aiming at manipulating the neural signals, the internal brain states, or behaviors. Controlling the patterns of neural activity facilitates the treatment of several neurological and neuropsychiatric disorders and restoring lost function. A variety of approaches, including neurostimulation, neuromodulation, and pharmacological interventions, can be applied to control neural dynamics. Previous studies have shown that neuromodulation can help restore normal function and reduce symptoms of neurological and psychiatric disorders (Edwards et al., 2017; Bradley et al., 2022; Sani et al., 2018; Scangos et al., 2021a). In addition to treating specific disorders, controlling neural dynamics can also help us better understand the underlying neural mechanisms of brain function and brain connection (Wagenaar et al., 2005; Bonifazi et al., 2009; Herz et al., 2018). This can lead to new insights into the causal relationship between neural dynamics and behaviors, and new treatments for a wide range of neurological and psychiatric disorders. In this section, we first summarize the control approaches in controlling neural dynamics ranging from open-loop control strategy to closed-loop control strategy. Then we discuss the recent studies about model-based control and data-driven control strategy in controlling neural dynamics. Finally, we introduce the applications of controlling neural dynamics.

4.1 Open-Loop and Closed-Loop Control Techniques in Controlling Neural Dynamics

4.1.1 Open-Loop Control Techniques in Controlling Neural Dynamics

Open-loop control involves applying a fixed or predetermined stimulating pattern of stimulation with empirical guidance to the nervous system, without taking into

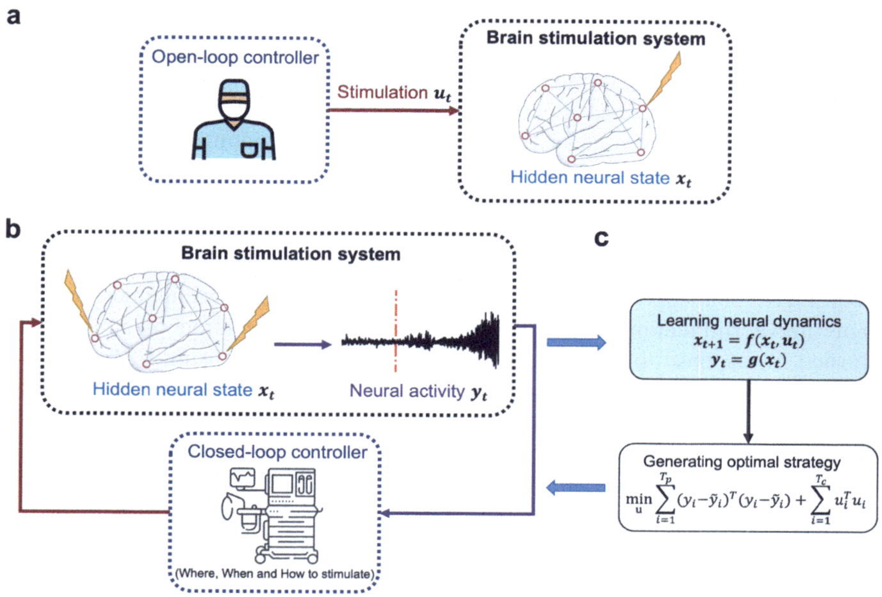

Fig. 4 Controlling neural dynamics. (**a**) Open-loop control strategy. (**b**) Closed-loop control strategy within (**c**) a model-based optimal control framework

account any feedback from the system being stimulated (Fig. 4a). The open-loop control strategies typically involve the direct application of external stimuli or perturbations to the neuronal network without feedback or adjustment based on the network's response. This approach is often used in neurostimulation techniques, such as deep brain stimulation (DBS) or spinal cord stimulation (SCS), where electrodes are implanted to directly stimulate neural circuits. In open-loop control, the stimulation parameters are preset and do not change in response to any changes in the neural activity being stimulated.

4.1.2 Closed-Loop Control Techniques in Controlling Neural Dynamics

Recently, the vast existing theory of control systems, especially closed-loop control theory, has been introduced in solving neurobiological control problems (Martínez et al., 2022; Liang et al., 2022). From a control-oriented point of view, closed-loop control involves real-time feedback signals from the system to adjust the control signal. It can be grouped into two types of closed-loop control strategies. One is the feedback signal to trigger the neural stimulation (Sani et al., 2018; Scangos et al., 2021a,b), and another type is about adjusting the control signal by minimizing the error between the reference signal and the observed signal (Liang et al., 2022; Chang et al., 2020). The first type of closed-loop control strategy requires the detection technique for triggering the stimulating device. The robustness and accuracy of the biomarkers in triggering the device are two important factors in neural treatment. The second type of closed-loop control strategy requires the solid control theory in developing a neural feedback control approach. For instance, an accurate model in capturing the neural dynamics advances the model-based optimal control problem (Fig. 4b).

Compared with the open-loop control strategy in controlling neural dynamics, closed-loop control, on the other hand, involves using real-time feedback from the nervous system to adjust the stimulation parameters and modulate neural activity accordingly. This approach is often used in neuromodulation techniques, such as vagus nerve stimulation (VNS) or TMS, where the stimulation parameters are adjusted in response to changes in neural activity measured using EEG or other techniques. Meanwhile, closed-loop control can provide more precise and adaptive modulation of neural activity compared to open-loop control. By incorporating real-time feedback from the nervous system, closed-loop control can adjust stimulation parameters in response to changes in neural activity, leading to more efficient and effective modulation of neural activity. Closed-loop control can also reduce side effects and optimize treatment outcomes by tailoring stimulation parameters to individual patients.

4.2 Model-Based and Data-Driven-Based Control in Controlling Neural Dynamics

Model-based control theory and data-driven control theory have been well developed in control theory. Model-based control involves using an accurate mathematical model of the system to design a control strategy, which was termed a two-step control. The model can be designed with the known principles or the identified model estimated with collected input–output data. Data-driven control, on the other hand, involves using machine learning techniques or optimization theory to develop a control strategy based on the collected data and it was termed one-step control. This approach is useful when the system is complex and not well understood, and there is a large amount of data available to analyze.

Recently, model-based control theory and data-driven-based control theory have been widely applied in controlling neural dynamics. For instance, the linear time-invariant model has been applied in modelling the internal brain state, and the optimal feedback controller (Linear-Quadratic Regulator, LQR) was developed for modulating the neural activity (Yang et al., 2021). In the designed linear time-invariant model, the control action was directly acted on the original state space. However, the linear time-invariant model can not accurately capture the long-time series of complex nonlinear dynamics, especially the complex neural dynamics. To conquer this problem with high model predictive accuracy and low model complexity, the Koopman operator theory would be a potential solution. It transformed the nonlinear dynamics from the original state space to another state space that satisfied the linear dynamics (Bevanda et al., 2021), and then the control action will be optimized in the new state space which advances the explicit solution of the linear optimal control problem. Liang et al. applied the Koopman operator theory equipped with model predictive control theory to implement the closed-loop optimal control framework. They validated the performance in seizure dynamics prediction in both synthetic seizure dynamics and real data of seizure dynamics, and seizure suppression effect in the well-known seizure dynamics model (Jansen–Rit model and Epileptor model) (Liang et al., 2022). Besides the model-based control strategy in controlling neural dynamics, a data-driven control framework has been developed in controlling complex networks (Baggio et al., 2021). The optimal control strategy in this study is constructed by using a finite set of random input–output data (Baggio et al., 2021).

4.3 Applications of Controlling the Neural Dynamics

Controlling neural dynamics has many potential applications in the diagnosis and treatment of neurological disorders, such as epilepsy, Parkinson's disease, and chronic pain, as well as in cognitive enhancement and brain–computer interfaces. In recent years, there has been a growing interest in closed-loop control methods, which involve real-time feedback from the neural system to adjust the stimulation parameters, as well as the use of advanced mathematical and machine learning techniques to develop more precise and personalized control strategies. In this section, we will list some representative applications in controlling neural dynamics.

4.3.1 Seizure Suppression

Seizure suppression is one of the most widely studied applications of controlling neural dynamics, with numerous studies investigating closed-loop electrical neurostimulation systems for the treatment of epilepsy (Chang et al., 2020; Takeuchi et al., 2021; Liang et al., 2022). These systems incorporate real-time feedback from neural recordings to adjust stimulation parameters and prevent or terminate seizures. Closed-loop systems have been shown to outperform traditional open-loop stimulation methods in terms of seizure control (Salam et al., 2015; Takeuchi et al., 2021). Experimental results report that closed-loop medial septum electrical stimulation can quickly and effectively terminate intrahippocampal seizures in a rat kindling model compared with the open-loop methods (Takeuchi et al., 2021). Liang et al. proposed a closed-loop seizure suppression framework rooted in the control theory. Owing to the ethical issue, direct validation of the framework's effectiveness was not feasible. Instead, the validations are only performed in well-known epilepsy dynamics models (Jansen–Rit model and Epileptor model) (Nakahira et al., 2021; Liang et al., 2022). More recently, an optimization-based parameter space search method has promoted the robustness of closed-loop electrical stimulation in cerebellar-directed epilepsy suppression (Stieve et al., 2023).

Despite the promising results of closed-loop neurostimulation and neuromodulation systems for seizure suppression, several challenges remain. The development of reliable and accurate seizure prediction algorithms is critical for the success of these systems. Controlling the complex brain network dynamics also requires hierarchical network models and more advanced control theory (Liang et al., 2022). In addition, issues such as electrode placement and implant longevity must be addressed in order to optimize the efficacy of these systems for long-term seizure control.

4.3.2 Parkinson's Treatment

Deep brain stimulation (DBS) consists in implanting a multi-contact lead into one or different parts of the deep brain, and it can deliver brief electrical pulses to the target regions. Recently, DBS approach has been validated as one of the most effective methods for the treatment of movement disorders, especially in the case of advanced Parkinson's disease (Gheisarnejad et al., 2019). And the closed-loop systems can provide better treatments than the traditional hand-crafted stimulated strategy (Rosin et al., 2011; Velisar et al., 2019). The classical proportional–integral–derivative (PID) feedback control methods have been explored to achieve an improved DBS algorithm in the treatment of Parkinson's disease (Gorzelic et al., 2013). The main motivation is to eliminate substantial decreases in the inhibition from the globus pallidus interna, which is the target region for DBS, to the thalamus. Robustness is increased in a fractional-order derivatives-based PID framework in the dampening oscillation of Parkinson's disease (Coronel-Escamilla et al., 2020).

4.3.3 In Vitro Neuron Modulation

Controlling neuronal dynamics in vitro involves manipulating the electrical and chemical activity of neurons in cultured brain tissue. This approach has been widely used to study the underlying mechanisms of neuronal activity and to develop new techniques for controlling and manipulating neural networks. For instance, researchers used closed-loop multi-electrode stimulation to control neuronal bursting in cortical cultures. This nonpharmacological method in controlling bursts of neuronal dynamics for in vitro neurons will be a critical tool for exploring how information processing of neuronal ensembles and has potential technical feasibility for the treatment of neural disorders (Wagenaar et al., 2005). Controlling in vitro microcircuits dynamics facilitates the exploration and identification of the engagement of network oscillation in the highly connected "hub" population that has strong directional inter-correlations (Bonifazi et al., 2009). More recently, the computational power of living neurons to create synthetic biological intelligence has been demonstrated in a simulated gameplay environment. The cultured neurons show the potential capability to self-organize neuronal dynamics in a goal-directed manner in response to sparse sensory information (Kagan et al., 2022).

References

Alamian, G., Pascarella, A., Lajnef, T., Knight, L., Walters, J., Singh, K. D., & Jerbi, K. (2020). Patient, interrupted: MEG oscillation dynamics reveal temporal dysconnectivity in schizophrenia. *NeuroImage: Clinical, 28*, 102485.

Amari, S.-I. (1977). Dynamics of pattern formation in lateral-inhibition type neural fields. *Biological Cybernetics, 27*(2), 77–87.

Baggio, G., Bassett, D. S., & Pasqualetti, F. (2021). Data-driven control of complex networks. *Nature Communications, 12*(1), 1429.

Banino, A., Barry, C., Uria, B., Blundell, C., Lillicrap, T., Mirowski, P., Pritzel, A., Chadwick, M. J., Degris, T., Modayil, J., et al. (2018) Vector-based navigation using grid-like representations in artificial agents. *Nature, 557*(7705), 429–433.

Barabási, A.-L., & Albert, R. (1999). Emergence of scaling in random networks. *Science, 286*(5439), 509–512.

Bassett, D. S., & Gazzaniga, M. S. (2011). Understanding complexity in the human brain. *Trends in Cognitive Sciences, 15*(5), 200–209.

Bassett, D. S., Wymbs, N. F., Puck Rombach, M., Porter, M. A., Mucha, P. J., & Grafton, S. T. (2013). Task-based core-periphery organization of human brain dynamics. *PLoS Computational Biology, 9*(9), e1003171.

Beggs, J. M., & Plenz, D. (2003). Neuronal avalanches in neocortical circuits. *Journal of Neuroscience, 23*(35), 11167–11177.

Bertschinger, N., & Natschläger, T. (2004). Real-time computation at the edge of chaos in recurrent neural networks. *Neural Computation, 16*(7), 1413–1436.

Bevanda, P., Sosnowski, S., & Hirche, S. (2021). Koopman operator dynamical models: Learning, analysis and control. *Annual Reviews in Control, 52*, 197–212.

Bieker, K., Peitz, S., Brunton, S. L., Kutz, J. N., & Dellnitz, M. (2020). Deep model predictive flow control with limited sensor data and online learning. *Theoretical and Computational Fluid Dynamics, 34*, 577–591.

Bikson, M., Grossman, P., Thomas, C., Zannou, A. L., Jiang, J., Adnan, T., Mourdoukoutas, A. P., Kronberg, G., Truong, D., Boggio, P., et al. (2016). Safety of transcranial direct current stimulation: evidence based update 2016. *Brain Stimulation, 9*(5), 641–661.

Bonifazi, P., Goldin, M., Picardo, M. A., Jorquera, I., Cattani, A., Bianconi, G., Represa, A., Ben-Ari, Y., & Cossart, R. (2009). GABAergic hub neurons orchestrate synchrony in developing hippocampal networks. *Science, 326*(5958), 1419–1424.

Bradley, C., Nydam, A. S., Dux, P. E., & Mattingley, J. B. (2022). State-dependent effects of neural stimulation on brain function and cognition. *Nature Reviews Neuroscience, 23*(8), 459–475.

Breakspear, M. (2017a). Dynamic models of large-scale brain activity. *Nature Neuroscience, 20*(3), 340–352.

Breakspear, M. (2017b). Dynamic models of large-scale brain activity. *Nature Neuroscience, 20*(3), 340–352.

Breakspear, M., Heitmann, S., & Daffertshofer, A. (2010). Generative models of cortical oscillations: Neurobiological implications of the Kuramoto model. *Frontiers in Human Neuroscience, 4*, 190.

Breakspear, M., Roberts, J. A., Terry, J. R., Rodrigues, S., Mahant, N., & Robinson, P. A. (2006). A unifying explanation of primary generalized seizures through nonlinear brain modeling and bifurcation analysis. *Cerebral Cortex, 16*(9), 1296–1313.

Breakspear, M., & Terry, J. R. (2002). Nonlinear interdependence in neural systems: Motivation, theory, and relevance. *International Journal of Neuroscience, 112*(10), 1263–1284.

Breakspear, M., Terry, J. R., & Friston, K. J. (2003). Modulation of excitatory synaptic coupling facilitates synchronization and complex dynamics in a biophysical model of neuronal dynamics. *Network: Computation in Neural Systems, 14*(4), 703.

Breakspear, M., Williams, L. M., & Stam, C. J. (2004). A novel method for the topographic analysis of neural activity reveals formation and dissolution of 'dynamic cell assemblies'. *Journal of Computational Neuroscience, 16*, 49–68.

Brunton, S. L., & Kutz, J. N. (2022). *Data-driven science and engineering: Machine learning, dynamical systems, and control*. Cambridge University Press.

Buonomano, D. V., & Maass, W. (2009). State-dependent computations: Spatiotemporal processing in cortical networks. *Nature Reviews Neuroscience, 10*(2), 113–125

Buzsaki, G., & Draguhn, A. (2004). Neuronal oscillations in cortical networks. *Science, 304*(5679), 1926–1929.

Byrne, Á., O'Dea, R. D., Forrester, M., Ross, J., & Coombes, S. (2020). Next-generation neural mass and field modeling. *Journal of Neurophysiology, 123*(2), 726–742.

Chang, S., Wei, X., Fei, S., Liu, C., Yi, G., Wang, J., et al. (2020). Model predictive control for seizure suppression based on nonlinear auto-regressive moving-average Volterra model. *IEEE Transactions on Neural Systems and Rehabilitation Engineering, 28*(10), 2173–2183.

Chialvo, D. R. (2010). Emergent complex neural dynamics. *Nature Physics, 6*(10), 744–750.

Cocchi, L., Gollo, L. L., Zalesky, A., & Breakspear, M. (2017). Criticality in the brain: A synthesis of neurobiology, models and cognition. *Progress in Neurobiology, 158*, 132–152.

Coffman, B. A., Clark, V. P., & Parasuraman, R. (2014). Battery powered thought: enhancement of attention, learning, and memory in healthy adults using transcranial direct current stimulation. *Neuroimage, 85*, 895–908.

Coombes, S. (2010). Large-scale neural dynamics: simple and complex. *NeuroImage, 52*(3), 731–739.

Coombes, S., Venkov, N. A., Shiau, L., Bojak, I., Liley, D. T. J., & Laing, C. R. (2007). Modeling electrocortical activity through improved local approximations of integral neural field equations. *Physical Review E, 76*(5), 051901.

Coronel-Escamilla, A., Gomez-Aguilar, J. F., Stamova, I., & Santamaria, F. (2020). Fractional order controllers increase the robustness of closed-loop deep brain stimulation systems. *Chaos, Solitons & Fractals, 140*, 110149.

Cui, Z., Stiso, J., Baum, G. L., Kim, J. Z., Roalf, D. R., Betzel, R. F., Gu, S., Lu, Z., Xia, C. H., He, X., et al. (2020) Optimization of energy state transition trajectory supports the development of executive function during youth. *Elife, 9*, e53060.

Deco, G., Jirsa, V. K., & McIntosh, A. R. (2011). Emerging concepts for the dynamical organization of resting-state activity in the brain. *Nature Reviews Neuroscience, 12*(1), 43–56.

Dosenbach, N. U. F., Fair, D. A., Cohen, A. L., Schlaggar, B. L., & Petersen, S. E. (2008). A dual-networks architecture of top-down control. *Trends in Cognitive Sciences, 12*(3), 99–105.

Edwards, C. A., Kouzani, A., Lee, K. H., & Ross, E. K. (2017). Neurostimulation devices for the treatment of neurologic disorders. In *Mayo Clinic Proceedings* (Vol. 92, pp. 1427–1444). Elsevier.

El-Guindy, A., Han, D., & Althoff, M. (2017). Estimating the region of attraction via forward reachable sets. In *2017 American Control Conference (ACC)* (pp. 1263–1270). IEEE.

Ermentrout, G. B., & Kopell, N. (1986). Parabolic bursting in an excitable system coupled with a slow oscillation. *SIAM Journal on Applied Mathematics, 46*(2), 233–253.

Freeman, W. J. (1991). The physiology of perception. *Scientific American, 264*(2), 78–87.

Friedman, N., Ito, S., Brinkman, B. A. W., Shimono, M., Lee DeVille, R. E., Dahmen, K. A., Beggs, J. M., & Butler, T. C. (2012). Universal critical dynamics in high resolution neuronal avalanche data. *Physical Review Letters, 108*(20), 208102.

Genkin, M., & Engel, T. A. (2020a). Moving beyond generalization to accurate interpretation of flexible models. *Nature Machine Intelligence, 2*(11), 674–683.

Genkin, M., & Engel, T. A. (2020b). Moving beyond generalization to accurate interpretation of flexible models. *Nature Machine Intelligence, 2*(11), 674–683.
Genkin, M., Hughes, O., & Engel, T. A. (2021). Learning non-stationary Langevin dynamics from stochastic observations of latent trajectories. *Nature Communications, 12*(1), 1–9.
Gheisarnejad, M., Faraji, B., Esfahani, Z., & Khooban, M.-H.. (2019). A close loop multi-area brain stimulation control for Parkinson's patients rehabilitation. *IEEE Sensors Journal, 20*(4), 2205–2213.
Gonçalves, P. J., Lueckmann, J.-M., Deistler, M., Nonnenmacher, M., Öcal, K., Bassetto, G., Chintaluri, C., Podlaski, W. F., Haddad, S. A., Vogels, T. P., Greenberg, D. S., & Macke, J. H. (2020). Training deep neural density estimators to identify mechanistic models of neural dynamics. *eLife, 9*, e56261.
Gorzelic, P., Schiff, S. J., & Sinha, A. (2013). Model-based rational feedback controller design for closed-loop deep brain stimulation of Parkinson's disease. *Journal of Neural Engineering, 10*(2), 026016.
Gu, S., Pasqualetti, F., Cieslak, M., Telesford, Q. K., Yu, A. B., Kahn, A. E., Medaglia, J. D., Vettel, J. M., Miller, M. B., Grafton, S. T., et al. (2015). Controllability of structural brain networks. *Nature Communications, 6*(1), 8414.
Haldeman, C., & Beggs, J. M. (2005). Critical branching captures activity in living neural networks and maximizes the number of metastable states. *Physical Review Letters, 94*(5), 058101.
Hashemi, M., Vattikonda, A. N., Jha, J., Sip, V., Woodman, M. M., Bartolomei, F., & Jirsa, V. K. (2023). Amortized Bayesian inference on generative dynamical network models of epilepsy using deep neural density estimators. *Neural Networks, 163*, 178–194.
Hazan, E., Singh, K., & Zhang, C. (2017). Learning linear dynamical systems via spectral filtering. In *Advances in Neural Information Processing Systems, 30*.
Heitmann, S., Boonstra, T., & Breakspear, M. (2013). A dendritic mechanism for decoding traveling waves: principles and applications to motor cortex. *PLoS Computational Biology, 9*(10), e1003260.
Herz, D. M., Little, S., Pedrosa, D. J., Tinkhauser, G., Cheeran, B., Foltynie, T., Bogacz, R., & Brown, P. (2018). Mechanisms underlying decision-making as revealed by deep-brain stimulation in patients with Parkinson's disease. *Current Biology, 28*(8), 1169–1178.
Ising, E. (1924). *Beitrag zur theorie des ferro-und paramagnetismus*. PhD Thesis, Grefe & Tiedemann Hamburg.
Izhikevich, E. M. (2000). Neural excitability, spiking and bursting. *International Journal of Bifurcation and Chaos, 10*(06), 1171–1266.
Jansen, B. H., & Rit, V. G. (1995). Electroencephalogram and visual evoked potential generation in a mathematical model of coupled cortical columns. *Biological Cybernetics, 73*(4), 357–366.
Jirsa, V. K., & Haken, H. (1996). Field theory of electromagnetic brain activity. *Physical Review Letters, 77*(5), 960.
Jirsa, V. K., Proix, T., Perdikis, D., Woodman, M. M., Wang, H., Gonzalez-Martinez, J., et al. (2017). The virtual epileptic patient: Individualized whole-brain models of epilepsy spread. *NeuroImage, 145*, 377–388.
Kagan, B. J., Kitchen, A. C., Tran, N. T., Habibollahi, F., Khajehnejad, M., Parker, B. J., Bhat, A., Rollo, B., Razi, A., & Friston, K. J. (2022). In vitro neurons learn and exhibit sentience when embodied in a simulated game-world. *Neuron, 110*(23), 3952–3969.
Kalman, R. E. (1960). On the general theory of control systems. In *Proceedings First International Conference on Automatic Control, Moscow, USSR* (pp. 481–492).
Kao, T.-C., & Hennequin, G. (2019). Neuroscience out of control: Control-theoretic perspectives on neural circuit dynamics. *Current Opinion in Neurobiology, 58*, 122–129.
Khambhati, A. N., Sizemore, A. E., Betzel, R. F., & Bassett, D. S. (2018). Modeling and interpreting mesoscale network dynamics. *NeuroImage, 180*, 337–349.
Kinouchi, O., & Copelli, M. (2006). Optimal dynamical range of excitable networks at criticality. *Nature Physics, 2*(5), 348–351.
Kleinman, M., Chandrasekaran, C., & Kao, J. C. (2019). Recurrent neural network models of multi-area computation underlying decision-making. *bioRxiv*, 798553.

Lai, Y.-C., & Tél, T. (2011). *Transient chaos: Complex dynamics on finite time scales* (Vol. 173). Springer Science & Business Media.

Lanzetti, N., Lian, Y. Z., Cortinovis, A., Dominguez, L., Mercangöz, M., & Jones, C. (2019). Recurrent neural network based MPC for process industries. In *2019 18th European Control Conference (ECC)* (pp. 1005–1010). IEEE.

Larter, R., Speelman, B., & Worth, R. M. (1999). A coupled ordinary differential equation lattice model for the simulation of epileptic seizures. *Chaos: An Interdisciplinary Journal of Nonlinear Science, 9*(3), 795–804.

Latham, P. E., Richmond, B. J., Nelson, P. G., & Nirenberg, S. (2000). Intrinsic dynamics in neuronal networks. I. theory. *Journal of Neurophysiology, 83*(2), 808–827.

Lee, W. H., Rodrigue, A., Glahn, D. C., Bassett, D. S., & Frangou, S. (2020). Heritability and cognitive relevance of structural brain controllability. *Cerebral Cortex, 30*(5), 3044–3054.

Levina, A., Herrmann, J. M., & Geisel, T. (2009). Phase transitions towards criticality in a neural system with adaptive interactions. *Physical Review Letters, 102*(11), 118110.

Liang, Z., Luo, Z., Liu, K., Qiu, J., & Liu, Q. (2022). Online learning Koopman operator for closed-loop electrical neurostimulation in epilepsy. *IEEE Journal of Biomedical and Health Informatics, 27*, 492–503.

Linkenkaer-Hansen, K., Nikouline, V. V., Palva, J. M., & Ilmoniemi, R. J. (2001). Long-range temporal correlations and scaling behavior in human brain oscillations. *Journal of Neuroscience, 21*(4), 1370–1377.

Litwin-Kumar, A., & Doiron, B. (2014). Formation and maintenance of neuronal assemblies through synaptic plasticity. *Nature Communications, 5*(1), 5319.

Lozano, A. M., & Lipsman, N. (2013). Probing and regulating dysfunctional circuits using deep brain stimulation. *Neuron, 77*(3), 406–424.

Luo, Z., Liang, Z., Xu, C., Zhou, C., & Liu, Q. (2022). Effective Brain Connectome: The whole-brain effective connectivity from neural perturbational inference. *arXiv preprint arXiv:2301.00148*.

Lygeros, J. (2004). On reachability and minimum cost optimal control. *Automatica, 40*(6), 917–927.

Lynn, C. W., & Bassett, D. S. (2019). The physics of brain network structure, function and control. *Nature Reviews Physics, 1*(5), 318–332.

Ma, Z., Turrigiano, G. G., Wessel, R., & Hengen, K. B. (2019). Cortical circuit dynamics are homeostatically tuned to criticality in vivo. *Neuron, 104*(4), 655–664.

Mante, V., Sussillo, D., Shenoy, K. V., & Newsome, W. T. (2013). Context-dependent computation by recurrent dynamics in prefrontal cortex. *Nature, 503*(7474), 78–84.

Martínez, S., García-Violini, D., Belluscio, M., Piriz, J., & Sánchez-Peña, R. (2022). Dynamical models in neuroscience from a closed-loop control perspective. *IEEE Reviews in Biomedical Engineering, 16*, 706–721.

McNamara, B., Ray, J. L., Arthurs, O. J., & Boniface, S. (2001). Transcranial magnetic stimulation for depression and other psychiatric disorders. *Psychological Medicine, 31*(7), 1141–1146.

Meisel, C., Schulze-Bonhage, A., Freestone, D., Cook, M. J., Achermann, P., & Plenz, D. (2015). Intrinsic excitability measures track antiepileptic drug action and uncover increasing/decreasing excitability over the wake/sleep cycle. *Proceedings of the National Academy of Sciences, 112*(47), 14694–14699.

Meisel, C., Storch, A., Hallmeyer-Elgner, S., Bullmore, E., & Gross, T. (2012). Failure of adaptive self-organized criticality during epileptic seizure attacks. *PLoS Computational Biology, 8*(1), e1002312.

Michaels, J. A., Dann, B., & Scherberger, H. (2016). Neural population dynamics during reaching are better explained by a dynamical system than representational tuning. *PLoS Computational Biology, 12*(11), e1005175.

Montez, T., Poil, S.-S., Jones, B. F., Manshanden, I., Verbunt, J. P. A., van Dijk, B. W., Brussaard, A. B., van Ooyen, A., Stam, C. J., Scheltens, P., et al. (2009) Altered temporal correlations in parietal alpha and prefrontal theta oscillations in early-stage Alzheimer disease. *Proceedings of the National Academy of Sciences, 106*(5), 1614–1619.

Muldoon, S. F., Pasqualetti, F., Gu, S., Cieslak, M., Grafton, S. T., Vettel, J. M., & Bassett, D. S. (2016). Stimulation-based control of dynamic brain networks. *PLoS Computational Biology, 12*(9), e1005076.

Muller, L., Chavane, F., Reynolds, J., & Sejnowski, T. J. (2018). Cortical travelling waves: Mechanisms and computational principles. *Nature Reviews Neuroscience, 19*(5), 255–268.

Muller, L., Reynaud, A., Chavane, F., & Destexhe, A. (2014). The stimulus-evoked population response in visual cortex of awake monkey is a propagating wave. *Nature Communications, 5*(1), 3675.

Nakahira, Y., Liu, Q., Sejnowski, T. J., & Doyle, J. C. (2021). Diversity-enabled sweet spots in layered architectures and speed–accuracy trade-offs in sensorimotor control. *Proceedings of the National Academy of Sciences, 118*(22), e1916367118.

O'Byrne, J., & Jerbi, K. (2022). How critical is brain criticality? *Trends in Neurosciences, 45*, 820–837.

Overview of TheVirtualBrain — TVB 2.7.2-17797 Documentation (2024). https://docs.thevirtualbrain.org/index.html

Pandarinath, C., Ames, K. C., Russo, A. A., Farshchian, A., Miller, L. E., Dyer, E. L., & Kao, E. L. (2018). Latent factors and dynamics in motor cortex and their application to brain–machine interfaces. *Journal of Neuroscience, 38*(44), 9390–9401.

Park, H.-J., & Friston, K. (2013). Structural and functional brain networks: From connections to cognition. *Science, 342*(6158), 1238411.

Phillips, A. J. K., & Robinson, P. A. (2007). A quantitative model of sleep-wake dynamics based on the physiology of the brainstem ascending arousal system. *Journal of Biological Rhythms, 22*(2), 167–179.

Pinotsis, D. A., & Friston, K. J. (2011). Neural fields, spectral responses and lateral connections. *Neuroimage, 55*(1), 39–48.

Pinto, L., Rajan, K., DePasquale, B., Thiberge, S. Y., Tank, D. W., & Brody, C. D. (2019). Task-dependent changes in the large-scale dynamics and necessity of cortical regions. *Neuron, 104*(4), 810–824.

Qin, S. J., & Badgwell, T. A. (2003). A survey of industrial model predictive control technology. *Control Engineering Practice, 11*(7), 733–764.

Rabinovich, M. I., Varona, P., Selverston, A. I., & Abarbanel, H. D. I. (2006). Dynamical principles in neuroscience. *Reviews of Modern Physics, 78*(4), 1213.

Remington, E. D., Narain, D., Hosseini, E. A., & Jazayeri, M. (2018). Flexible sensorimotor computations through rapid reconfiguration of cortical dynamics. *Neuron, 98*(5), 1005–1019.

Rennie, C. J., Robinson, P. A., & Wright, J. J. (2002). Unified neurophysical model of EEG spectra and evoked potentials. *Biological Cybernetics, 86*(6), 457–471.

Robinson, P. A., Rennie, C. J., & Rowe, D. L. (2002). Dynamics of large-scale brain activity in normal arousal states and epileptic seizures. *Physical Review E, 65*(4), 041924.

Robinson, P. A., Rennie, C. J., & Wright, J. J. (1997). Propagation and stability of waves of electrical activity in the cerebral cortex. *Physical Review E, 56*(1), 826.

Rosin, B., Slovik, M., Mitelman, R., Rivlin-Etzion, M., Haber, S. N., Israel, Z., Vaadia, E., & Bergman, H. (2011). Closed-loop deep brain stimulation is superior in ameliorating parkinsonism. *Neuron, 72*(2), 370–384.

Rubino, D., Robbins, K. A., & Hatsopoulos, N. G. (2006). Propagating waves mediate information transfer in the motor cortex. *Nature Neuroscience, 9*(12), 1549–1557.

Rungger, M., & Tabuada, P. (2017). Computing robust controlled invariant sets of linear systems. *IEEE Transactions on Automatic Control, 62*(7), 3665–3670.

Salam, M. T., Velazquez, J. L. P., & Genov, R. (2015). Seizure suppression efficacy of closed-loop versus open-loop deep brain stimulation in a rodent model of epilepsy. *IEEE Transactions on Neural Systems and Rehabilitation Engineering, 24*(6), 710–719.

Samsonovich, A., & McNaughton, B. L. (1997). Path integration and cognitive mapping in a continuous attractor neural network model. *Journal of Neuroscience, 17*(15), 5900–5920.

Sani, O. G., Yang, Y., Lee, M. B., Dawes, H. E., Chang, E. F., & Shanechi, M. M. (2018). Mood variations decoded from multi-site intracranial human brain activity. *Nature Biotechnology, 36*(10), 954–961.

Scangos, K. W., Khambhati, A. N., Daly, P. M., Makhoul, G. S., Sugrue, L. P., Zamanian, H., Liu, T. X., Rao, V. R., Sellers, K. K., Dawes, H. E., et al. (2021a). Closed-loop neuromodulation in an individual with treatment-resistant depression. *Nature Medicine, 27*(10), 1696–1700.

Scangos, K. W., Makhoul, G. S., Sugrue, L. P., Chang, E. F., & Krystal, A. D. (2021b). State-dependent responses to intracranial brain stimulation in a patient with depression. *Nature Medicine, 27*(2), 229–231.

Scheid, B. H., Ashourvan, A., Stiso, J., Davis, K. A., Mikhail, F., Pasqualetti, F., Litt, B., & Bassett, D. S. (2021). Time-evolving controllability of effective connectivity networks during seizure progression. *Proceedings of the National Academy of Sciences, 118*(5), e2006436118.

Schneidman, E., Berry, M. J., Segev, R., & Bialek, W. (2006). Weak pairwise correlations imply strongly correlated network states in a neural population. *Nature, 440*(7087), 1007–1012.

Shenoy, K. V., & Kao, J. C. (2021). Measurement, manipulation and modeling of brain-wide neural population dynamics. *Nature Communications, 12*(1), 633.

Shenoy, K. V., Sahani, M., & Churchland, M. M. (2013). Cortical control of arm movements: A dynamical systems perspective. *Annual Review of Neuroscience, 36*, 337–359.

Shew, W. L., Clawson, W. P., Pobst, J., Karimipanah, Y., Wright, N. C., & Wessel, R. (2015). Adaptation to sensory input tunes visual cortex to criticality. *Nature Physics, 11*(8), 659–663.

Shew, W. L., Yang, H., Petermann, T., Roy, R., & Plenz, D. (2009). Neuronal avalanches imply maximum dynamic range in cortical networks at criticality. *Journal of Neuroscience, 29*(49), 15595–15600.

Shine, J. M., Müller, E. J., Munn, B., Cabral, J., Moran, R. J., & Breakspear, M. (2021). Computational models link cellular mechanisms of neuromodulation to large-scale neural dynamics. *Nature Neuroscience, 24*(6), 765–776.

Shriki, O., Alstott, J., Carver, F., Holroyd, T., Henson, R. N. A., Smith, M. L., Coppola, R., Bullmore, E.,& Plenz, D. (2013). Neuronal avalanches in the resting MEG of the human brain. *Journal of Neuroscience, 33*(16), 7079–7090.

Singh, M. F., Cole, M. W., Braver, T. S., & Ching, S. (2022). Developing control-theoretic objectives for large-scale brain dynamics and cognitive enhancement. *Annual Reviews in Control, 54*, 363–376.

Song, H. F., Yang, G. R., & Wang, X.-J. (2016). Training excitatory-inhibitory recurrent neural networks for cognitive tasks: a simple and flexible framework. *PLoS Computational Biology, 12*(2), e1004792.

Stanley, H. E. (1971). *Phase Transitions and Critical Phenomena* (Vol. 7). Clarendon Press.

Stieve, B. J., Richner, T. J., Krook-Magnuson, C., Netoff, T. I., & Krook-Magnuson, E. (2023). Optimization of closed-loop electrical stimulation enables robust cerebellar-directed seizure control. *Brain, 146*(1), 91–108.

Sussillo, D., & Abbott, L. F. (2009). Generating coherent patterns of activity from chaotic neural networks. *Neuron, 63*(4), 544–557.

Tagliazucchi, E., Balenzuela, P., Fraiman, D., & Chialvo, D. R. (2012). Criticality in large-scale brain fMRI dynamics unveiled by a novel point process analysis. *Frontiers in Physiology, 3*, 15.

Takeuchi, Y., Harangozó, M., Pedraza, L., Földi, T., Kozák, G., Li, Q., & Berényi, A. (2021). Closed-loop stimulation of the medial septum terminates epileptic seizures. *Brain, 144*(3), 885–908.

Todorov, E. (2004). Optimality principles in sensorimotor control. *Nature Neuroscience, 7*(9), 907–915.

Tu, T., Paisley, J., Haufe, S., & Sajda, P. (2019). A state-space model for inferring effective connectivity of latent neural dynamics from simultaneous EEG/fMRI. In: *Advances in Neural Information Processing Systems, 32*.

Uthamacumaran, A. (2021). A review of dynamical systems approaches for the detection of chaotic attractors in cancer networks. *Patterns, 2*(4), 100226.

Van Vreeswijk, C., & Sompolinsky, H. (1996). Chaos in neuronal networks with balanced excitatory and inhibitory activity. *Science, 274*(5293), 1724–1726.
Velisar, A., Syrkin-Nikolau, J., Blumenfeld, Z., Trager, M. H., Afzal, M. F., Prabhakar, V., & Bronte-Stewart, H. (2019). Dual threshold neural closed loop deep brain stimulation in Parkinson disease patients. *Brain Stimulation, 12*(4), 868–876.
Vyas, S., Golub, M. D., Sussillo, D., & Shenoy, K. V. (2020). Computation through neural population dynamics. *Annual Review of Neuroscience, 43*, 249–275.
Wagenaar, D. A., Madhavan, R., Pine, J., & Potter, S. M. (2005). Controlling bursting in cortical cultures with closed-loop multi-electrode stimulation. *Journal of Neuroscience, 25*(3), 680–688.
Wang, X.-J.. (2001). Synaptic reverberation underlying mnemonic persistent activity. *Trends in Neurosciences, 24*(8), 455–463.
Wang, R., Liu, M., Cheng, X., Ying, W., Hildebrandt, A., & Zhou, C. (2021). Segregation, integration, and balance of large-scale resting brain networks configure different cognitive abilities. *Proceedings of the National Academy of Sciences, 118*(23), e2022288118.
Watts, D. J., & Strogatz, S. H. (1998). Collective dynamics of 'small-world' networks. *Nature, 393*(6684), 440–442.
Wilson, H. R., & Cowan, J. D. (1972). Excitatory and inhibitory interactions in localized populations of model neurons. *Biophysical Journal, 12*(1), 1–24.
Wong, K.-F., & Wang, X.-J. (2006). A recurrent network mechanism of time integration in perceptual decisions. *Journal of Neuroscience, 26*(4), 1314–1328.
Yang, G. R., Joglekar, M. R., Song, H. F., Newsome, W. T., & Wang, X.-J. (2019). Task representations in neural networks trained to perform many cognitive tasks. *Nature Neuroscience, 22*(2), 297–306.
Yang, Y., Qiao, S., Sani, O. G., Sedillo, J. I., Ferrentino, B., Pesaran, B., & Shanechi, M. M. (2021). Modelling and prediction of the dynamic responses of large-scale brain networks during direct electrical stimulation. *Nature Biomedical Engineering, 5*(4), 324–345.

Part II
Cutting-Edge Technologies

Adeno-associated Virus-Mediated Gene Delivery Across the Blood-Brain Barrier

Yizheng Yao and Fengfeng Bei

Abstract Recombinant adeno-associated viruses (AAVs) have emerged as a popular tool for gene therapy in the central nervous system (CNS). Given the dense vasculature in the CNS, systemic administration is an appealing approach for achieving a broad distribution of AAV vectors across the CNS. However, the blood-brain barrier (BBB) is a major obstacle that blocks the entry of AAV vectors into the brain and spinal cord. Thus, there is a great need to develop novel AAV vector technology with enhanced BBB penetration. In this chapter, we briefly summarize AAV biology, possible mechanisms for AAV vectors to overcome the BBB and further engineering strategies, and current clinical trials using systemic AAV gene therapy for CNS diseases.

Keywords Adeno-associated viruses · Central nervous system · Blood-brain barrier · Gene therapy

1 Introduction

Gene therapy has tremendous potential as a "one-time" treatment for debilitating disorders in the central nervous system (CNS) (Deverman et al., 2018). AAVs

Y. Yao
Jiangsu Key Laboratory of Neuropsychiatric Diseases and Institute of Neuroscience, Soochow University, Suzhou, China

Clinical Research Center of Neurological Disease, The Second Affiliated Hospital of Soochow University, Suzhou, China

Department of Neurosurgery, Brigham and Women's Hospital, Harvard Medical School, Boston, MA, USA
e-mail: yzyao28@suda.edu.cn

F. Bei (✉)
Department of Neurosurgery, Brigham and Women's Hospital, Harvard Medical School, Boston, MA, USA
e-mail: fbei@bwh.harvard.edu

A. C.-H. Yu et al. (eds.), *Systems Neuroscience*, Advances in Neurobiology 41,
https://doi.org/10.1007/978-3-031-69188-1_4

are widely used as a delivery tool for gene therapy because of their favorable attributes including their long-term expression in both dividing and non-dividing cells, low pathogenicity, and non-integration properties (Hudry & Vandenberghe, 2019). However, improving the efficiency of gene delivery remains a major hurdle for AAV-mediated gene therapy. To achieve global CNS transduction, two administration routes are usually exploited: (1) systemic route, which could leverage the abundant vasculature in the CNS for vector distribution, but would require vectors to overcome the blood-brain barrier (BBB) to enter the CNS (Pardridge, 2005); (2) intra-cerebrospinal fluid (intra-CSF) route, which could bypass the obstacle of the BBB, but presents difficulties for AAVs to diffuse into deep CNS regions. AAV serotype 9, applied in the FDA-approved drug Zolgensma for treating spinal muscular atrophy (Mendell et al., 2017), is now regarded as the "gold standard" of AAV vectors for global CNS delivery. Due to limited efficiency, the high systemic dosage of AAV9 is often used to reach the therapeutic threshold in the CNS, which in turn leads to an increased risk of off-target toxicity and severe immune response, let alone the high manufacturing cost. Thus, there have been considerable efforts aiming to develop new AAV vectors with enhanced BBB penetration and CNS transduction. Here, we provide a brief review of BBB-penetrant AAVs. We will first introduce an overview of AAV biology. Then, we will review the current understanding of potential mechanisms on how AAV penetrates the BBB and transduces the CNS. Subsequently, we will discuss engineering approaches to enhancing AAV's ability to cross the BBB. Finally, we will summarize current clinical efforts in using systemic AAV-mediated gene therapy to treat CNS diseases and provide our perspective on future development.

2 AAV Biology

2.1 Wild-Type AAVs

Wild-type AAV (wtAAV) was first reported in 1965 as DNA-containing contamination particles of adenovirus isolation from cultured rhesus monkey kidney cells (Atchison et al., 1965).

Later, AAV was classified as a member of the *Dependovirus* genus in the parvovirus family, with defective replication unless in the presence of helper viruses (such as adenoviruses, herpes viruses, or vaccinia virus). AAV virion is non-enveloped with a 20–25 nm diameter capsid. It contains a single-stranded, ~4.7 kb long DNA genome, comprising two in-*cis* open reading frames encoding Rep and Cap genes flanked by two 145 bp hairpin-like inverted terminal repeats (ITRs) (Fig. 1).

The AAV Rep gene encodes four proteins Rep78, Rep68, Rep52, and Rep40 that are responsible for viral replication and packaging (Fig. 1). Expression of Rep78 and Rep68 is controlled by the p5 promoter while expression of Rep52 and Rep40

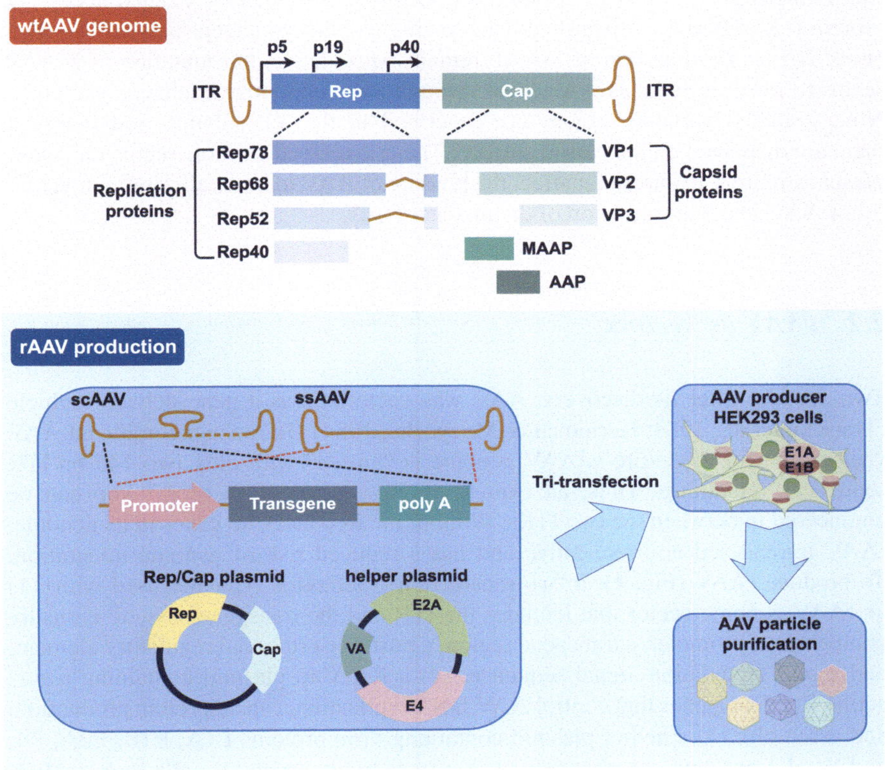

Fig. 1 wtAAV genome components and a common recombinant AAV (rAAV) production strategy. (The figure was created with Figdraw: https://www.figdraw.com/)

is controlled by the p19 promoter. Rep68 and Rep40 are splice variants of Rep78 and Rep52, respectively. Importantly, homologous recognition sequences, which Rep78/Rep68 proteins bind to, are found within the Rep-binding element in the ITR and AAVS1 locus of human chromosome 19, which could result in site-specific integration of AAV DNA into host genome (Weitzman et al., 1994; Balague et al., 1997). Such specific integration into the AAVS1 site by wtAAV reduces the chance of random mutagenesis in host genomes. Unloading the Rep gene, a strategy used in designing recombinant AAV (rAAV) vectors, further reduces the possibility of genome integration.

The AAV Cap gene encodes three viral capsid proteins (VP1, VP2, and VP3) and the assembly activating protein (AAP) under the p40 promoter. AAV capsid is a 60-mer icosahedral structure, assembled with 60 VP subunits of VP1, VP2, and VP3 capsid proteins in a molar ratio of 1:1:10. During capsid production, the AAP is a pivotal scaffolding protein that facilitates VP nucleation by assembling and connecting the C termini of VP proteins (Naumer et al., 2012). In addition, a frameshifted gene in the VP1 region encodes the membrane-associated accessory protein (MAAP) that is identified using a single codon mutagenesis study (Ogden et al., 2019). The function of MAAP remains speculative, but mutation of MAAP seems to result in reduced AAV yield through the competitive exclusion principle. Nine so-called variable regions are described in the VP subunit, and many of them are displayed on the capsid surface. These displayed variable regions together present structure features that affect the binding of AAV to host cells, immunogenicity of AAV, and determination of distinct serotypes.

2.2 *rAAV as Vectors*

Two decades after its discovery, AAV was vectorized as a gene delivery vehicle (Tratschin et al., 1984; Hermonat & Muzyczka, 1984). To make recombinant AAV vectors, almost the entire wtAAV genome is "gutted" with only two 145-bp ITR sequences remaining. Thus, an expression cassette of close to 4.7 kbp can be engineered in between the two ITRs. Without the Rep and Cap genes in its genome, rAAV is rendered non-replicative and has a reduced risk of genome integration. To produce rAAV (Fig. 1), a "tri-transfection" method is typically used with: (1) an rAAV genome vector that includes the ITR and the transgene cassette typically comprising a promoter, a transgene region, a posttranscriptional regulatory element, and a polyadenylation signal sequence; (2) a Rep/Cap plasmid containing *trans*-acting Rep/Cap genes that control AAV DNA replication, capsid protein production, and assembly; (3) a helper plasmid containing viral proteins E1A, E1B, E2A, E4, and VA. E1 proteins are often provided in trans by the producer cells such as HEK 293 cells. Helper viruses can also be applied to produce rAAVs. Unless specified, AAVs discussed herein refer to recombinant AAVs.

Upon cellular uptake, AAV travels to the nucleus of the host cell. AAV genome is released after uncaging and remains in an episomal form. AAV-mediated transgene expression is subject to both in-*cis* regulation of the expression cassette and the host cell environment. For designing gene therapy, two regulation elements are often used:

2.2.1 ITR

wtAAV contains a single-stranded DNA genome. After reaching the nucleus, it initiates replication of the original genome by binding Rep78/Rep68 to a Rep-binding element in ITR, which introduces a site-specific cleavage at the terminal resolution site (*trs*) to linearize and generate a dimeric genome. The synthesis of the second strand represents a rate-limiting step in determining rAAV transduction efficiency. However, this has been circumvented by the development of the so-called self-complementary AAV (scAAV) (McCarty et al., 2003). The alteration in scAAV to dislodge the *trs* cleavage site in one ITR avoids Rep-mediated nicking and allows the viral genome to instantly fold into double-stranded DNA before the encapsidation (McCarty et al., 2001). Although scAAV displays 10–100-fold more efficient transduction over ssAAV (McCarty et al., 2001), it is worth noticing that (1) the cargo capacity of scAAV is about half the size of ssAAV with a maximum accommodation up to ~2.2 kb within ITRs (McCarty, 2008); (2) scAAV may induce stronger innate and adaptive immune responses (Martino et al., 2011). Of note, it was reported that packing viral genome with scAAV demonstrated significantly increased CNS transduction over ssAAV after systemic administration, with expanded astrocyte tropism (Mathiesen et al., 2020; Rincon et al., 2018).

2.2.2 Promoter

The promoter is a vital *cis*-acting transcriptional regulatory component that can control transgene expression in tissue and cell-type-specific manners. Promoter selection and optimization are therefore an essential part of AAV cassette design. To achieve high levels of overall transgene expression, ubiquitous promoters, such as the cytomegalovirus (CMV), the chicken beta-acting (CBA), and cytomegalovirus immediate-early enhancer/chicken β-actin/rabbit β-globin (CAG) promoter, are often used in AAV gene therapy (Au et al., 2021). On the other hand, ubiquitous promoters will induce expression in undesirable cell types and increase the risk of immune-mediated toxicity. Thus, tissue or cell-type-specific promoters may be preferred to allow more restricted gene expression and reduce off-target transduction. Several cell-type-specific promoters have been applied in controlling desired expression in the CNS, such as neuron-specific hSyn promoter (Kugler et al., 2003), astrocyte-specific GfABC1D promoter (Lee et al., 2008), oligodendrocyte-specific MAG promoter (von Jonquieres et al., 2016), and microglia-specific F4/80 promoter (Ussher & Taylor, 2010). For targeted transduction in peripheral organs, examples of tissue-specific promoters include the TBG or ApoE/hAAT promoters with liver specificity (Yan et al., 2012; Manno et al., 2006) and the MCK promoter with muscle specificity (Larochelle et al., 1997).

Two optional elements have been devised to augment the efficiency of the promoter: enhancer element upstream of the core promoter and intron element between the core promoter and the ORF region. Incorporating a strong distal enhancer element can increase transgene expression level while not compromising the tissue specificity of the promoter (Powell et al., 2015; Muller et al., 2006). Muller and colleagues reported that the addition of CMV enhancer upstream of a cardiac muscle promoter led to a 50-fold enhancement of gene expression in the heart (Muller et al., 2006). Another strategy is to use a chimeric intron to improve gene expression by improving mRNA stability (Parra et al., 2011). For example, the minute virus of mice (MVM) intron was reported to significantly improve FIX expression in the liver (Wu et al., 2008).

2.3 AAV Serotypes and CNS Applications

A novel viral serotype is defined as a new type of virus showing no cross-reaction with sera-specific neutralization antibodies for existing serotypes (Wu et al., 2006). To date, 13 AAV serotypes comprising over 150 naturally occurring variants have been isolated from humans, non-human primates, and other animal species (Srivastava, 2016). Different AAV serotypes present distinct transduction profiles and tissue tropism (Wu et al., 2008; Colon-Thillet et al., 2021; Gao et al., 2003). Developing successful gene therapy in the CNS would require careful selection of the right AAV serotype and the most appropriate route of administration.

For intraparenchymal administration, AAV1, AAV2, AAV5, AAV8, AAV9, and AAV-DJ are commonly used to target local populations. Of these, AAV2 and AAV-DJ show minimal diffusion to adjacent parenchymal tissues, which would be ideal for a precise small region targeting (Burger et al., 2004; Davidson et al., 2000; Grimm et al., 2008). Other AAV serotypes may also be used for local delivery, as all known AAV serotypes appear to exhibit distinct but significantly overlapping tropisms and distribution properties in the CNS.

Intra-cerebrospinal fluid (CSF) administration of AAVs can be achieved via intracerebroventricular (ICV) injection, intra-cisterna magna injection (ICM), or intrathecal (IT) lumbar puncture (Miyanohara et al., 2016). Although broad CNS transduction can be achieved in neonates, intra-CSF administered AAV vectors in adults preferentially transduce cells close to the CSF circulation pathway and show limited diffusion into deep brain parenchymal regions. A comparative study demonstrated that AAV9 and AAVrh10 provided the highest level of neurotropism following the intra-CSF delivery (Swain et al., 2014; Cearley & Wolfe, 2006). IT delivery is generally applied to transduce spinal neurons and dorsal root ganglions (Schuster et al., 2014).

In contrast to other routes, systemic delivery of AAV has the potential to achieve widespread transgene expression given the vast vascular network in the CNS. However, the BBB is an obstacle that needs to be overcome. To date, it has been discovered that only a few naturally occurring AAVs can cross the BBB with limited efficiency. One example is AAV9, generally regarded as the "gold standard" for the global CNS delivery (Foust et al., 2009; Mendell et al., 2017). More on BBB-penetrant AAVs will be discussed further.

3 Potential Mechanisms for AAV to Cross the Blood-Brain Barrier

The BBB presents a formidable obstacle that impedes the passage of pathogens, all macromolecules, and the majority of small molecules to enter the CNS (Abbott et al., 2006, 2010). While it shields the brain from the circulating blood, the BBB hampers the systemic delivery of therapeutic drugs. While much research has been

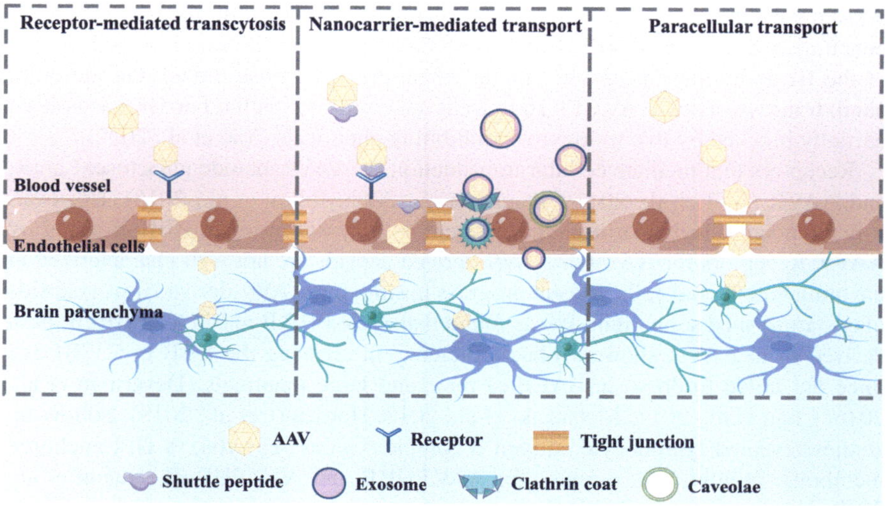

Fig. 2 Potential mechanisms for AAVs to cross the BBB. (The figure was created with Figdraw: https://www.figdraw.com/)

focused on molecular interaction between AAV capsids and host receptors and the intracellular pathway of AAV transduction, little is known about how AAV interacts with the BBB. More informative in vivo studies are particularly lacking. Here, we will list possible mechanisms on how circulating AAV may cross the BBB (Fig. 2):

3.1 Receptor-Mediated Transcytosis

To deliver particles through the BBB in a noninvasive way, three mechanisms could be potentially exploited: adsorptive-mediated transcytosis (AMT), carrier-mediated transcytosis, and receptor-mediated transcytosis (RMT) (Terstappen et al., 2021). Among them, receptor-mediated transcytosis is the most extensively investigated. However, receptors and cellular pathways responsible for AAV trafficking across the BBB are not well understood.

As a widely studied serotype in BBB-crossing, AAV9 has served as a prototype for exploring the mechanism of AAV-BBB interaction. Using an in vitro transduction and transcytosis assay of primary human brain microvascular endothelial cells (BMVEC), Ramirez and colleagues (Merkel et al., 2017) demonstrated that AAV9 preferably transcytosed the endothelial barrier through an active, cell-mediated process, with little transduction of the endothelial cells. On the contrary, AAV2, which has no BBB-crossing ability, was found to transduce BMVEC. Also, they provided evidence that transcytosis of AAV9 did not affect the integrity and barrier homeostasis of the BBB. Similar results were reported for another AAV9-

derived capsid, namely AAV.CPP.16 (Yao et al., 2022). Despite its enhanced BBB penetration as compared with AAV9, AAV.CPP16 did not compromise the integrity of the BBB in either an in vitro model or an in vivo mouse model. On the other hand, transcytosis by AAV.CPP.16 to transverse the endothelial barrier was at least partially blocked by two transcytosis-inhibiting chemicals (Yao et al., 2022).

Receptors that mediate cellular transduction by AAV9 include galactose, LamR, and AAVR (Bell et al., 2011; Akache et al., 2006; Pillay et al., 2016). However, it is unknown whether or how these receptors play any roles in BBB-crossing by AAV9. Receptors for AAV9 or AAV9-derived capsids are not well characterized in the brain endothelial cells. Recent progress in studying AAV9-derived novel capsids may start to shed some light. It was found that AAV-PHP·B and AAV-PHP.eB, both derived from AAV9, showed robust efficiency in crossing the BBB in C57BL/6 J mice but failed to do so in BALB/cJ mice and large mammals (Deverman et al., 2016; Chan et al., 2017; Matsuzaki et al., 2018; Hordeaux et al., 2018). Follow-up studies revealed lymphocyte antigen 6 complex, locus A (Ly6a), a GPI-anchored membrane protein, as a receptor for AAV.PHP.B and AAV.PHP.eB (Huang et al., 2019; Hordeaux et al., 2019). Ly6a is highly expressed in brain endothelial cells of C57BL/6 J mice but not in BALB/cJ mice or primates, which correlates with the strain and species specificity of AAV-PHP·B and AAV-PHP.eB. These findings suggest that GPI-anchored membrane proteins may play a role in AAV transcytosis and CNS biodistribution.

In a recent study, Kay and colleagues established an in vitro transwell model of human BBB for selecting BBB-crossing AAVs and further applied a siRNA screen approach to identify receptors that participate in the AAV transcytosis (Song et al., 2022). They reported that scavenger receptors (MSR1, SCARB1, and SCARA3), insulin-like growth factor receptors (IGF1R and IGF2R), antibody transporting receptors (FCGRT and TMEM30A), as well as LEPR, LRP8, CNR1, EPHA2, PTAFR, and CCR2 may facilitate AAV transcytosis. On the contrary, INSR, AGER, AAVR, M6PR, TFRC, and LRP2 receptors showed antagonistic effects on transcytosis. More study is warranted to elucidate the roles of these, as well as other unrevealed molecules, in mediating the BBB-crossing of AAVs.

3.2 Nanocarrier-Mediated Transport

CNS-targeting ligands such as short peptides and antibodies have been extensively studied for shuttling drugs into the CNS (Terstappen et al., 2021). Among them, short peptides may be particularly interesting to be combined with AAVs in enhancing CNS delivery. Such "Shuttle peptides" can facilitate drug transport in a specific way of receptor-mediated transport or a non-specific way of AMT. The less-studied AMT pathway requires the attachment of positively charged BBB shuttle peptides to the brain endothelial cells (Herve et al., 2008; Lombardo et al., 2020). AAV may also exploit the ligand-receptor interaction of shuttle peptides for better CNS penetration, for example, using previously discovered "BBB receptors" such as transferrin, insulin receptors, and low-density lipoproteins (Fu & McCarty, 2016). So far, some successes have been reported using THR peptide, LAH4 peptide, and PB5-3 peptide, which can facilitate certain AAV serotypes to cross the BBB (more discussion in Sect. 4.3).

Nanoparticles may be another way of facilitating CNS penetration by AAV, at least in theory. Such nanoparticles include exosomes, liposomes, solid lipid nanoparticles, and polymeric particles (Kamaly et al., 2016; Fonseca-Santos et al., 2015). For example, exosomes may help bypass the BBB using clathrin and caveolae-mediated mechanism (Tian et al., 2014; Nanbo et al., 2013). However, little success has been reported so far.

3.3 Paracellular Transport

The BBB is composed of densely packed brain microvascular endothelial cells that form intercellular tight junctions (TJs) and block paracellular transport of molecular weight over 500 Da into the CNS (Abbott et al., 2010; Cardoso et al., 2010). However, the TJs may be compromised in disease conditions or artificially disrupted in a transient and noninvasive manner. Such change to the TJs could increase the influx of drugs into the CNS via the paracellular transport (Sweeney et al., 2018, 2019).

Disease-associated breakdown of TJs occurs in a wide spectrum of neurological disorders including Alzheimer's disease (AD), Parkinson's disease, multiple sclerosis (MS), and traumatic brain injuries (Sweeney et al., 2019; Yamazaki & Kanekiyo, 2017; Kimura & Harashima, 2020). For example, Rozemuller and colleagues demonstrated that β-amyloid accumulated capillaries contributed to disrupted BBB function by loss of TJ proteins including claudin-5, occluding, and ZO-1 (Carrano et al., 2012). TJ abnormality was also found in MS lesions and is associated with BBB leakage (Kirk et al., 2003). It would be interesting to see whether and to what extent disease-associated TJ damage could allow AAVs to cross the BBB via paracellular transport.

Strategies have also been developed to artificially induce transient breakdown of the TJs and open the BBB for drug entry into the CNS. Such strategies include co-administration of chemical mannitol and focused ultrasound with microbubbles (FUS-MB). More details will be discussed in Sect. 4.2.

4 Methods to Enhance AAV Penetration Across the BBB

4.1 BBB-Crossing AAV Capsids

AAV9 is the first naturally discovered serotype that can overcome the BBB, especially in young mammals (Foust et al., 2009; Saraiva et al., 2016). Systemic administration of AAV9 can successfully transduce astrocytes, as well as some neuronal cell types in the CNS in animal models (Bevan et al., 2011; Duque et al., 2009). Importantly, AAV9, as in Zolgensma®, was approved by the FDA for treating patients with spinal muscular atrophy, a genetic disease preferentially causing degeneration of the motor neurons (Mendell et al., 2017). In addition to AAV9, AAVrh.8, and AAVrh.10, a new clade of AAVs isolated from rhesus macaques, were subsequently identified as CNS-tropic in adult mice following i.v. injection (Yang et al., 2014). AAVrh.8 was shown to better transduce neurons, astrocytes, and oligodendrocytes in various areas of the CNS as compared to AAV9 and AAVrh.10. AAVrh.10 also targets the spinal motor neurons of adult marmosets and is reported to be superior to AAV9 in neonatal mice (Tanguy et al., 2015; Zhang et al., 2011). Furthermore, AAVHSC7, AAVHSC15, and AAVHSC17 were demonstrated to produce widespread vector distribution in glial cells of the CNS, with the highest levels of expression in the pons and lateral geniculate nuclei (Ellsworth et al., 2019).

To increase BBB-crossing efficiency, new AAV capsids have been developed through directed evolution and rational design. To date, several AAV9 variants with enhanced BBB-crossing ability have been reported. Deverman et al. established a directed evolution platform of Cre recombination-based AAV targeted evolution (CREATE) and employed 7-mer insertion into threefold protrusion between AA588/589 of VP1 for screening of AAV capsid variants (Deverman et al., 2016). After discovering AAV-PHP.B using CREATE, the same group identified an enhanced version, namely AAV-PHP.eB, with even more robust CNS transduction efficiency in mice (Chan et al., 2017). However, it is now known that the effect of AAV.PHP.B/eB is limited to certain strains of mice. The capsids fail to translate into primates due to strain/species-specific expression of their receptor (Hordeaux et al., 2019). Other examples of AAV screening platforms include the "iTransduce" and the "TRACER" platforms that yielded the AAV variants with over 50-fold higher transduction of astrocytes and neurons than parental AAV9 in C57BL/6 J and BALB/c mice (Hanlon et al., 2019; Nonnenmacher et al., 2021). Despite such progress in mice, more effort is still required to translate such engineered AAVs from rodents to primates.

Our laboratory recently applied a rational design approach to engineer BBB-penetrant AAV variants (Yao et al., 2022). Cell-penetrating peptides (CPPs) were displayed on the AAV capsid, and the resulting variants were screened in both mice and non-human primates (NHPs) for their brain transduction after intravenous administration. We identified one variant, namely AAV.CPP.16, demonstrated 6- to 249-fold enhancement across four mouse strains and approximately five-fold enhancement in NHPs, as compared to the AAV9 parent vector. The superior neurotropism of AAV.CPP.16 is retained in both young and adult NHPs. We also showed that AAV.CPP.16 could be applied for the development of systemic gene therapies delivering both secreted and non-secreted antitumor payloads for glioblastoma. Thus, AAV.CPP.16 could be a translatable vector for developing systemic gene therapy for a variety of CNS disorders including glioblastoma.

4.2 Transient Opening of the BBB

Another approach to improve AAV penetration into the CNS is to induce transient BBB opening allowing AAV to access the CNS via paracellular transport. Here, two strategies for inducing transient BBB disruption will be discussed: the use of chemical compounds and focused ultrasound with microbubbles (FUS-MB).

Several chemical compounds, such as mannitol (Louboutin et al., 2010), Borneol (Zhang et al., 2013), and agonists of adenosine receptors (Carman et al., 2011; Jacobson & Gao, 2006), have been used as enhancer for drug delivery to overcome the BBB. Among them, mannitol has been well-reported for facilitating the delivery of viral vectors. Intravascular administration of mannitol temporarily increases vascular osmotic pressure and thereby osmotically shrinks the BBB cells to enhance the CNS delivery (Garg et al., 2015). As the mannitol-mediated BBB disruption is temporary and reversible, the BBB permeability peaks 5 min after intra-arterial infusion and lasts for 20–30 min (Ikeda et al., 2002; Rapoport, 2001). Fu and colleagues first reported an intravenous infusion of 25% mannitol solution 15–20 min before AAV2 vector administration increased up to tenfold more vector distribution in the mouse brain (McCarty et al., 2009). Moreover, intra-arterial infusion of mannitol was shown to increase the BBB crossing efficiency of AAVrh10 (Foley et al., 2014). These studies demonstrated that a minimum threshold of BBB disruption by mannitol seems to be sufficient to enhance CNS targeting by AAV vectors. To test the application of mannitol-enhanced AAV delivery in CNS gene therapy, a single intravenous infusion of rAAV2-CMV-hNaGlu vector was performed after mannitol administration. Such treatment was reported to be therapeutically beneficial in a mouse model of mucopolysaccharidosis (MPS) IIIB, which is a lysosomal storage disease with α-N-acetylglucosaminidase (NaGlu) gene deficiency (McCarty et al., 2009). Of note, this study showed that optimal transduction by AAV2 vectors in the brain was achieved when AAV injection was performed 8 min after mannitol administration. Thus, it seems that the effect of mannitol on the CNS distribution of AAV vectors is dependent on the dose, infusion route as well as timing of viral injection.

Another approach to achieve more localized transient disruption of BBB is by using focused ultrasound in combination with systemic infusion of microbubbles (FUS-MB). Magnetic resonance-guided FUS-MB has been successfully applied to noninvasively induce temporary and localized BBB opening in different animal models including mice, rats, pigs, and NHPs (Tung et al., 2011; Choi et al., 2007; Hynynen et al., 2003; Xie et al., 2008). FUS waves converge energy to a targeted region of the brain and cause oscillation of the intravenously injected microbubbles, resulting in controlled BBB opening through disruption of the tight junctions between the endothelial cells (Fan et al., 2017; Hynynen et al., 2005). Several studies showed that intravenously infused AAV1, AAV2, or chimeric AAV1/2 vectors successfully penetrated into directed brain regions after FUS-MB treatment, despite that some of the vectors don't naturally cross the BBB (Hsu et al., 2013; Stavarache et al., 2018; Alonso et al., 2013). Strikingly, 50–75% of the striatal neurons on average were transduced by AAVs when combined with FUS-MB. It is worth noticing that mild inflammation was found for the first 2 days after BBB disruption and AAV administration, while no evidence of inflammation was observed over longer terms (Stavarache et al., 2018). Recently, a comprehensive study compared the efficiency of AAV-mediated gene delivery to the brain using AAV1, AAV2, AAV5, AAV8, AAV9, and AAVrg in combination with FUS-MB. It was found that AAV1, AAV8, and AAV9 overall resulted in the highest transgene expression, although the transduction level can vary depending on brain regions and cell types (Kofoed et al., 2022).

4.3 Nanocarriers for AAV

Nanocarriers such as short peptides and exosomes can be used to boost systemic AAV delivery to the CNS. For example, the THR peptide (Sequence: THRPPMWSPVWP) that specifically binds to the transferrin receptor 1 (TfR1) was identified as a BBB shuttle peptide (Zhang et al., 2018). THR peptide can increase the ability of the AAV9 vector to cross the BBB and improve brain transduction in a dose-dependent manner in C57BL/6 mice. The mechanistic study revealed that THR peptide increased the transcytosis of AAV8 in hCMEC/D3 endothelial cells, resulting in an enhanced brain penetration ability. A slow blood clearance of the AAV vector in circulation was observed after THR peptide conjugation, which could also contribute to improved brain penetration by the AAV (Zhang et al., 2018). Meng and colleagues also screened for CPPs with CNS enhancement and identified LAH4 that can increase the CNS tropism of AAV9 using both in vitro and *vivo* models (Meng et al., 2021). PB5-3 (Sequence: QFAALPVRAHYG) is another example of shuttle peptides for AAVs (Zhang et al., 2022). PB5-3 peptide can specifically bind to AAV9, but not to other serotypes tested, and enhance its transduction efficiency in different brain cells in mice. In an animal model of mucopolysaccharidosis type I (MPS I), the AAV9-PB5-3 complex achieved more than two-fold higher IDUA expression and better brain correction. Thus, BBB shuttle peptides may be applied with AAVs for the systemic treatment of CNS diseases.

Exosome is another type of nanocarrier for potential co-application with AAVs. During AAV production, it was observed that a fraction of AAV capsids were wrapped in macrovesicles/exosomes (Maguire et al., 2012). This so-called exo-AAV showed improved transduction efficiency in multiple cultured cell lines and more resistance to anti-AAV neutralizing antibody as compared to "naked" AAVs (Gyorgy et al., 2014). Furthermore, a simple approach to concentrate exo-AAV from media via ultracentrifugation yields high-titer AAVs capable of broad CNS transduction after systemic injection in mice. Further study reported that exosome-coupled AAV9 outperformed AAV9 alone in achieving higher percentages of astrocytes and neuron transduction (Hudry et al., 2016). Similarly, exo-AAV8 exhibits higher transgene delivery to the brain than conventional AAV8, despite less efficiency than exo-AAV9. Reasons for enhanced brain penetration by exo-AAV have not been fully explored.

5 Systemic AAV Gene Therapy in the CNS

The lack of effective AAV delivery tools has significantly hampered the clinical development of systemic AAV gene therapy for CNS diseases. Here in this section, we highlight recent progress in clinical testing of such therapies using AAV9 and/or AAVrh10 (Table 1).

As discussed earlier, AAV9-based Zolgensma® is the first successfully approved treatment in the field of systemic gene therapy for CNS disease. In addition to the high manufacturing cost, the high therapeutic dose set by Zolgensma (1.1×10^{14} vg/kg of AAV9) could cause capsid-related immune toxicity, especially for the liver. Despite recommended co-administration of immunomodulatory medicine, two fatalities have been reported due to acute liver failure after intravenous delivery of Zolgensma (Philippidis, 2022). In both cases, fatal toxicity occurred 5–6 weeks after infusion of AAV and 1–10 days after the initiation of corticosteroid tapering. Developing more efficient AAV vectors may lower the therapeutic dose, which in turn reduces the risk of dose-associated toxicity.

In addition, clinical effort has been made to try to use BBB-penetrant AAV vectors for treating lysosomal storage diseases (LSDs).LSDs are a group of diseases that are caused by a deficiency of lysosomal enzymes and typically cause both peripheral pathologies and progressive neurodegeneration in the CNS (Patil & Maegawa, 2013). Enzyme replacement therapy is the mainstay treatment for several LSDs. However, intravenously administered enzymes are unable to penetrate the BBB and thus fail to treat the CNS pathology (Eisengart et al., 2018). The development of a systemic AAV platform that targets both the peripherical tissues and the CNS may help address the unmet clinical need. Currently, a phase I/II trial is active for treating MPS IIIA patients by using a single intravenous injection of scAAV9 to deliver replacement gene of N-sulphoglucosamine sulphohydrolase (SGSH) with multiple dose cohorts from 0.5×10^{12} to 3×10^{13} vg/kg (NCT02716246). Of note, another clinical trial using rAAV9-CMV-hNAGLU vector for MPS IIIB treatment, with the highest dose cohort at 1×10^{13} vg/kg (NCT03315182), was terminated due to drug supply and business reasons. Furthermore, AAV9-mediated gene transfer of human Beta-galactosidase enzyme to type I and type II of GM1 gangliosidosis is actively recruiting (NCT03952637). Overall, clinical studies on systemic AAV gene therapy in the CNS have been few, as compared with other routes such as intrathecal and intra-cerebroventricular administration. Reducing the viral dose without compromising the therapeutic benefit would open more opportunities for developing clinical programs.

6 Perspectives

Most CNS disorders affect large volumes of CNS tissues. Treating these diseases requires broad CNS targeting by therapeutics. Systemic administration of drugs

Table 1 Clinical trials of AAV-based systemic gene therapy for CNS diseases

Disease	Serotype	Route	Dose	Promoter	Transgene	Prophylactic administration of immunosuppressants	Exclude NAb	Exclude immunosuppressive patients	Phase	Sponsor	Identifier
Spinal Muscular Astrophy	scAAV9	IV	6.7e+13vg/kg; 1.1e+14vg/kg	CBA	Human survival motor neuron (SMN)	Corticosteroids	Y	Y	Completed	Novartis Gene Therapies	NCT03461289
Spinal Muscular Astrophy	scAAV9	IV	6.7e+13vg/kg; 2e+14vg/kg	CBA	Human survival motor neuron (SMN)	Y	Y	Y	Completed	Novartis Gene Therapies	NCT02122952
Spinal Muscular Astrophy	AAV9	IV	N/A	N/A	Human survival motor neuron (SMN)	N/A	Y	Y	Recruiting	Hangzhou Jiayin Biotech Ltd	NCT05614531
Canavan Disease	AAV9	IV and ICV	N/A	CB6	Aspartoacylase (AspA)	Rituximab, Sirolimus	Y	Y	Available	Barry J. Byrne	NCT05317780
Canavan Disease	AAV9	IV	N/A	N/A	Aspartoacylase (AspA)	N/A	Y	Y	Recruiting	Aspa Therapeutics	NCT04998396
Krabbe Disease	AAVrh10	IV	N/A	CBA	Human galactocerebrosidase (GALC)	N/A	N/A	N/A	Recruiting	Forge Biologics, Inc	NCT04693598

GM1 Gangliosidosis	AAV9	IV	1.5e+13vg/kg; 4.5e+13vg/kg	N/A	Beta-galactosidase (β-gal)	Rituximab, Sirolimus, Methylprednisolone, Prednisone	Y	Y	Recruiting	National Human Genome Research Institute (NHGRI)	NCT03952637
Pompe Disease	AAV9	IV	1.2e+14vg/kg; 2.4e+14vg/kg	N/A	Human acid alpha-glucosidase (GAA)	N/A	Y	N/A	Recruiting	Seventh Medical Center of PLA General Hospital	NCT05567627

Y yes, *N/A* not applicable, *IV* intravenous, *ICV* intracerebroventricular with BBB-crossing properties is the "ideal" route for treatment given the extensive vasculature in the CNS tissue. However, the challenge of developing BBB-crossing drugs is longstanding. With now proven clinical success, AAV has become a promising delivery tool for developing systemic gene therapy in the CNS. Significant progress has been made in engineering efficient BBB-penetrant AAV capsids primarily in mice. However, most of such AAV capsids so far have failed to translate into primates. Future studies on mechanistic understanding of these new AAVs, despite their limited translatability, would guide the development of effective BBB-penetrant AAVs in humans. Combining neurotropic AAVs with other modalities such as FUS-MB may enhance CNS targeting efficiency. The unmet need for a safe and effective BBB-penetrant gene delivery vector is substantial and a breakthrough is still to be made with AAV.

References

Abbott, N. J., Ronnback, L., & Hansson, E. (2006). Astrocyte-endothelial interactions at the blood-brain barrier. *Nature Reviews Neuroscience, 7*, 41–53.

Abbott, N. J., Patabendige, A. A., Dolman, D. E., Yusof, S. R., & Begley, D. J. (2010). Structure and function of the blood-brain barrier. *Neurobiology of Disease, 37*, 13–25.

Akache, B., Grimm, D., Pandey, K., Yant, S. R., Xu, H., & Kay, M. A. (2006). The 37/67-kilodalton laminin receptor is a receptor for adeno-associated virus serotypes 8, 2, 3, and 9. *Journal of Virology, 80*, 9831–9836.

Alonso, A., Reinz, E., Leuchs, B., Kleinschmidt, J., Fatar, M., Geers, B., Lentacker, I., Hennerici, M. G., De Smedt, S. C., & Meairs, S. (2013). Focal delivery of AAV2/1-transgenes into the rat brain by localized ultrasound-induced BBB opening. *Molecular Therapy – Nucleic Acids, 2*, e73.

Atchison, R. W., Casto, B. C., & Hammon, W. M. (1965). Adenovirus-associated defective virus particles. *Science, 149*, 754–756.

Au, H. K. E., Isalan, M., & Mielcarek, M. (2021). Gene therapy advances: A meta-analysis of AAV usage in clinical settings. *Frontiers in Medicine(Lausanne), 8*, 809118.

Balague, C., Kalla, M., & Zhang, W. W. (1997). Adeno-associated virus Rep78 protein and terminal repeats enhance integration of DNA sequences into the cellular genome. *Journal of Virology, 71*, 3299–3306.

Bell, C. L., Vandenberghe, L. H., Bell, P., Limberis, M. P., Gao, G. P., Van Vliet, K., Agbandje-Mckenna, M., & Wilson, J. M. (2011). The AAV9 receptor and its modification to improve in vivo lung gene transfer in mice. *The Journal of Clinical Investigation, 121*, 2427–2435.

Bevan, A. K., Duque, S., Foust, K. D., Morales, P. R., Braun, L., Schmelzer, L., Chan, C. M., Mccrate, M., Chicoine, L. G., Coley, B. D., Porensky, P. N., Kolb, S. J., Mendell, J. R., Burghes, A. H., & Kaspar, B. K. (2011). Systemic gene delivery in large species for targeting spinal cord, brain, and peripheral tissues for pediatric disorders. *Molecular Therapy, 19*, 1971–1980.

Burger, C., Gorbatyuk, O. S., Velardo, M. J., Peden, C. S., Williams, P., Zolotukhin, S., Reier, P. J., Mandel, R. J., & Muzyczka, N. (2004). Recombinant AAV viral vectors pseudotyped with viral capsids from serotypes 1, 2, and 5 display differential efficiency and cell tropism after delivery to different regions of the central nervous system. *Molecular Therapy, 10*, 302–317.

Cardoso, F. L., Brites, D., & Brito, M. A. (2010). Looking at the blood-brain barrier: Molecular anatomy and possible investigation approaches. *Brain Research Reviews, 64*, 328–363.

Carman, A. J., Mills, J. H., Krenz, A., Kim, D. G., & Bynoe, M. S. (2011). Adenosine receptor signaling modulates permeability of the blood-brain barrier. *The Journal of Neuroscience, 31*, 13272–13280.

Carrano, A., Hoozemans, J. J., Van Der Vies, S. M., Van Horssen, J., De Vries, H. E., & Rozemuller, A. J. (2012). Neuroinflammation and blood-brain barrier changes in capillary amyloid angiopathy. *Neurodegenerative Diseases, 10*, 329–331.

Cearley, C. N., & Wolfe, J. H. (2006). Transduction characteristics of adeno-associated virus vectors expressing cap serotypes 7, 8, 9, and Rh10 in the mouse brain. *Molecular Therapy, 13*, 528–537.

Chan, K. Y., Jang, M. J., Yoo, B. B., Greenbaum, A., Ravi, N., Wu, W. L., Sanchez-Guardado, L., Lois, C., Mazmanian, S. K., Deverman, B. E., & Gradinaru, V. (2017). Engineered AAVs for efficient noninvasive gene delivery to the central and peripheral nervous systems. *Nature Neuroscience, 20*, 1172–1179.

Choi, J. J., Pernot, M., Brown, T. R., Small, S. A., & Konofagou, E. E. (2007). Spatio-temporal analysis of molecular delivery through the blood-brain barrier using focused ultrasound. *Physics in Medicine and Biology, 52*, 5509–5530.

Colon-Thillet, R., Jerome, K. R., & Stone, D. (2021). Optimization of AAV vectors to target persistent viral reservoirs. *Virology Journal, 18*, 85.

Davidson, B. L., Stein, C. S., Heth, J. A., Martins, I., Kotin, R. M., Derksen, T. A., Zabner, J., Ghodsi, A., & Chiorini, J. A. (2000). Recombinant adeno-associated virus type 2, 4, and 5 vectors: Transduction of variant cell types and regions in the mammalian central nervous system. *Proceedings of the National Academy of Sciences of the United States of America, 97*, 3428–3432.

Deverman, B. E., Pravdo, P. L., Simpson, B. P., Kumar, S. R., Chan, K. Y., Banerjee, A., Wu, W. L., Yang, B., Huber, N., Pasca, S. P., & Gradinaru, V. (2016). Cre-dependent selection yields AAV variants for widespread gene transfer to the adult brain. *Nature Biotechnology, 34*, 204–209.

Deverman, B. E., Ravina, B. M., Bankiewicz, K. S., Paul, S. M., & Sah, D. W. Y. (2018). Gene therapy for neurological disorders: Progress and prospects. *Nature Reviews Drug Discovery, 17*, 641–659.

Duque, S., Joussemet, B., Riviere, C., Marais, T., Dubreil, L., Douar, A. M., Fyfe, J., Moullier, P., Colle, M. A., & Barkats, M. (2009). Intravenous administration of self-complementary AAV9 enables transgene delivery to adult motor neurons. *Molecular Therapy, 17*, 1187–1196.

Eisengart, J. B., Rudser, K. D., Xue, Y., Orchard, P., Miller, W., Lund, T., Van Der Ploeg, A., Mercer, J., Jones, S., Mengel, K. E., Gokce, S., Guffon, N., Giugliani, R., De Souza, C. F. M., Shapiro, E. G., & Whitley, C. B. (2018). Long-term outcomes of systemic therapies for hurler syndrome: An international multicenter comparison. *Genetics in Medicine, 20*, 1423–1429.

Ellsworth, J. L., Gingras, J., Smith, L. J., Rubin, H., Seabrook, T. A., Patel, K., Zapata, N., Olivieri, K., O'callaghan, M., Chlipala, E., Morales, P., & Seymour, A. (2019). Clade F AAVHSCs cross the blood brain barrier and transduce the central nervous system in addition to peripheral tissues following intravenous administration in nonhuman primates. *PLoS One, 14*, e0225582.

Fan, C. H., Lin, C. Y., Liu, H. L., & Yeh, C. K. (2017). Ultrasound targeted CNS gene delivery for Parkinson's disease treatment. *Journal of Controlled Release, 261*, 246–262.

Foley, C. P., Rubin, D. G., Santillan, A., Sondhi, D., Dyke, J. P., Crystal, R. G., Gobin, Y. P., & Ballon, D. J. (2014). Intra-arterial delivery of AAV vectors to the mouse brain after mannitol mediated blood brain barrier disruption. *Journal of Controlled Release, 196*, 71–78.

Fonseca-Santos, B., Gremiao, M. P., & Chorilli, M. (2015). Nanotechnology-based drug delivery systems for the treatment of Alzheimer's disease. *International Journal of Nanomedicine, 10*, 4981–5003.

Foust, K. D., Nurre, E., Montgomery, C. L., Hernandez, A., Chan, C. M., & Kaspar, B. K. (2009). Intravascular AAV9 preferentially targets neonatal neurons and adult astrocytes. *Nature Biotechnology, 27*, 59–65.

Fu, H., & Mccarty, D. M. (2016). Crossing the blood-brain-barrier with viral vectors. *Current Opinion in Virology, 21*, 87–92.

Gao, G., Alvira, M. R., Somanathan, S., Lu, Y., Vandenberghe, L. H., Rux, J. J., Calcedo, R., Sanmiguel, J., Abbas, Z., & Wilson, J. M. (2003). Adeno-associated viruses undergo substantial evolution in primates during natural infections. *Proceedings of the National Academy of Sciences of the United States of America, 100*, 6081–6086.

Garg, P., Pandey, S., Seonwoo, H., Yeom, S., Choung, Y. H., Cho, C. S., Choung, P. H., & Hoon Chung, J. (2015). Hyperosmotic polydixylitol for crossing the blood brain barrier and efficient nucleic acid delivery. *Chemical Communications (Cambridges), 51*, 3645–3648.

Grimm, D., Lee, J. S., Wang, L., Desai, T., Akache, B., Storm, T. A., & Kay, M. A. (2008). In vitro and in vivo gene therapy vector evolution via multispecies interbreeding and retargeting of adeno-associated viruses. *Journal of Virology, 82*, 5887–5911.

Gyorgy, B., Fitzpatrick, Z., Crommentuijn, M. H., Mu, D., & Maguire, C. A. (2014). Naturally enveloped AAV vectors for shielding neutralizing antibodies and robust gene delivery in vivo. *Biomaterials, 35*, 7598–7609.

Hanlon, K. S., Meltzer, J. C., Buzhdygan, T., Cheng, M. J., Sena-Esteves, M., Bennett, R. E., Sullivan, T. P., Razmpour, R., Gong, Y., Ng, C., Nammour, J., Maiz, D., Dujardin, S., Ramirez, S. H., Hudry, E., & Maguire, C. A. (2019). Selection of an efficient AAV vector for robust CNS transgene expression. *Molecular Therapy – Methods & Clinical Development, 15*, 320–332.

Hermonat, P. L., & Muzyczka, N. (1984). Use of adeno-associated virus as a mammalian DNA cloning vector: Transduction of neomycin resistance into mammalian tissue culture cells. *Proceedings of the National Academy of Sciences of the United States of America, 81*, 6466–6470.

Herve, F., Ghinea, N., & Scherrmann, J. M. (2008). CNS delivery via adsorptive transcytosis. *The AAPS Journal, 10*, 455–472.

Hordeaux, J., Wang, Q., Katz, N., Buza, E. L., Bell, P., & Wilson, J. M. (2018). The neurotropic properties of AAV-PHP.B are limited to C57BL/6J mice. *Molecular Therapy, 26*, 664–668.

Hordeaux, J., Yuan, Y., Clark, P. M., Wang, Q., Martino, R. A., Sims, J. J., Bell, P., Raymond, A., Stanford, W. L., & Wilson, J. M. (2019). The GPI-linked protein LY6A drives AAV-PHP.B transport across the blood-brain barrier. *Molecular Therapy, 27*, 912–921.

Hsu, P. H., Wei, K. C., Huang, C. Y., Wen, C. J., Yen, T. C., Liu, C. L., Lin, Y. T., Chen, J. C., Shen, C. R., & Liu, H. L. (2013). Noninvasive and targeted gene delivery into the brain using microbubble-facilitated focused ultrasound. *PLoS One, 8*, e57682.

Huang, Q., Chan, K. Y., Tobey, I. G., Chan, Y. A., Poterba, T., Boutros, C. L., Balazs, A. B., Daneman, R., Bloom, J. M., Seed, C., & Deverman, B. E. (2019). Delivering genes across the blood-brain barrier: LY6A, a novel cellular receptor for AAV-PHP.B capsids. *PLoS One, 14*, e0225206.

Hudry, E., & Vandenberghe, L. H. (2019). Therapeutic AAV gene transfer to the nervous system: A clinical reality. *Neuron, 101*, 839–862.

Hudry, E., Martin, C., Gandhi, S., Gyorgy, B., Scheffer, D. I., Mu, D., Merkel, S. F., Mingozzi, F., Fitzpatrick, Z., Dimant, H., Masek, M., Ragan, T., Tan, S., Brisson, A. R., Ramirez, S. H., Hyman, B. T., & Maguire, C. A. (2016). Exosome-associated AAV vector as a robust and convenient neuroscience tool. *Gene Therapy, 23*, 380–392.

Hynynen, K., Mcdannold, N., Martin, H., Jolesz, F. A., & Vykhodtseva, N. (2003). The threshold for brain damage in rabbits induced by bursts of ultrasound in the presence of an ultrasound contrast agent (Optison). *Ultrasound in Medicine & Biology, 29*, 473–481.

Hynynen, K., Mcdannold, N., Sheikov, N. A., Jolesz, F. A., & Vykhodtseva, N. (2005). Local and reversible blood-brain barrier disruption by noninvasive focused ultrasound at frequencies suitable for trans-skull sonications. *NeuroImage, 24*, 12–20.

Ikeda, M., Bhattacharjee, A. K., Kondoh, T., Nagashima, T., & Tamaki, N. (2002). Synergistic effect of cold mannitol and Na(+)/ca(2+) exchange blocker on blood-brain barrier opening. *Biochemical and Biophysical Research Communications, 291*, 669–674.

Jacobson, K. A., & Gao, Z. G. (2006). Adenosine receptors as therapeutic targets. *Nature Reviews Drug Discovery, 5*, 247–264.

Kamaly, N., Yameen, B., Wu, J., & Farokhzad, O. C. (2016). Degradable controlled-release polymers and polymeric nanoparticles: Mechanisms of controlling drug release. *Chemical Reviews, 116*, 2602–2663.

Kimura, S., & Harashima, H. (2020). Current status and challenges associated with CNS-targeted gene delivery across the BBB. *Pharmaceutics, 12*, 1216.

Kirk, J., Plumb, J., Mirakhur, M., & Mcquaid, S. (2003). Tight junctional abnormality in multiple sclerosis white matter affects all calibres of vessel and is associated with blood-brain barrier leakage and active demyelination. *The Journal of Pathology, 201*, 319–327.

Kofoed, R. H., Dibia, C. L., Noseworthy, K., Xhima, K., Vacaresse, N., Hynynen, K., & Aubert, I. (2022). Efficacy of gene delivery to the brain using AAV and ultrasound depends on serotypes and brain areas. *Journal of Controlled Release, 351*, 667–680.

Kugler, S., Kilic, E., & Bahr, M. (2003). Human synapsin 1 gene promoter confers highly neuron-specific long-term transgene expression from an adenoviral vector in the adult rat brain depending on the transduced area. *Gene Therapy, 10*, 337–347.

Larochelle, N., Lochmuller, H., Zhao, J., Jani, A., Hallauer, P., Hastings, K. E., Massie, B., Prescott, S., Petrof, B. J., Karpati, G., & Nalbantoglu, J. (1997). Efficient muscle-specific transgene expression after adenovirus-mediated gene transfer in mice using a 1.35 kb muscle creatine kinase promoter/enhancer. *Gene Therapy, 4*, 465–472.

Lee, Y., Messing, A., Su, M., & Brenner, M. (2008). GFAP promoter elements required for region-specific and astrocyte-specific expression. *Glia, 56*, 481–493.

Lombardo, S. M., Schneider, M., Tureli, A. E., & Gunday Tureli, N. (2020). Key for crossing the BBB with nanoparticles: The rational design. *Beilstein Journal of Nanotechnology, 11*, 866–883.

Louboutin, J. P., Chekmasova, A. A., Marusich, E., Chowdhury, J. R., & Strayer, D. S. (2010). Efficient CNS gene delivery by intravenous injection. *Nature Methods, 7*, 905–907.

Maguire, C. A., Balaj, L., Sivaraman, S., Crommentuijn, M. H., Ericsson, M., Mincheva-Nilsson, L., Baranov, V., Gianni, D., Tannous, B. A., Sena-Esteves, M., Breakefield, X. O., & Skog, J. (2012). Microvesicle-associated AAV vector as a novel gene delivery system. *Molecular Therapy, 20*, 960–971.

Manno, C. S., Pierce, G. F., Arruda, V. R., Glader, B., Ragni, M., Rasko, J. J., Ozelo, M. C., Hoots, K., Blatt, P., Konkle, B., Dake, M., Kaye, R., Razavi, M., Zajko, A., Zehnder, J., Rustagi, P. K., Nakai, H., Chew, A., Leonard, D., Wright, J. F., Lessard, R. R., Sommer, J. M., Tigges, M., Sabatino, D., Luk, A., Jiang, H., Mingozzi, F., Couto, L., Ertl, H. C., High, K. A., & Kay, M. A. (2006). Successful transduction of liver in hemophilia by AAV-factor IX and limitations imposed by the host immune response. *Nature Medicine, 12*, 342–347.

Martino, A. T., Suzuki, M., Markusic, D. M., Zolotukhin, I., Ryals, R. C., Moghimi, B., Ertl, H. C., Muruve, D. A., Lee, B., & Herzog, R. W. (2011). The genome of self-complementary adeno-associated viral vectors increases toll-like receptor 9-dependent innate immune responses in the liver. *Blood, 117*, 6459–6468.

Mathiesen, S. N., Lock, J. L., Schoderboeck, L., Abraham, W. C., & Hughes, S. M. (2020). Cns transduction benefits of AAV-PHP.eB over AAV9 are dependent on administration route and mouse strain. *Molecular Therapy – Methods & Clinical Development, 19*, 447–458.

Matsuzaki, Y., Konno, A., Mochizuki, R., Shinohara, Y., Nitta, K., Okada, Y., & Hirai, H. (2018). Intravenous administration of the adeno-associated virus-PHP.B capsid fails to upregulate transduction efficiency in the marmoset brain. *Neuroscience Letters, 665*, 182–188.

Mccarty, D. M. (2008). Self-complementary AAV vectors; advances and applications. *Molecular Therapy, 16*, 1648–1656.

Mccarty, D. M., Monahan, P. E., & Samulski, R. J. (2001). Self-complementary recombinant adeno-associated virus (scAAV) vectors promote efficient transduction independently of DNA synthesis. *Gene Therapy, 8*, 1248–1254.

Mccarty, D. M., Fu, H., Monahan, P. E., Toulson, C. E., Naik, P., & Samulski, R. J. (2003). Adeno-associated virus terminal repeat (TR) mutant generates self-complementary vectors to overcome the rate-limiting step to transduction in vivo. *Gene Therapy, 10*, 2112–2118.

Mccarty, D. M., Dirosario, J., Gulaid, K., Muenzer, J., & Fu, H. (2009). Mannitol-facilitated CNS entry of RAAV2 vector significantly delayed the neurological disease progression in MPS IIIB mice. *Gene Therapy, 16*, 1340–1352.

Mendell, J. R., Al-Zaidy, S., Shell, R., Arnold, W. D., Rodino-Klapac, L. R., Prior, T. W., Lowes, L., Alfano, L., Berry, K., Church, K., Kissel, J. T., Nagendran, S., L'italien, J., Sproule, D. M., Wells, C., Cardenas, J. A., Heitzer, M. D., Kaspar, A., Corcoran, S., Braun, L., Likhite, S., Miranda, C., Meyer, K., Foust, K. D., Burghes, A. H. M., & Kaspar, B. K. (2017). Single-dose gene-replacement therapy for spinal muscular atrophy. *The New England Journal of Medicine, 377*, 1713–1722.

Meng, Y., Sun, D., Qin, Y., Dong, X., Luo, G., & Liu, Y. (2021). Cell-penetrating peptides enhance the transduction of adeno-associated virus serotype 9 in the central nervous system. *Molecular Therapy – Methods & Clinical Development, 21*, 28–41.

Merkel, S. F., Andrews, A. M., Lutton, E. M., Mu, D., Hudry, E., Hyman, B. T., Maguire, C. A., & Ramirez, S. H. (2017). Trafficking of adeno-associated virus vectors across a model of the blood-brain barrier; a comparative study of transcytosis and transduction using primary human brain endothelial cells. *Journal of Neurochemistry, 140*, 216–230.

Miyanohara, A., Kamizato, K., Juhas, S., Juhasova, J., Navarro, M., Marsala, S., Lukacova, N., Hruska-Plochan, M., Curtis, E., Gabel, B., Ciacci, J., Ahrens, E. T., Kaspar, B. K., Cleveland, D., & Marsala, M. (2016). Potent spinal parenchymal AAV9-mediated gene delivery by subpial injection in adult rats and pigs. *Molecular Therapy – Methods & Clinical Development, 3*, 16046.

Muller, O. J., Leuchs, B., Pleger, S. T., Grimm, D., Franz, W. M., Katus, H. A., & Kleinschmidt, J. A. (2006). Improved cardiac gene transfer by transcriptional and transductional targeting of adeno-associated viral vectors. *Cardiovascular Research, 70*, 70–78.

Nanbo, A., Kawanishi, E., Yoshida, R., & Yoshiyama, H. (2013). Exosomes derived from Epstein-Barr virus-infected cells are internalized via caveola-dependent endocytosis and promote phenotypic modulation in target cells. *Journal of Virology, 87*, 10334–10347.

Naumer, M., Sonntag, F., Schmidt, K., Nieto, K., Panke, C., Davey, N. E., Popa-Wagner, R., & Kleinschmidt, J. A. (2012). Properties of the adeno-associated virus assembly-activating protein. *Journal of Virology, 86*, 13038–13048.

Nonnenmacher, M., Wang, W., Child, M. A., Ren, X. Q., Huang, C., Ren, A. Z., Tocci, J., Chen, Q., Bittner, K., Tyson, K., Pande, N., Chung, C. H., Paul, S. M., & Hou, J. (2021). Rapid evolution of blood-brain-barrier-penetrating AAV capsids by RNA-driven biopanning. *Molecular Therapy – Methods & Clinical Development, 20*, 366–378.

Ogden, P. J., Kelsic, E. D., Sinai, S., & Church, G. M. (2019). Comprehensive AAV capsid fitness landscape reveals a viral gene and enables machine-guided design. *Science, 366*, 1139–1143.

Pardridge, W. M. (2005). The blood-brain barrier: Bottleneck in brain drug development. *NeuroRx, 2*, 3–14.

Parra, G., Bradnam, K., Rose, A. B., & Korf, I. (2011). Comparative and functional analysis of intron-mediated enhancement signals reveals conserved features among plants. *Nucleic Acids Research, 39*, 5328–5337.

Patil, S. A., & Maegawa, G. H. (2013). Developing therapeutic approaches for metachromatic leukodystrophy. *Drug Design, Development and Therapy, 7*, 729–745.

Philippidis, A. (2022). Novartis confirms deaths of two patients treated with gene therapy Zolgensma. *Human Gene Therapy, 33*, 842–844.

Pillay, S., Meyer, N. L., Puschnik, A. S., Davulcu, O., Diep, J., Ishikawa, Y., Jae, L. T., Wosen, J. E., Nagamine, C. M., Chapman, M. S., & Carette, J. E. (2016). An essential receptor for adeno-associated virus infection. *Nature, 530*, 108–112.

Powell, S. K., Rivera-Soto, R., & Gray, S. J. (2015). Viral expression cassette elements to enhance transgene target specificity and expression in gene therapy. *Discovery Medicine, 19*, 49–57.

Rapoport, S. I. (2001). Advances in osmotic opening of the blood-brain barrier to enhance CNS chemotherapy. *Expert Opinion on Investigational Drugs, 10*, 1809–1818.

Rincon, M. Y., De Vin, F., Duque, S. I., Fripont, S., Castaldo, S. A., Bouhuijzen-Wenger, J., & Holt, M. G. (2018). Widespread transduction of astrocytes and neurons in the mouse central nervous

system after systemic delivery of a self-complementary AAV-PHP.B vector. *Gene Therapy, 25*, 83–92.
Saraiva, J., Nobre, R. J., & Pereira De Almeida, L. (2016). Gene therapy for the CNS using AAVs: The impact of systemic delivery by AAV9. *Journal of Controlled Release, 241*, 94–109.
Schuster, D. J., Dykstra, J. A., Riedl, M. S., Kitto, K. F., Belur, L. R., Mcivor, R. S., Elde, R. P., Fairbanks, C. A., & Vulchanova, L. (2014). Biodistribution of adeno-associated virus serotype 9 (AAV9) vector after intrathecal and intravenous delivery in mouse. *Frontiers in Neuroanatomy, 8*, 42.
Song, R., Pekrun, K., Khan, T. A., Zhang, F., Pasca, S. P., & Kay, M. A. (2022). Selection of rAAV vectors that cross the human blood-brain barrier and target the central nervous system using a transwell model. *Molecular Therapy – Methods & Clinical Development, 27*, 73–88.
Srivastava, A. (2016). In vivo tissue-tropism of adeno-associated viral vectors. *Current Opinion in Virology, 21*, 75–80.
Stavarache, M. A., Petersen, N., Jurgens, E. M., Milstein, E. R., Rosenfeld, Z. B., Ballon, D. J., & Kaplitt, M. G. (2018). Safe and stable noninvasive focal gene delivery to the mammalian brain following focused ultrasound. *Journal of Neurosurgery, 130*, 989–998.
Swain, G. P., Prociuk, M., Bagel, J. H., O'donnell, P., Berger, K., Drobatz, K., Gurda, B. L., Haskins, M. E., Sands, M. S., & Vite, C. H. (2014). Adeno-associated virus serotypes 9 and rh10 mediate strong neuronal transduction of the dog brain. *Gene Therapy, 21*, 28–36.
Sweeney, M. D., Sagare, A. P., & Zlokovic, B. V. (2018). Blood-brain barrier breakdown in Alzheimer disease and other neurodegenerative disorders. *Nature Reviews Neurology, 14*, 133–150.
Sweeney, M. D., Zhao, Z., Montagne, A., Nelson, A. R., & Zlokovic, B. V. (2019). Blood-brain barrier: From physiology to disease and back. *Physiological Reviews, 99*, 21–78.
Tanguy, Y., Biferi, M. G., Besse, A., Astord, S., Cohen-Tannoudji, M., Marais, T., & Barkats, M. (2015). Systemic AAVrh10 provides higher transgene expression than AAV9 in the brain and the spinal cord of neonatal mice. *Frontiers in Molecular Neuroscience, 8*, 36.
Terstappen, G. C., Meyer, A. H., Bell, R. D., & Zhang, W. (2021). Strategies for delivering therapeutics across the blood-brain barrier. *Nature Reviews Drug Discovery, 20*, 362–383.
Tian, T., Zhu, Y. L., Zhou, Y. Y., Liang, G. F., Wang, Y. Y., Hu, F. H., & Xiao, Z. D. (2014). Exosome uptake through clathrin-mediated endocytosis and macropinocytosis and mediating miR-21 delivery. *The Journal of Biological Chemistry, 289*, 22258–22267.
Tratschin, J. D., West, M. H., Sandbank, T., & Carter, B. J. (1984). A human parvovirus, adeno-associated virus, as a eucaryotic vector: Transient expression and encapsidation of the procaryotic gene for chloramphenicol acetyltransferase. *Molecular and Cellular Biology, 4*, 2072–2081.
Tung, Y. S., Marquet, F., Teichert, T., Ferrera, V., & Konofagou, E. E. (2011). Feasibility of noninvasive cavitation-guided blood-brain barrier opening using focused ultrasound and microbubbles in nonhuman primates. *Applied Physics Letters, 98*, 163704.
Ussher, J. E., & Taylor, J. A. (2010). Optimized transduction of human monocyte-derived dendritic cells by recombinant adeno-associated virus serotype 6. *Human Gene Therapy, 21*, 1675–1686.
Von Jonquieres, G., Frohlich, D., Klugmann, C. B., Wen, X., Harasta, A. E., Ramkumar, R., Spencer, Z. H., Housley, G. D., & Klugmann, M. (2016). Recombinant human myelin-associated glycoprotein promoter drives selective AAV-mediated transgene expression in oligodendrocytes. *Frontiers in Molecular Neuroscience, 9*, 13.
Weitzman, M. D., Kyostio, S. R., Kotin, R. M., & Owens, R. A. (1994). Adeno-associated virus (AAV) rep proteins mediate complex formation between AAV DNA and its integration site in human DNA. *Proceedings of the National Academy of Sciences of the United States of America, 91*, 5808–5812.
Wu, Z., Asokan, A., & Samulski, R. J. (2006). Adeno-associated virus serotypes: Vector toolkit for human gene therapy. *Molecular Therapy, 14*, 316–327.
Wu, Z., Sun, J., Zhang, T., Yin, C., Yin, F., Van Dyke, T., Samulski, R. J., & Monahan, P. E. (2008). Optimization of self-complementary AAV vectors for liver-directed expression results in sustained correction of hemophilia B at low vector dose. *Molecular Therapy, 16*, 280–289.

Xie, F., Boska, M. D., Lof, J., Uberti, M. G., Tsutsui, J. M., & Porter, T. R. (2008). Effects of transcranial ultrasound and intravenous microbubbles on blood brain barrier permeability in a large animal model. *Ultrasound in Medicine & Biology, 34*, 2028–2034.

Yamazaki, Y., & Kanekiyo, T. (2017). Blood-brain barrier dysfunction and the pathogenesis of Alzheimer's disease. *International Journal of Molecular Sciences, 18*, 1965.

Yan, Z., Yan, H., & Ou, H. (2012). Human thyroxine binding globulin (TBG) promoter directs efficient and sustaining transgene expression in liver-specific pattern. *Gene, 506*, 289–294.

Yang, B., Li, S., Wang, H., Guo, Y., Gessler, D. J., Cao, C., Su, Q., Kramer, J., Zhong, L., Ahmed, S. S., Zhang, H., He, R., Desrosiers, R. C., Brown, R., Xu, Z., & Gao, G. (2014). Global CNS transduction of adult mice by intravenously delivered rAAVrh.8 and rAAVrh.10 and nonhuman primates by rAAVrh.10. *Molecular Therapy, 22*, 1299–1309.

Yao, Y., Wang, J., Liu, Y., Qu, Y., Wang, K., Zhang, Y., Chang, Y., Yang, Z., Wan, J., Liu, J., Nakashima, H., Lawler, S. E., Chiocca, E. A., Cho, C. F., & Bei, F. (2022). Variants of the adeno-associated virus serotype 9 with enhanced penetration of the blood-brain barrier in rodents and primates. *Nature Biomedical Engineering, 6*, 1257–1271.

Zhang, H., Yang, B., Mu, X., Ahmed, S. S., Su, Q., He, R., Wang, H., Mueller, C., Sena-Esteves, M., Brown, R., Xu, Z., & Gao, G. (2011). Several rAAV vectors efficiently cross the blood-brain barrier and transduce neurons and astrocytes in the neonatal mouse central nervous system. *Molecular Therapy, 19*, 1440–1448.

Zhang, L., Han, L., Qin, J., Lu, W., & Wang, J. (2013). The use of borneol as an enhancer for targeting aprotinin-conjugated PEG-PLGA nanoparticles to the brain. *Pharmaceutical Research, 30*, 2560–2572.

Zhang, X., He, T., Chai, Z., Samulski, R. J., & Li, C. (2018). Blood-brain barrier shuttle peptides enhance AAV transduction in the brain after systemic administration. *Biomaterials, 176*, 71–83.

Zhang, X., Chai, Z., Lee Dobbins, A., Itano, M. S., Askew, C., Miao, Z., Niu, H., Samulski, R. J., & Li, C. (2022). Customized blood-brain barrier shuttle peptide to increase AAV9 vector crossing the BBB and augment transduction in the brain. *Biomaterials, 281*, 121340.

Virus-Based Neural Circuit Tracing

Fuqiang Xu and Qing Liu

Abstract Neural circuits provide an anatomical basis for functional networks. Therefore, dissecting the structure of neural circuits is an indispensable prerequisite to understanding how the brain functions. Knowing how the neural circuits organize and function under physiological conditions and their progressive alterations under pathophysiological conditions are key to understanding the underlying circuit mechanism of diseases, thus finding cures for the diseases. Recombinant neurotropic viruses are important tools for neural circuit tracing with many advantages over non-viral tracers: they allow for anterograde, retrograde, and trans-synaptic delivery of tracers in a cell-type-specific, circuit-selective manner. We herein summarize the recent developments in the viral tools for neural circuit tracing, discuss the key principles of using viral tools in neuroscience research, and highlight innovations for developing and optimizing viral tools for neural circuit tracing across diverse animal species, including nonhuman primates.

Graphical Abstract

F. Xu (✉) · Q. Liu
The Brain Cognition and Brain Disease Institute (BCBDI), Shenzhen Institute of Advanced Technology, Chinese Academy of Sciences, Shenzhen-Hong Kong Institute of Brain Science-Shenzhen Fundamental Research Institutions, Shenzhen, Guangdong, China
e-mail: fq.xu@siat.ac.cn; qing.liu1@siat.ac.cn

A. C.-H. Yu et al. (eds.), *Systems Neuroscience*, Advances in Neurobiology 41,
https://doi.org/10.1007/978-3-031-69188-1_5

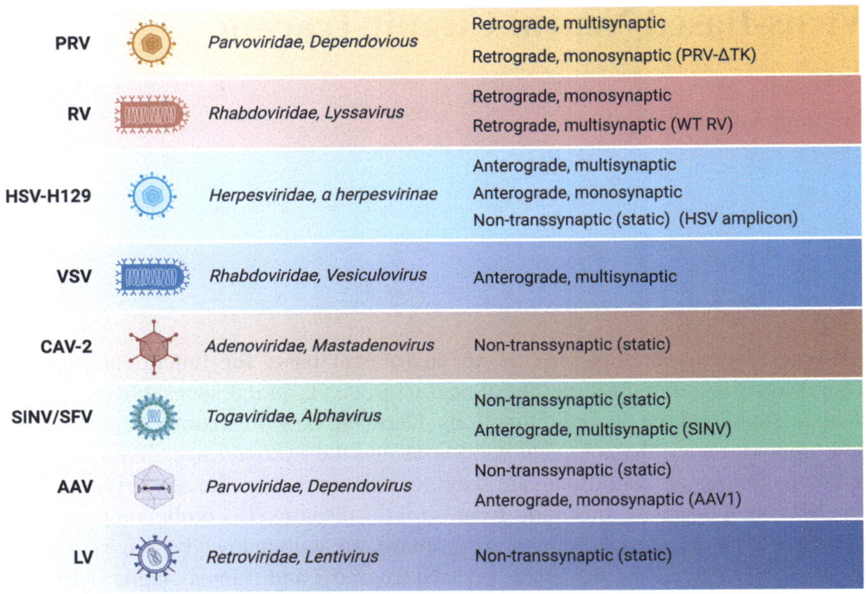

The widely used viral vectors for neural circuit tracing

Keywords Neural circuit · Neurotropic viruses · Tracer · Trans-synaptic · Non-trans-synaptic

The human brain is the most mysterious and complex structure in the universe. It consists of nearly 10^{11} neurons connected by about 10^{14} synapses. This complicated network lays a solid structural foundation for higher brain functions such as emotion, cognition, learning, and memory. Dissecting how neurons connect, mapping the fine structure of neural circuits, and recognizing their function are fundamental to our understanding of how the brain works. Over the past decades, many powerful labeling, imaging, and measuring techniques have been developed to decipher neural circuits. Classical tracers including cholera toxin B subunit (CTB), horseradish peroxidase (HRP), and biotinylated dextran amine (BDA) were widely used to determine neural connections between different brain regions in experimental animals. However, these tracers have the following disadvantages: lack of cell-type-specificity, variability of trans-synaptic efficiency, and the inability of trans-multisynaptic spreading. It is difficult to study the neural circuits formed by the synaptic connections of multiple brain regions and different types of neurons by using these tracers. Scientists have been working hard to develop new tracers with cell-type-specific, spread direction–/trans-synaptic-controllable, and visualization-feasible properties. Recently, virus-based tracers have provided new solutions to these problems and achieved a precise understanding of neural circuits, not only visualizing the structure by encoding fluorescent proteins but also monitoring and manipulating the function by carrying optogenetic (ChR2 or eNpHR), chemogenetic (hM3Dq or hM4Di), neurotransmitter probes (iGluSnFR or dLight), Ca^{2+} sensors (GCaMP-X), recombination-based gene expression systems (Cre or Flp), or genome editing systems (CRISPR-Cas9), etc.

Neurotropic virus-based tracers have mostly been modified from the natural strains. They can infect nerve cells and spread across synapses along neural circuits, which makes them ideal tools for mapping neural circuits. Neural circuit tracing using viral tools has become an increasingly popular method in neuroscience research. These tools allow researchers to label and manipulate specific populations of neurons or defined projections within a circuit and investigate neural connectivity and function. The commonly used viral tools are pseudorabies virus (PRV), rabies virus (RV), herpes simplex virus 1 (HSV-1), vesicular stomatitis virus (VSV), canine adenoviruses (CAV), Sindbis virus (SINV), and Semliki Forest virus (SFV) (Nassi et al., 2015; Xu et al., 2020; Lin et al., 2020; Soudais et al., 2001; Shi et al., 2022; Jia et al., 2019a). These viruses have high infectivity and specificity in tracing neural circuits, but with strong immunogenicity and high toxicity to nerve cells, which lead to neuronal pathological changes and even the death of infected experimental animals. Moreover, they are potentially dangerous to the environment and laboratory personnel and thus require operation in Biosafety Level 2 facilities. It is necessary to be cautious when using neurotropic viruses in research and develop viral tools with much lower immunogenicity. In addition, some recombinant viral vectors can serve as helpers for the expression of exogenous genes, such as adeno-associated virus (AAV) and lentivirus (LV) (Wang et al., 2019; Parr-Brownlie et al., 2015). According to whether spread trans-synaptically and the transmission direction (anterograde: from the cell body through the axon to the terminals, retrograde: from the axon terminals back to the cell body or from the cell body to dendrites), viral tools are classified into non-trans-synaptic (static), anterograde trans-multisynaptic, retrograde trans-multisynaptic, anterograde trans-monosynaptic, and retrograde trans-monosynaptic ones (Liu et al., 2022). The widely used viral tools will be introduced in the following paragraphs.

1 PRV

PRV is a powerful viral tool in neuroscience research. It is a member of the alpha-herpesvirus subfamily, 200–250 nm in diameter, with a complicated genome of ~150 kb. It is a spherical double-stranded DNA virus with an envelope, and large exogenous genes can be inserted after molecular modification. PRV strains used include Becker and Bartha strains. The Becker strain is rarely used in neural circuit research due to its significant toxicity and non-direction-specific transmission. The Bartha strain is less toxic and transfers in a retrograde direction through neural circuits. It has been widely used to trans-multisynaptically trace neural circuits in the central nervous system and the connections from the peripheral organs to the central nervous system (Ekstrand et al., 2008), which has attracted the attention of many researchers in this field, and has been used for identifying a novel corticopontine circuit for initiation of urination (Yao et al., 2018), tracing the brain origin of splenic nerve-dependent immunostimulatory neural activity (Zhang et al., 2020), defining premotor brainstem neurons that may control gut-sympathetic activity (Muller et al., 2020), etc. Previously, it was reported that PRV mainly infects pigs and rodents, but is not capable of causing human infections. Recently, however, it has been reported that wild-type PRV infects the human central nervous system and induces encephalitis (Liu et al., 2021), so personal protection should be emphasized for experimental use.

Although PRV is already the most commonly used trans-synaptic tracer at present, the weak fluorescent signal after PRV infection needs to be amplified by immunohistochemistry, which is a time-consuming and tedious process. To address this issue, PRV531 and PRV724 with robust EGFP and mRuby3 expression have been developed, which can be used as efficient retrograde trans-multisynaptic tracers (Jia et al., 2019b). Because these PRVs infect many different types of neurons, the lack of cell-type specificity of originally infected neurons limits the precise elucidation of neural circuits. Therefore, PRVs that can be retrogradely transmitted from selected neurons were developed. For example, Ba2001 is a Cre-dependent retrograde trans-multisynaptic tracer encoding GFP and the thymidine kinase (TK) essential for viral replication. It only replicates in neurons expressing the Cre recombinase and in neurons synaptically connected with the originally infected ones; the inputs of neuropeptide Y-expressing neurons in the arcuate nucleus have been traced with this method (DeFalco et al., 2001). However, high-titer PRV-Ba2001 may lead to infection of neurons in the absence of Cre. Then, PRV-Introvert, a nonleaky PRV by inserting a synthetic intron into the PRV *tk* gene and then inverting one of the exons with *FLEx,* was developed, in combination with translating ribosome affinity purification (TRAP), has been used to anatomically map and molecularly profile multisynaptic inputs to mesolimbic dopaminergic neurons (Pomeranz et al., 2017). PRVB180 expressing TK fused to GFP in a Cre recombinase-dependent manner combined with single-cell transcriptomics has been used to uncover the molecular identities of upstream neurons of the paraventricular nucleus of the hypothalamus (PVN) in corticotropin-releasing hormone (CRH) -Cre mice (Hanchate et al., 2020). Moreover, Ba2017, a novel recombinant PRV derived from attenuated Bartha strain with improved genetic stability and fluorescent protein expression, was generated (Enquist & Card, 2003). It has been successfully applied in identifying and characterizing a subpopulation of oxytocin neurons in the PVN that is multisynaptically connected to pancreatic islet β cells (Papazoglou et al., 2022). An *IE180*-null recombinant PRV has been engineered to express site-specific Cre recombinase (PRV-hSyn-Cre). It can infect neurons efficiently and robustly in vivo and activate transgene expression from Cre-dependent vectors in local and retrogradely projecting neurons, which can be used for targeted gene expression in the brain, facilitating functional dissection of neural circuits in vivo (Oyibo et al., 2014).

2 RV

RV is another highly effective tracer for labeling retrograde trans-multisynaptic circuits. It is an enveloped negative-sense single-stranded RNA virus. It is ~75 nm in diameter and 180 nm in length. It is bullet-shaped with a genome of ~12 kb. Wild-type RV (WT RV) is by far the most common causative agent of rabies and is mostly transmitted by the bite of an infected dog. It infects peripheral nerve terminals at the site of the bite and proceeds through the central nervous system. It is highly neurotropic that can infect neurons and spread through synaptically connected circuits. RV has been modified to be safer by deletion of the envelope glycoprotein gene (RV-ΔG) and then used as a tracer for retrograde infection of projection neurons through their axon terminals, targeted infection of genetically specified neurons, or trans-monosynaptic tracing of direct input neurons after further modifications. The recombinant RV-ΔG can express exogenous genes with a high abundance in a short time, which can be used to depict the fine structure of neurons projecting to the injection site and monitor Ca^{2+} activity in specific input circuits (Wickersham et al., 2007a; Tang et al., 2020). Moreover, the RV-ΔG-based retrograde trans-monosynaptic tracing system is a unique and powerful tool to identify the presynaptic inputs to genetically defined neurons across the whole brain. The RV-ΔG pseudotyped with EnvA, the avian sarcoma, and leukosis virus A protein, cannot infect mammalian neurons. It can only infect cells that are specifically engineered to express the avian cell-surface molecule tumor virus A (TVA), the receptor of EnvA. This receptor-ligand pair (TVA-EnvA) allows specific infection of target neurons. The RV-ΔG is replication-deficient and so cannot spread beyond the initially infected neurons unless the glycoprotein (G) is supplied. Then the virus can produce infectious progeny, which spread from these neurons to the presynaptic neurons directly connected to them. The retrograde trans-synaptic system based on the modified SAD-B19 RV strain has been widely used to map the input network of specific types of neurons (Wickersham et al., 2007b), not only in normal mice but also in disease model mice. For example, using the rabies viral monosynaptic tracing method, it has been found that excitatory neural connections from lateral hypothalamus to hypothalamic oxytocin neurons are drastically strengthened when male mice become fathers (Inada et al., 2022). Both local and global circuit connectivity to hippocampal CA1 excitatory neurons are altered in AD model mice compared with control mice (Ye et al., 2022). Since input network analysis at single-neuron resolution is a powerful method for understanding fundamental principles of neural circuits, an efficient RV-ΔG tracing method from projection-defined single neurons using TVA950, a transmembrane isoform of TVA receptors, was established, which has a much-improved success rate than RV-ΔG tracing combined with single-cell electroporation. Presynaptic neurons were trans-synaptically labeled from 80% of the TVA950-expressing single starter cells that survived after infection with EnvA-pseudotyped RV-ΔG in the adult mouse brain (Masaki et al., 2022). However, this tracing system can label only a fraction of the actual input neurons due to the low trans-synaptic efficiency, and the high toxicity of the tracing system makes it difficult to carry functional elements for monitoring and modulating neuronal activity. Therefore, there is an urgent need to improve its trans-synaptic efficiency and reduce its toxicity. Much effort has been made by researchers, but the overall effect has not been great.

A codon-optimized chimeric glycoprotein (oG) can greatly improve the trans-synaptic efficiency of SAD-B19-ΔG, which has been applied in input network mapping (Kim et al., 2016). However, the self-inactivating RV (SiR) based on SAD-B19-ΔG with high expectations has not been widely used (Ciabatti et al., 2017). Due to the low expression efficiency of the nontoxic G/L gene deletion RV (RV-ΔGL), it can only be applied to monitor and manipulate neural activity by using Cre or Flp recombinase in combination with functional probes (Chatterjee et al., 2018). It is necessary to use appropriate viral vectors or produce transgenic animals carrying L ("large protein" gene) to achieve trans-synaptic spread, which has not been realized yet. Moreover, the trans-synaptic efficiency is significantly enhanced by a two-plasmid co-packaging system for helper AAV, but the underlying mechanism is still not clear (Sun et al., 2020). Therefore, there is still a requirement to develop retrograde trans-synaptic tracing systems with high efficiency and low toxicity. Then the oG sequence was reshuffled with a positive codon pair bias score to further optimize the glycoprotein and generate the reshuffled oG (ooG), with which the tracing efficiency increased by up to 10 times (Jia et al., 2021). AAV-PHP.eB-assisted retrograde tracing systems (PARTS) based on RV-ΔG were developed. Through the transduction of the TVA receptor using AAV-PHP.eB to the whole brain in C57BL/6 J mice, EnvA-pseudotyped SAD-B19-ΔG can label brain-wide upstream neurons of a defined brain area. Then, cell-type-specific PARTS (cPARTS) was constructed and combined with Cre transgenic mice, it can retrogradely label genetically defined projection neurons across the whole brain (Han et al., 2022a).

Compared with the SAD-B19 strain, the retrograde trans-synaptic ability of the modified CVS-N2c strain is significantly enhanced and the toxicity to host neurons is significantly reduced, which allows functional effectors such as Ca^{2+} sensors and optogenetic actuators to be carried in functional circuit dissection (Reardon et al., 2016; Zhu et al., 2020). However, the difficulties in preparation and low titers limit its application. Then, an improved production system for the rapid rescue and preparation of high-titer CVS-N2c-ΔG virus was developed, the trans-monosynaptic efficiency of oG-mediated CVS-N2c-ΔG was 2–3 times higher than that of oG-mediated SAD-B19-ΔG, the time window for neural activity monitoring was 3 weeks and can express sufficient recombinases for efficient transgene recombination (Lin et al., 2023).

3 HSV 1 Strain H129

HSV is a large DNA virus that can infect neurons and establish lifelong latency. HSV can be used for both anterograde and retrograde tracing of neural circuits, as well as for the transgene expression in specific neuronal populations. HSV1 strain

H129, originally isolated from the brain of a patient suffering from acute HSV encephalitis, is widely used as an anterograde trans-synaptic neural circuit tracer (Zemanick et al., 1991; Dum et al., 2009). However, there are some shortcomings of H129, including low brightness and nonspecific retrograde labeling due to axon terminal absorption. H129 has been reported to be capable of infecting innervating neurons through retrograde uptake and showing temporally delayed retrograde trans-neuronal spread. H306, a *TK* and *ICP34.5* dual-deleted H129 recombinant, is replication-deficient in non-dividing postmitotic neurons. It can infect upstream neurons efficiently through axon terminal uptake (Su et al., 2019). Therefore, the neural networks traced by replication-competent, trans-multisynaptic H129 might be from two types of starter cells. One is neurons local to the injection site (true starter site), and the other is the retrogradely infected neurons (ectopic starter sites), in which H129 can replicate and spread anterogradely through their axon collaterals. As a result, when using H129 to dissect output neural networks, the issue of axon terminal uptake should be carefully considered.

In addition, animals will die within 3–5 days following intracranial H129 tracer infection, when the fluorescent protein is not fully expressed. Immunohistochemical staining is necessary to amplify the fluorescence signals for visualization of the output neural circuits. Applying a Trojan horse-like strategy, HSV integrating AAV replication elements was generated to enhance exogenous gene expression with low immunogenicity and improved fluorescent brightness (Su et al., 2020a). This approach may be useful as a general enhanced expression strategy for HSV-based tracing tools or gene-delivery vectors. However, the lack of cell-type-specificity of initially infected neurons limits the precise elucidation of output neural networks. Therefore, H129ΔTK-TT, a Cre-dependent anterograde trans-synaptic tracer derived from H129, has been developed, which Cre-dependently expresses TK and tdTomato. It only replicates in neurons expressing the Cre recombinase and in neurons synaptically connected with the initially infected cells. It has been injected into mice lumbar 5 dorsal root ganglia (DRG), together with immunosuppressant Bortezomib to facilitate the labeling of ascending neural pathways from specific types of DRG neurons for processing pain and itch (Chen et al., 2022).

Furthermore, in combination with the retargeting oncolytic virus strategy, a highly specific H129 anterograde tracing system has been developed by modifying the *gD* gene essential for HSV to infect cells (Su et al., 2020b). The recombinant H129 exhibits a new infection tropism by replacing the receptor recognition domain with a single-chain antibody targeting the Her2 receptor. The Her2 receptor and gD protein are pre-expressed in a specific type of neurons, and then the retargeted H129 with *gD* gene deletion is injected to infect the cell bodies of corresponding starter neurons. By specifically targeting the Her2 receptors, which are not endogenously expressed in neurons, the axon terminal infection is avoided, and specificity of the tracer is achieved. With helper viruses encoding truncated Her2 and wild-type gD proteins, the rigorous direct output of defined neurons can be elucidated. Moreover, some other anterograde monosynaptic tracer systems were also developed based on H129. For example, one consists of an H129 amplicon (H129Amp) tracer and an H129-dTK-T2-pacFlox helper, which assists in trans-synaptic transmission of the H129Amp tracer (Xiong et al., 2022). Another is composed of H129-

dgK-G4 with the H129 envelope glycoprotein K (gK) gene knockout and AAV helper compensating gK (Yang et al., 2022). These viruses enriched anterograde monosynaptic tracing tools and their applications contribute to revealing the output neural networks by combining anatomical tracing with other functional analysis.

4 VSV

VSV can be used as a trans-multisynaptic tracer, which spreads in the anterograde direction across synapses, owing to the fact that the replication-competent form of the virus does not cause serious diseases in mammals. VSV belongs to the genus *Vesiculovirus* in the family Rhabdoivirida. It is a bullet-shaped enveloped negative-sense RNA virus with a genome of ~11 kb, which contains genes encoding the nucleocapsid (N) protein, phosphoprotein (P), matrix (M) protein, G protein, and large polymerase (L). The G protein enables VSV to infect most mammalian cell types, and the N protein tightly encapsulates the RNA genome, forming a nuclease-resistant nucleocapsid. VSV is known for its fast replication and assembly, high efficiency of infection, and exogenous gene expression. These properties make it suitable for anterograde trans-multisynaptic tracing output networks efficiently (Lin et al., 2020). Moreover, a chimeric gene coding for ferritin and EGFP was integrated into VSV, this rVSV-Ferritin-EGFP can be used as a bimodal imaging tool for detecting synaptically connected neural circuits with both ex vivo magnetic resonance imaging (MRI) and fluorescent imaging. It combines neurotropic virus tracing with MRI to visualize the global network for a given type of neurons or their function in the brain (Zheng et al., 2019).

The VSV G protein can be pseudotyped with glycoproteins from other viruses to achieve different labeling characteristics and trans-synaptic transmission direction (Beier et al., 2011; Mundell et al., 2015). However, the high neurotoxicity in laboratory animals hinders its application in long-term studies of the structure and function of neural circuits. To overcome these drawbacks, a recombinant VSV with the seventh amino acid of the replication-related N protein mutation (VSV-NR7A) was developed. It displays attenuated cytotoxicity and more efficient anterograde trans-synaptic capability (Lin et al., 2020). To disambiguate between monosynaptic and multisynaptic connections, a tracer with restricted anterograde spread (TRAS) based on replication-incompetent VSV (VSV-ΔG) was developed. The VSV-ΔG cannot spread on its own and is complemented by providing replication-incompetent LV coated with VSV glycoprotein (Ma et al., 2019), which can be used for detecting novel efferent targets and output neural network changes underlying neuropsychiatric disorders. However, the high level of cytotoxicity in both the starter cells and downstream target cells hinders its application. Then, the second generation of TRAS utilizing a mutant variant of VSV-ΔG [VSV (M51R)-ΔG] with reduced cytotoxicity was generated (TRAS-M51R) incorporating with Cre-Lox combinatory gene expression system. This thus can be used to label the direct projection targets of the initially infected neurons in a Cre-dependent manner and image their function for multiple days post-infection (Kler et al., 2021).

5 CAV-2

CAV consists of two predominant serotypes, CAV-1 and CAV-2. They are genetically distinct from each other. CAV-1 is the causative agent of infectious canine hepatitis, while CAV-2 is of infectious canine respiratory disease. CAV-2 is a form of adenovirus and a non-enveloped double-stranded DNA virus. It has already been used in numerous studies in neuroscience research due to its advantages, including relatively low immunogenicity, preferential transduction of neurons, long-term transgene expression in the mammalian brain, and efficient retrograde transport from the axon terminal to the cell body. Its retrograde property relies on the coxsackievirus and adenovirus receptor (CAR), which is a cell adhesion molecule responsible for the attachment, internalization, and axonal transport of CAV. CARs are expressed by maturing and mature neurons and are located on the cell bodies and projections (Wehbi et al., 2020). Therefore, CAV-2 transduces neurons at the injection site as well as projection neurons innervating the injection site. It is transported from axon terminals to distant cell bodies and drives the gene expression there, thus useful for targeting neurons based on their projections. For example, CAV-2 vectors containing the PRS promoter and expressing ChR2 allow efficient transduction of locus coeruleus (LC) neurons in rats (both local and retrogradely labeled from the spinal cord) and enable reliable optogenetic activation (Li et al., 2016). CAV-2 can not only be effective for transducing genes into multiple regions of the central nervous system but also label motoneurons following injection into extraocular muscles in primates (Bohlen et al., 2019). However, the modest transgene expression levels, virus-related toxicity to the nervous system, and not easily scalable production limit its applications. Thus, the development of a nontoxic, readily manufactured CAV-2 vector that affords flexible packaging of different transgenes and supports long-term, high-level transgene expression is still urgently needed.

CAV-2 expressing Cre recombinase combined with another vector containing Cre-inducible expression of effectors can be used to monitor or perturb neurons that innervate a distal region. For example, a combination of CAV2-Cre and anterograde AAV enables the selective expression of AAV-carrying effectors in neurons projecting from the AAV injection site to the CAV-2 injection site. This strategy is especially valuable for studying reciprocal neural pathways (Carter et al., 2013). Furthermore, CAV has a high carrying capacity allowing a variety of genes such as large promoters, optogenetic elements, indicators of neural activity, and recombinases in one vector. This can be used to selectively label neurons in a projection- and/or cell-type-specific manner, facilitating monitoring and manipulating neural activity simultaneously (Lavoie & Liu, 2020).

6 SINV and SFV

SINV and SFV are closely related members of the *Alphavirus* genus of the Togaviridae family. They are small, enveloped, positive-strand RNA viruses with ~12 kb genomes encoding four non-structural proteins (nsPs 1–4) and five structural proteins (capsid, E3, E2, 6 K, and E1), which are translated from a highly active subgenomic promoter (26S promoter). Vectors based on these two viruses have been engineered to express exogenous genes at high levels both in vitro and in vivo (Bredenbeek et al., 1993; Kebschull et al., 2016).

SINV was first isolated from febrile patients and identified as a human pathogen in Uganda in 1961. It has high infection efficiency in neurons (Furuta et al., 2001) and can be optimized as a tool for neuroscience research. Recent studies have found that SINV with non-structural and capsid regions from the Toto1101 strain and structural regions from the neuroadapted Sindbis virus (NSV) strain can infect neurons in the peripheral and central nervous systems of zebrafish larvae after intravenous inoculation (Passoni et al., 2017). Moreover, a new anterograde trans-synaptic tracer based on SINV has been developed, which serves as a supplement to the viral tools for anterograde neural circuit tracing (Shi et al., 2022).

SFV was first isolated from mosquitoes captured in the Semliki Forest in Uganda in 1942. Wild-type SFV mainly infects neurons and is suitable for the rapid labeling of neurons, but not suitable for long-term transgene expression, which limits its application in the functional investigation of neural circuits. Currently, a mutant SFV virus-like particle has been developed that efficiently expresses GFP with single-round infection, which can be used for rapid and sparse labeling of neurons, thus delineating the detailed morphology of neurons as well as SINV (Jia et al., 2019a; Kuramoto, 2019). In addition, it was found that the infection tropism of SFV strain A7 (74) is temperature-dependent. Contrary to conventional SFV, it exclusively transduces glial cells at 37 °C, but mostly infects neurons at 31 °C (Ehrengruber et al., 2003), indicating that the SFV can potentially be developed into viral tools targeting different types of cells in the nervous system.

The production of high-titer SINV and SFV is rapid and generally requires Biosafety Level 2 facilities. They are usually produced by replicons and helper RNAs. The replicons are self-replicating with structural protein-encoding genes replaced by heterologous genes, and the helper RNAs provide structural proteins for the replicons (Ehrengruber et al., 2011). The infectious particles can be packaged by co-transfection of replicons with helper RNAs into target cells, which can be used to decipher the fine structure of neurons (Guo et al., 2019). Besides, the plasmid containing a promoter to initiate viral genome transcription can produce viral particle progeny, which can be utilized to dissect neural circuits trans-synaptically. Wild-type SINV and SFV vectors allowing fast and efficient transgene expression are cytotoxic; therefore, mutant vectors with reduced cytotoxicity have been developed. For example, mutant strains of SINV, SFV-SINrep ($nsP2S^{726}$), and SFV (PD) show attenuated cytotoxicity in primary cultured neurons, dissociated neurons, and brain slices, respectively (Kim et al., 2004; Lundstrom et al., 2003).

SINV and SFV have gradually become potent viral tools for depicting the fine structure of neurons and non-neuronal cells, and vectors for tracing neural circuits with reduced cytotoxicity in rodents and non-human primates in vivo are under development.

7 Recombinant AAV (rAAV)

rAAV is the most commonly used viral tool in biomedicine (Wang et al., 2019). It was developed as a gene transfer vector from the wild-type AAV, which belongs to the Parvoviridae family and the genus *Dependovirus*. rAAV not only has an excellent safety profile but also inherits many advantages of wild-type AAV, such as low immunogenicity, multiple serotypes, a wide range of infected host cells, and the ability to mediate long-term stable expression of foreign genes in vivo. It can be engineered to express fluorescent proteins or other genetic markers for labeling specific neuronal populations and manipulating or monitoring neuronal activity. AAV is the simplest type of replication-defective virus found so far and is dependent on the presence of adenovirus, HSV, and other auxiliary viruses to produce progeny. The genome of AAV is a single-stranded DNA molecule ~4.7 kb long. The AAV replicase protein (Rep) and capsid protein (Cap*)* genes are flanked by inverted terminal repeat (ITR) sequences. The virus particle does not have an envelope and is an icosahedron with a diameter of ~20 nm. The rAAV genome only conserves the AAV-derived ITRs, and all viral DNA between the ITRs is replaced by the gene of interest. The AAV ITRs serve as cis-elements, while the AAV *Rep*, *Cap,* and helper genes are provided in trans for viral replication and packaging (Ayuso et al., 2010).

AAV capsid engineering and scaling-up production have attracted much attention in academic and industrial circles. To date, 13 serotypes (AAV1–13) and >100 variants have been identified from hosts. Different serotypes differ in their capsid structure and display variable transduction efficiency, tissue and cell tropism, and immunological profile (Srivastava, 2016). In AAV capsid engineering, the traditional methods for AAV mutant library construction are mainly based on DNA shuffling, the peptide display library, and random point mutations (Grimm & Zolotukhin, 2015), and now computer-aided rational design combined with directed evolution has become a hot topic for optimizing the capsid of AAV vectors (Li & Samulski, 2020). A machine learning strategy has also been applied for the generation of improved AAV vectors (Wu et al., 2019; Bryant et al., 2021). Moreover, a single baculovirus expression vector (BEV)-derived OneBac system suitable for scaling up the production of different serotypes of rAAV has been generated. About 1×10^{15} vector genomes of purified rAAV can be obtained from 5 L of suspension-cultured Sf9 cells by using this new OneBac system, which can meet the needs of primate research and clinical therapy (Wu et al., 2021).

In the field of neuroscience, rAAVs are increasingly popular tools. Different serotypes of rAAV display different tropisms in the central nervous system and can transduce nerve cells both in vivo and in vitro (Watakabe et al., 2015; Castle et al., 2016). They can be either anterograde or retrograde, and non-trans-synaptic or trans-synaptic, which are serotype-, concentration-, and host-species-dependent (Salegio et al., 2013). Recently, AAV13 that can transduce neurons locally and label rigorously anterograde in the brain was found. Then, sparse labeling and functional monitoring of local neurons with AAV13-based single-viral and dual-viral strategies in C57BL/6 mice or transgenic mice expressing Cre-recombinase were achieved (Han et al., 2022b). By using anterograde and retrograde AAV tracing with optogenetic stimulations, a fine-grained anatomical and functional map between the anterior cortex and medulla for skilled forelimb movements in mice has been revealed (Yang et al., 2023). In general, most AAV serotypes prefer to enter the cell bodies of neurons at the injection site, and subsequent cellular trafficking may result in the spread of transgenic products throughout the whole neurons. This property may be useful for labeling neural projections and manipulating projection targets. Because host cells infected with rAAVs are unable to produce new viral particles, rAAVs cannot efficiently transmit trans-synaptically between neurons. For example, rAAV1 has been shown to exhibit anterograde trans-synaptic spread (Zingg et al., 2017), and high-titer AAV1-Cre can infect presynaptic neurons and drive Cre-dependent expression in postsynaptic neurons specifically, thus allowing the tracing and manipulation of direct input-defined neurons. An avian AAV (A3V) has been shown to spread trans-multi synaptically in the anterograde direction and display no extrasynaptic leakage (Ito et al., 2021). Some engineered AAV variants (AAV2-retro and AAV9-retro) are preferentially taken up by axon terminals at the injection site, then transported retrogradely from the axon to the cell body, which is suitable for labeling projection neurons with defined outputs. Some AAV serotypes, such as AAV9, are able to cross the blood-brain barrier (BBB) to transduce brain cells, but the efficiency is limited. By using the AAV capsid engineering strategy, a series of AAV variants (AAV-PHP.B (Deverman et al., 2016), AAV-PHP.eB and AAV-PHP.S (Chan et al., 2017), AAV-F (Hanlon et al., 2019), and AAV CAP-B10 (Goertsen et al., 2022)) have been developed with enhanced BBB penetration and broader host tropism; these are ideal viral tools for brain-wide neural cell transduction after intravenous delivery.

8 LV

LV vectors can be used to express transgenes in specific neuronal populations or for anterograde and retrograde tracing of neural circuits. LVs are spherical, enveloped viruses ~80–100 nm in diameter. Their genome is single-stranded RNA with *gag*, *pol*, and *env* genes, which encode polyprotein components of the capsid, the enzymes reverse transcriptase, protease, integrase, and envelope glycoproteins, respectively. The viral RNA genome is reverse-transcribed into DNA for integration into the host genome. LV vectors based on the human immunodeficiency virus 1 (HIV-1) are the most widely used in neuroscience research. They can mediate stable and efficient in vivo gene transfer into differentiated neurons and infect dividing and non-dividing cells, including neurons and glia in the adult mammalian brain. They have a packaging capacity of up to 9 kb, which makes them attractive gene-delivery vehicles for gene therapy, especially for the treatment of neurological and neurodegenerative disorders (Azzouz et al., 2004).

The tropism of LV vectors is largely defined by the viral envelope glycoproteins, which are essential for vector transduction. A variety of glycoproteins have been used to pseudotype LV vectors, typically the envelope glycoproteins from VSV (VSV-G) and RV (RV-G) (Mazarakis et al., 2001)· The LV vectors packaged with VSV-G can transduce neurons with anterograde transport of the expressed proteins throughout the cell bodies and axons and low efficiency of retrograde gene transfer. However, the LV vectors pseudotyped with RV-G show improved retrograde axonal transport property. Moreover, LV vectors pseudotyped with a fusion envelope glycoprotein type B (FuG-B) show highly efficient retrograde transport (HiRet), and those pseudotyped with a fusion glycoprotein type C (FuG-C) exhibit neuron-specific retrograde gene transfer (NeuRet) (Kobayashi et al., 2018). These viral vectors have been applied for targeting neuronal populations innervating the injection site and thus can be used as retrograde tracers for anatomical mapping and the functional analysis of specific projection pathways.

9 Cautions

Structure is the basis of function, and function is the embodiment of structure. The viral tools for neural circuit tracing have revolutionized our ability to investigate the organization and function of neural circuits in the brain. The choice of viral tools in dissecting neural circuits is complex and depends on the experimental design required to solve a specific research question or problem.

Currently, the common viral tools for anterograde tracing include AAV, HSV, and VSV. AAV is well known for its low cytotoxicity but limited trans-synaptic capability. The H129 and VSV have been used as anterograde trans-multisynaptic tracers in neural circuits for decades. However, their high toxicity leads to a short time window for examination. The anterograde trans-monosynaptic tracers based on replication-incompetent H129 or VSV have been developed to dissect the direct output networks. PRV, RV-B2c, RV-N2c, RV-CVS, HSV, CAV-2, and AAV2-retro can infect axon terminals and be used to map the input networks of specific brain regions. Systematic comparisons have shown that they have different retrograde labeling efficiency and selectivity (cell type or cortical layer) (Sun et al., 2019). AAV2-retro mainly targets cortical areas, whereas RV targets both cortical and subcortical areas, and the retrograde labeling efficiency of N2cG-enveloped RV is ten times higher than that of B19G-enveloped RV. Compared with AAV2-retro, CAV-2 poorly infects neurons in basal pontine nuclei and therefore does not induce significant cortical expression of GFP in Rosa26-LSL-H2B-eGFP mice (Tervo et al., 2016). These provide valuable information for the selection of appropriate viral tools in research. In addition, retrograde and anterograde trans-monosynaptic tracing can be achieved by combining a replication-incompetent viral tracer with a helper virus. However, the variable overlap of the two viruses due to sequential injections of the helper virus and viral tracer may potentially result in inconsistent results between animals and inefficient tracing of the direct input or output neural networks.

Moreover, it is noteworthy that many functional studies have been performed on freely-moving animals for a long time, so viral vectors that do not disturb neural activity are needed. Replication-defective viruses such as AAV vectors, which are less toxic to nerve cells, are mainly used for the functional analysis of neural circuits. The viral tools that can trace neural circuits trans-multisynaptically are currently less virulent after modifications, but they are still unsuitable for long-term functional research. Fortunately, trans-monosynaptic viral vectors are much less toxic to nerve cells and animals, and tools based on these vectors for long-term functional dissection are under development. The complexity of neural circuits poses an unparalleled challenge for developing and improving viral tools in many ways including attenuating toxicity, increasing specificity, enhancing sensitivity, and then combining structural and functional elements and being compatible with omics and other technologies. The development of new strategies and technologies and the endless demand of neuroscience research for new viral tools will promote rapid improvements in this field, thus accelerating progress in neuroscience research.

References

Ayuso, E., Mingozzi, F., & Bosch, F. (2010). Production, purification and characterization of adeno-associated vectors. *Current Gene Therapy, 10*(6), 423–436. https://doi.org/10.2174/156652310793797685

Azzouz, M., Ralph, G. S., Storkebaum, E., Walmsley, L. E., Mitrophanous, K. A., Kingsman, S. M., Carmeliet, P., & Mazarakis, N. D. (2004). VEGF delivery with retrogradely transported lentivector prolongs survival in a mouse ALS model. *Nature, 429*(6990), 413–417. https://doi.org/10.1038/nature02544

Beier, K. T., Saunders, A., Oldenburg, I. A., Miyamichi, K., Akhtar, N., Luo, L., Whelan, S. P., Sabatini, B., & Cepko, C. L. (2011). Anterograde or retrograde transsynaptic labeling of CNS neurons with vesicular stomatitis virus vectors. *Proceedings of the National Academy of Sciences of the United States of America, 108*(37), 15414–15419. https://doi.org/10.1073/pnas.1110854108

Bohlen, M. O., El-Nahal, H. G., & Sommer, M. A. (2019). Transduction of craniofacial motoneurons following intramuscular injections of canine adenovirus type-2 (CAV-2) in rhesus macaques. *Frontiers in Neuroanatomy, 13*, 84. https://doi.org/10.3389/fnana.2019.00084

Bredenbeek, P. J., Frolov, I., Rice, C. M., & Schlesinger, S. (1993). Sindbis virus expression vectors: Packaging of RNA replicons by using defective helper RNAs. *Journal of Virology, 67*(11), 6439–6446. https://doi.org/10.1128/JVI.67.11.6439-6446.1993

Bryant, D. H., Bashir, A., Sinai, S., Jain, N. K., Ogden, P. J., Riley, P. F., Church, G. M., Colwell, L. J., & Kelsic, E. D. (2021). Deep diversification of an AAV capsid protein by machine learning. *Nature Biotechnology, 39*(6), 691–696. https://doi.org/10.1038/s41587-020-00793-4

Carter, M. E., Soden, M. E., Zweifel, L. S., & Palmiter, R. D. (2013). Genetic identification of a neural circuit that suppresses appetite. *Nature, 503*(7474), 111–114. https://doi.org/10.1038/nature12596

Castle, M. J., Turunen, H. T., Vandenberghe, L. H., & Wolfe, J. H. (2016). Controlling AAV tropism in the nervous system with natural and engineered capsids. *Methods in Molecular Biology, 1382*, 133–149. https://doi.org/10.1007/978-1-4939-3271-9_10

Chan, K. Y., Jang, M. J., Yoo, B. B., Greenbaum, A., Ravi, N., Wu, W. L., Sanchez-Guardado, L., Lois, C., Mazmanian, S. K., Deverman, B. E., et al. (2017). Engineered AAVs for efficient noninvasive gene delivery to the central and peripheral nervous systems. *Nature Neuroscience, 20*(8), 1172–1179. https://doi.org/10.1038/nn.4593

Chatterjee, S., Sullivan, H. A., MacLennan, B. J., Xu, R., Hou, Y., Lavin, T. K., Lea, N. E., Michalski, J. E., Babcock, K. R., Dietrich, S., et al. (2018). Nontoxic, double-deletion-mutant rabies viral vectors for retrograde targeting of projection neurons. *Nature Neuroscience, 21*(4), 638–646. https://doi.org/10.1038/s41593-018-0091-7

Chen, Y., Song, Y., Wang, H., Zhang, Y., Hu, X., Wang, K., Lu, Y., Zhang, Z., Li, S., Li, A., et al. (2022). Distinct neural networks derived from galanin-containing nociceptors and neurotensin-expressing pruriceptors. *Proceedings of the National Academy of Sciences of the United States of America, 119*(33), e2118501119. https://doi.org/10.1073/pnas.2118501119

Ciabatti, E., Gonzalez-Rueda, A., Mariotti, L., Morgese, F., & Tripodi, M. (2017). Life-long genetic and functional access to neural circuits using self-inactivating rabies virus. *Cell, 170*(2), 382–392 e314. https://doi.org/10.1016/j.cell.2017.06.014

DeFalco, J., Tomishima, M., Liu, H., Zhao, C., Cai, X., Marth, J. D., Enquist, L., & Friedman, J. M. (2001). Virus-assisted mapping of neural inputs to a feeding center in the hypothalamus. *Science, 291*(5513), 2608–2613. https://doi.org/10.1126/science.1056602

Deverman, B. E., Pravdo, P. L., Simpson, B. P., Kumar, S. R., Chan, K. Y., Banerjee, A., Wu, W. L., Yang, B., Huber, N., Pasca, S. P., et al. (2016). Cre-dependent selection yields AAV variants for widespread gene transfer to the adult brain. *Nature Biotechnology, 34*(2), 204–209. https://doi.org/10.1038/nbt.3440

Dum, R. P., Levinthal, D. J., & Strick, P. L. (2009). The spinothalamic system targets motor and sensory areas in the cerebral cortex of monkeys. *The Journal of Neuroscience, 29*(45), 14223–14235. https://doi.org/10.1523/JNEUROSCI.3398-09.2009

Ehrengruber, M. U., Renggli, M., Raineteau, O., Hennou, S., Vaha-Koskela, M. J., Hinkkanen, A. E., & Lundstrom, K. (2003). Semliki Forest virus A7(74) transduces hippocampal neurons and glial cells in a temperature-dependent dual manner. *Journal of Neurovirology, 9*(1), 16–28. https://doi.org/10.1080/13550280390173346

Ehrengruber, M. U., Schlesinger, S., & Lundstrom, K. (2011). Alphaviruses: Semliki forest virus and Sindbis virus vectors for gene transfer into neurons. Current Protocols in Neuroscience Chapter 4:Unit 4 22. https://doi.org/10.1002/0471142301.ns0422s57

Ekstrand, M. I., Enquist, L. W., & Pomeranz, L. E. (2008). The alpha-herpesviruses: Molecular pathfinders in nervous system circuits. *Trends in Molecular Medicine, 14*(3), 134–140. https://doi.org/10.1016/j.molmed.2007.12.008

Enquist, L. W., & Card, J. P. (2003). Recent advances in the use of neurotropic viruses for circuit analysis. *Current Opinion in Neurobiology, 13*(5), 603–606. https://doi.org/10.1016/j.conb.2003.08.001

Furuta, T., Tomioka, R., Taki, K., Nakamura, K., Tamamaki, N., & Kaneko, T. (2001). In vivo transduction of central neurons using recombinant Sindbis virus: Golgi-like labeling of dendrites and axons with membrane-targeted fluorescent proteins. *The Journal of Histochemistry and Cytochemistry, 49*(12), 1497–1508. https://doi.org/10.1177/002215540104901203

Goertsen, D., Flytzanis, N. C., Goeden, N., Chuapoco, M. R., Cummins, A., Chen, Y., Fan, Y., Zhang, Q., Sharma, J., Duan, Y., et al. (2022). AAV capsid variants with brain-wide transgene expression and decreased liver targeting after intravenous delivery in mouse and marmoset. *Nature Neuroscience, 25*(1), 106–115. https://doi.org/10.1038/s41593-021-00969-4

Grimm, D., & Zolotukhin, S. (2015). E Pluribus Unum: 50 years of research, millions of viruses, and one goal–tailored acceleration of AAV evolution. *Molecular Therapy, 23*(12), 1819–1831. https://doi.org/10.1038/mt.2015.173

Guo, B., Chen, J., Chen, Q., Ren, K., Feng, D., Mao, H., Yao, H., Yang, J., Liu, H., Liu, Y., et al. (2019). Anterior cingulate cortex dysfunction underlies social deficits in Shank3 mutant mice. *Nature Neuroscience, 22*(8), 1223–1234. https://doi.org/10.1038/s41593-019-0445-9

Han, Z., Luo, N., Kou, J., Li, L., Xu, Z., Wei, S., Wu, Y., Wang, J., Ye, C., Lin, K., et al. (2022a). Brain-wide TVA compensation allows rabies virus to retrograde target cell-type-specific projection neurons. *Molecular Brain, 15*(1), 13. https://doi.org/10.1186/s13041-022-00898-8

Han, Z., Luo, N., Wu, Y., Kou, J., Ma, W., Yang, X., Cai, Y., Ma, L., Han, L., Wang, X., et al. (2022b). AAV13 enables precise targeting of local neural populations. *International Journal of Molecular Sciences, 23*(21). https://doi.org/10.3390/ijms232112806

Hanchate, N. K., Lee, E. J., Ellis, A., Kondoh, K., Kuang, D., Basom, R., Trapnell, C., & Buck, L. B. (2020). Connect-seq to superimpose molecular on anatomical neural circuit maps. *Proceedings of the National Academy of Sciences of the United States of America, 117*(8), 4375–4384. https://doi.org/10.1073/pnas.1912176117

Hanlon, K. S., Meltzer, J. C., Buzhdygan, T., Cheng, M. J., Sena-Esteves, M., Bennett, R. E., Sullivan, T. P., Razmpour, R., Gong, Y., Ng, C., et al. (2019). Selection of an efficient AAV vector for robust CNS transgene expression. *Molecular Therapy - Methods & Clinical Development, 15*, 320–332. https://doi.org/10.1016/j.omtm.2019.10.007

Inada, K., Hagihara, M., Tsujimoto, K., Abe, T., Konno, A., Hirai, H., Kiyonari, H., & Miyamichi, K. (2022). Plasticity of neural connections underlying oxytocin-mediated parental behaviors of male mice. *Neuron, 110*(12), 2009–2023 e2005. https://doi.org/10.1016/j.neuron.2022.03.033

Ito, T., Ono, M., Matsui, R., Watanabe, D., & Ohmori, H. (2021). Avian adeno-associated virus as an anterograde transsynaptic vector. *Journal of Neuroscience Methods, 359*, 109221. https://doi.org/10.1016/j.jneumeth.2021.109221

Jia, F., Zhu, X., Lv, P., Hu, L., Liu, Q., Jin, S., & Xu, F. (2019a). Rapid and sparse labeling of neurons based on the mutant virus-like particle of Semliki Forest virus. *Neuroscience Bulletin, 35*(3), 378–388. https://doi.org/10.1007/s12264-019-00362-z

Jia, F., Lv, P., Miao, H., Shi, X., Mei, H., Li, L., Xu, X., Tao, S., & Xu, F. (2019b). Optimization of the fluorescent protein expression level based on pseudorabies virus Bartha strain for neural circuit tracing. *Frontiers in Neuroanatomy, 13*, 63. https://doi.org/10.3389/fnana.2019.00063

Jia, F., Li, L., Liu, H., Lv, P., Shi, X., Wu, Y., Ling, C., & Xu, F. (2021). Development of a rabies virus-based retrograde tracer with high trans-monosynaptic efficiency by reshuffling glycoprotein. *Molecular Brain, 14*(1), 109. https://doi.org/10.1186/s13041-021-00821-7

Kebschull, J. M., Garcia da Silva, P., & Zador, A. M. (2016). A new defective helper RNA to produce recombinant sindbis virus that infects neurons but does not propagate. *Frontiers in Neuroanatomy, 10*, 56. https://doi.org/10.3389/fnana.2016.00056

Kim, J., Dittgen, T., Nimmerjahn, A., Waters, J., Pawlak, V., Helmchen, F., Schlesinger, S., Seeburg, P. H., & Osten, P. (2004). Sindbis vector SINrep(nsP2S726): A tool for rapid heterologous expression with attenuated cytotoxicity in neurons. *Journal of Neuroscience Methods, 133*(1–2), 81–90. https://doi.org/10.1016/j.jneumeth.2003.09.029

Kim, E. J., Jacobs, M. W., Ito-Cole, T., & Callaway, E. M. (2016). Improved monosynaptic neural circuit tracing using engineered rabies virus glycoproteins. *Cell Reports, 15*(4), 692–699. https://doi.org/10.1016/j.celrep.2016.03.067

Kler, S., Ma, M., Narayan, S., Ahrens, M. B., & Pan, Y. A. (2021). Cre-dependent anterograde transsynaptic labeling and functional imaging in zebrafish using VSV with reduced cytotoxicity. *Frontiers in Neuroanatomy, 15*, 758350. https://doi.org/10.3389/fnana.2021.758350

Kobayashi, K., Kato, S., & Kobayashi, K. (2018). Genetic manipulation of specific neural circuits by use of a viral vector system. *Journal of Neural Transmission (Vienna), 125*(1), 67–75. https://doi.org/10.1007/s00702-016-1674-7

Kuramoto, E. (2019). Method for labeling and reconstruction of single neurons using Sindbis virus vectors. *Journal of Chemical Neuroanatomy, 100*, 101648. https://doi.org/10.1016/j.jchemneu.2019.05.002

Lavoie, A., & Liu, B. H. (2020). Canine adenovirus 2: A natural choice for brain circuit dissection. *Frontiers in Molecular Neuroscience, 13*, 9. https://doi.org/10.3389/fnmol.2020.00009

Li, C., & Samulski, R. J. (2020). Engineering adeno-associated virus vectors for gene therapy. *Nature Reviews Genetics, 21*(4), 255–272. https://doi.org/10.1038/s41576-019-0205-4

Li, Y., Hickey, L., Perrins, R., Werlen, E., Patel, A. A., Hirschberg, S., Jones, M. W., Salinas, S., Kremer, E. J., & Pickering, A. E. (2016). Retrograde optogenetic characterization of the pontospinal module of the locus coeruleus with a canine adenoviral vector. *Brain Research, 1641*(Pt B), 274–290. https://doi.org/10.1016/j.brainres.2016.02.023

Lin, K., Zhong, X., Ying, M., Li, L., Tao, S., Zhu, X., He, X., & Xu, F. (2020). A mutant vesicular stomatitis virus with reduced cytotoxicity and enhanced anterograde trans-synaptic efficiency. *Molecular Brain, 13*(1), 45. https://doi.org/10.1186/s13041-020-00588-3

Lin, K. Z., Li, L., Ma, W. Y., Yang, X., Han, Z. P., Luo, N. S., Wang, J., & Xu, F. Q. (2023). A rabies virus-based toolkit for efficient retrograde labeling and monosynaptic tracing. *Neural Regeneration Research, 18*(8), 1827–1833. https://doi.org/10.4103/1673-5374.358618

Liu, Q., Wang, X., Xie, C., Ding, S., Yang, H., Guo, S., Li, J., Qin, L., Ban, F., Wang, D., et al. (2021). A novel human acute encephalitis caused by pseudorabies virus variant strain. *Clinical Infectious Diseases, 73*(11), e3690–e3700. https://doi.org/10.1093/cid/ciaa987

Liu, Q., Wu, Y., Wang, H., Jia, F., & Xu, F. (2022). Viral tools for neural circuit tracing. *Neuroscience Bulletin, 38*(12), 1508–1518. https://doi.org/10.1007/s12264-022-00949-z

Lundstrom, K., Abenavoli, A., Malgaroli, A., & Ehrengruber, M. U. (2003). Novel Semliki Forest virus vectors with reduced cytotoxicity and temperature sensitivity for long-term enhancement of transgene expression. *Molecular Therapy, 7*(2), 202–209. https://doi.org/10.1016/s1525-0016(02)00056-4

Ma, M., Kler, S., & Pan, Y. A. (2019). Structural neural connectivity analysis in zebrafish with restricted anterograde transneuronal viral labeling and quantitative brain mapping. *Frontiers in Neural Circuits, 13*, 85. https://doi.org/10.3389/fncir.2019.00085

Masaki, Y., Yamaguchi, M., Takeuchi, R. F., & Osakada, F. (2022). Monosynaptic rabies virus tracing from projection-targeted single neurons. *Neuroscience Research, 178*, 20–32. https://doi.org/10.1016/j.neures.2022.01.007

Mazarakis, N. D., Azzouz, M., Rohll, J. B., Ellard, F. M., Wilkes, F. J., Olsen, A. L., Carter, E. E., Barber, R. D., Baban, D. F., Kingsman, S. M., et al. (2001). Rabies virus glycoprotein pseudotyping of lentiviral vectors enables retrograde axonal transport and access to the nervous system after peripheral delivery. *Human Molecular Genetics, 10*(19), 2109–2121. https://doi.org/10.1093/hmg/10.19.2109

Muller, P. A., Schneeberger, M., Matheis, F., Wang, P., Kerner, Z., Ilanges, A., Pellegrino, K., Del Marmol, J., Castro, T. B. R., Furuichi, M., et al. (2020). Microbiota modulate sympathetic neurons via a gut-brain circuit. *Nature, 583*(7816), 441–446. https://doi.org/10.1038/s41586-020-2474-7

Mundell, N. A., Beier, K. T., Pan, Y. A., Lapan, S. W., Goz Ayturk, D., Berezovskii, V. K., Wark, A. R., Drokhlyansky, E., Bielecki, J., Born, R. T., et al. (2015). Vesicular stomatitis virus enables gene transfer and transsynaptic tracing in a wide range of organisms. *The Journal of Comparative Neurology, 523*(11), 1639–1663. https://doi.org/10.1002/cne.23761

Nassi, J. J., Cepko, C. L., Born, R. T., & Beier, K. T. (2015). Neuroanatomy goes viral! *Frontiers in Neuroanatomy, 9*, 80. https://doi.org/10.3389/fnana.2015.00080

Oyibo, H. K., Znamenskiy, P., Oviedo, H. V., Enquist, L. W., & Zador, A. M. (2014). Long-term Cre-mediated retrograde tagging of neurons using a novel recombinant pseudorabies virus. *Frontiers in Neuroanatomy, 8*, 86. https://doi.org/10.3389/fnana.2014.00086

Papazoglou, I., Lee, J. H., Cui, Z., Li, C., Fulgenzi, G., Bahn, Y. J., Staniszewska-Goraczniak, H. M., Pinol, R. A., Hogue, I. B., Enquist, L. W., et al. (2022). A distinct hypothalamus-to-beta cell circuit modulates insulin secretion. *Cell Metabolism, 34*(2), 285–298 e287. https://doi.org/10.1016/j.cmet.2021.12.020

Parr-Brownlie, L. C., Bosch-Bouju, C., Schoderboeck, L., Sizemore, R. J., Abraham, W. C., & Hughes, S. M. (2015). Lentiviral vectors as tools to understand central nervous system biology in mammalian model organisms. *Frontiers in Molecular Neuroscience, 8*, 14. https://doi.org/10.3389/fnmol.2015.00014

Passoni, G., Langevin, C., Palha, N., Mounce, B. C., Briolat, V., Affaticati, P., De Job, E., Joly, J. S., Vignuzzi, M., Saleh, M. C., et al. (2017). Imaging of viral neuroinvasion in the zebrafish reveals that Sindbis and chikungunya viruses favour different entry routes. *Disease Models & Mechanisms, 10*(7), 847–857. https://doi.org/10.1242/dmm.029231

Pomeranz, L. E., Ekstrand, M. I., Latcha, K. N., Smith, G. A., Enquist, L. W., & Friedman, J. M. (2017). Gene expression profiling with cre-conditional pseudorabies virus reveals a subset of midbrain neurons that participate in reward circuitry. *The Journal of Neuroscience, 37*(15), 4128–4144. https://doi.org/10.1523/JNEUROSCI.3193-16.2017

Reardon, T. R., Murray, A. J., Turi, G. F., Wirblich, C., Croce, K. R., Schnell, M. J., Jessell, T. M., & Losonczy, A. (2016). Rabies virus CVS-N2c(DeltaG) strain enhances retrograde synaptic transfer and neuronal viability. *Neuron, 89*(4), 711–724. https://doi.org/10.1016/j.neuron.2016.01.004

Salegio, E. A., Samaranch, L., Kells, A. P., Mittermeyer, G., San Sebastian, W., Zhou, S., Beyer, J., Forsayeth, J., & Bankiewicz, K. S. (2013). Axonal transport of adeno-associated viral vectors is serotype-dependent. *Gene Therapy, 20*(3), 348–352. https://doi.org/10.1038/gt.2012.27

Shi, X. W., Jia, F., Lyu, P., & Xu, F. Q. (2022). A new anterograde trans-synaptic tracer based on Sindbis virus. *Neural Regeneration Research, 17*(12), 2761–2764. https://doi.org/10.4103/1673-5374.339495

Soudais, C., Laplace-Builhe, C., Kissa, K., & Kremer, E. J. (2001). Preferential transduction of neurons by canine adenovirus vectors and their efficient retrograde transport in vivo. *The FASEB Journal, 15*(12), 2283–2285. https://doi.org/10.1096/fj.01-0321fje

Srivastava, A. (2016). In vivo tissue-tropism of adeno-associated viral vectors. *Current Opinion in Virology, 21*, 75–80. https://doi.org/10.1016/j.coviro.2016.08.003

Su, P., Wang, H., Xia, J., Zhong, X., Hu, L., Li, Y., Li, Y., Ying, M., & Xu, F. (2019). Evaluation of retrograde labeling profiles of HSV1 H129 anterograde tracer. *Journal of Chemical Neuroanatomy, 100*, 101662. https://doi.org/10.1016/j.jchemneu.2019.101662

Su, P., Ying, M., Han, Z., Xia, J., Jin, S., Li, Y., Wang, H., & Xu, F. (2020a). High-brightness anterograde transneuronal HSV1 H129 tracer modified using a Trojan horse-like strategy. *Molecular Brain, 13*(1), 5. https://doi.org/10.1186/s13041-020-0544-2

Su, P., Ying, M., Xia, J., Li, Y., Wu, Y., Wang, H., & Xu, F. (2020b). Rigorous anterograde trans-monosynaptic tracing of genetic defined neurons with retargeted HSV1 H129. bioRxiv:2020.2012.2001.407312. https://doi.org/10.1101/2020.12.01.407312

Sun, L., Tang, Y., Yan, K., Yu, J., Zou, Y., Xu, W., Xiao, K., Zhang, Z., Li, W., Wu, B., et al. (2019). Differences in neurotropism and neurotoxicity among retrograde viral tracers. *Molecular Neurodegeneration, 14*(1), 8. https://doi.org/10.1186/s13024-019-0308-6

Sun, P., Jin, S., Tao, S., Wang, J., Li, A., Li, N., Wu, Y., Kuang, J., Liu, Y., Wang, L., et al. (2020). Highly efficient and super-bright neurocircuit tracing using vector mixing-based virus cocktail. bioRxiv:705772. https://doi.org/10.1101/705772

Tang, Y., Li, L., Sun, L., Yu, J., Hu, Z., Lian, K., Cao, G., & Dai, J. (2020). In vivo two-photon calcium imaging in dendrites of rabies virus-labeled V1 corticothalamic neurons. *Neuroscience Bulletin, 36*(5), 545–553. https://doi.org/10.1007/s12264-019-00452-y

Tervo, D. G., Hwang, B. Y., Viswanathan, S., Gaj, T., Lavzin, M., Ritola, K. D., Lindo, S., Michael, S., Kuleshova, E., Ojala, D., et al. (2016). A designer AAV variant permits efficient retrograde access to projection neurons. *Neuron, 92*(2), 372–382. https://doi.org/10.1016/j.neuron.2016.09.021

Wang, D., Tai, P. W. L., & Gao, G. (2019). Adeno-associated virus vector as a platform for gene therapy delivery. *Nature Reviews Drug Discovery, 18*(5), 358–378. https://doi.org/10.1038/s41573-019-0012-9

Watakabe, A., Ohtsuka, M., Kinoshita, M., Takaji, M., Isa, K., Mizukami, H., Ozawa, K., Isa, T., & Yamamori, T. (2015). Comparative analyses of adeno-associated viral vector serotypes 1, 2, 5, 8 and 9 in marmoset, mouse and macaque cerebral cortex. *Neuroscience Research, 93*, 144–157. https://doi.org/10.1016/j.neures.2014.09.002

Wehbi, A., Kremer, E. J., & Dopeso-Reyes, I. G. (2020). Location of the cell adhesion molecule "coxsackievirus and adenovirus receptor" in the adult mouse brain. *Frontiers in Neuroanatomy, 14*, 28. https://doi.org/10.3389/fnana.2020.00028

Wickersham, I. R., Finke, S., Conzelmann, K. K., & Callaway, E. M. (2007a). Retrograde neuronal tracing with a deletion-mutant rabies virus. *Nature Methods, 4*(1), 47–49. https://doi.org/10.1038/nmeth999

Wickersham, I. R., Lyon, D. C., Barnard, R. J., Mori, T., Finke, S., Conzelmann, K. K., Young, J. A., & Callaway, E. M. (2007b). Monosynaptic restriction of transsynaptic tracing from single, genetically targeted neurons. *Neuron, 53*(5), 639–647. https://doi.org/10.1016/j.neuron.2007.01.033

Wu, Z., Kan, S. B. J., Lewis, R. D., Wittmann, B. J., & Arnold, F. H. (2019). Machine learning-assisted directed protein evolution with combinatorial libraries. *Proceedings of the National Academy of Sciences of the United States of America, 116*(18), 8852–8858. https://doi.org/10.1073/pnas.1901979116

Wu, Y., Han, Z., Duan, M., Jiang, L., Tian, T., Jin, D., Wang, Q., & Xu, F. (2021). Popularizing recombinant baculovirus-derived OneBac system for laboratory production of all recombinant adeno-associated virus vector serotypes. *Current Gene Therapy, 21*(2), 167–176. https://doi.org/10.2174/1566523221666210118111657

Xiong, F., Yang, H., Song, Y. G., Qin, H. B., Zhang, Q. Y., Huang, X., Jing, W., Deng, M., Liu, Y., Liu, Z., et al. (2022). An HSV-1-H129 amplicon tracer system for rapid and efficient monosynaptic anterograde neural circuit tracing. *Nature Communications, 13*(1), 7645. https://doi.org/10.1038/s41467-022-35355-6

Xu, X., Holmes, T. C., Luo, M. H., Beier, K. T., Horwitz, G. D., Zhao, F., Zeng, W., Hui, M., Semler, B. L., & Sandri-Goldin, R. M. (2020). Viral vectors for neural circuit mapping and recent advances in trans-synaptic anterograde tracers. *Neuron, 107*(6), 1029–1047. https://doi.org/10.1016/j.neuron.2020.07.010

Yang, H., Xiong, F., Qin, H. B., Yu, Q. T., Sun, J. Y., Zhao, H. W., Li, D., Zhou, Y., Zhang, F. K., Zhu, X. W., et al. (2022). A novel H129-based anterograde monosynaptic tracer exhibits features of strong labeling intensity, high tracing efficiency, and reduced retrograde labeling. *Molecular Neurodegeneration, 17*(1), 6. https://doi.org/10.1186/s13024-021-00508-6

Yang, W., Kanodia, H., & Arber, S. (2023). Structural and functional map for forelimb movement phases between cortex and medulla. *Cell, 186*(1), 162–177 e118. https://doi.org/10.1016/j.cell.2022.12.009

Yao, J., Zhang, Q., Liao, X., Li, Q., Liang, S., Li, X., Zhang, Y., Li, X., Wang, H., Qin, H., et al. (2018). A corticopontine circuit for initiation of urination. *Nature Neuroscience, 21*(11), 1541–1550. https://doi.org/10.1038/s41593-018-0256-4

Ye, Q., Gast, G., Su, X., Saito, T., Saido, T. C., Holmes, T. C., & Xu, X. (2022). Hippocampal neural circuit connectivity alterations in an Alzheimer's disease mouse model revealed by monosynaptic rabies virus tracing. *Neurobiology of Disease, 172*, 105820. https://doi.org/10.1016/j.nbd.2022.105820

Zemanick, M. C., Strick, P. L., & Dix, R. D. (1991). Direction of transneuronal transport of herpes simplex virus 1 in the primate motor system is strain-dependent. *Proceedings of the National Academy of Sciences of the United States of America, 88*(18), 8048–8051. https://doi.org/10.1073/pnas.88.18.8048

Zhang, X., Lei, B., Yuan, Y., Zhang, L., Hu, L., Jin, S., Kang, B., Liao, X., Sun, W., Xu, F., et al. (2020). Brain control of humoral immune responses amenable to behavioural modulation. *Nature, 581*(7807), 204–208. https://doi.org/10.1038/s41586-020-2235-7

Zheng, N., Su, P., Liu, Y., Wang, H., Nie, B., Fang, X., Xu, Y., Lin, K., Lv, P., He, X., et al. (2019). Detection of neural connections with ex vivo MRI using a ferritin-encoding trans-synaptic virus. *NeuroImage, 197*, 133–142. https://doi.org/10.1016/j.neuroimage.2019.04.039

Zhu, X., Lin, K., Liu, Q., Yue, X., Mi, H., Huang, X., He, X., Wu, R., Zheng, D., Wei, D., et al. (2020). Rabies virus pseudotyped with CVS-N2C glycoprotein as a powerful tool for retrograde neuronal network tracing. *Neuroscience Bulletin, 36*(3), 202–216. https://doi.org/10.1007/s12264-019-00423-3

Zingg, B., Chou, X. L., Zhang, Z. G., Mesik, L., Liang, F., Tao, H. W., & Zhang, L. I. (2017). AAV-mediated anterograde transsynaptic tagging: Mapping corticocollicular input-defined neural pathways for defense behaviors. *Neuron, 93*(1), 33–47. https://doi.org/10.1016/j.neuron.2016.11.045

Spatial Omics: Navigating Neuroscience Research into the New Era

Pengfei Guo and Yanxiang Deng

Abstract The human brain's complexity is underpinned by billions of neurons and trillions of synapses, necessitating coordinated activities across diverse cell types. Conventional techniques like in situ hybridization and immunohistochemistry, while valuable, face limitations in resolution and comprehensiveness when analyzing neuron types. Advances in spatial omics technologies, especially those integrating transcriptomics and proteomics, have revolutionized our understanding of brain tissue organization. These technologies, such as FISH-based, in situ sequencing-based (ISS), and next-generation sequencing (NGS)-based methods, provide detailed spatial context, overcoming previous limitations. FISH techniques, including smFISH and its variants like seqFISH and MERFISH, offer high-resolution spatial gene expression data. ISS approaches leverage padlock probes and rolling circle amplification to yield spatial transcriptome information. NGS-based methods, such as spatial transcriptomics and spatial-epigenomics, integrate spatial barcodes with single-cell sequencing, enabling comprehensive profiling of gene expression and epigenetic states in tissues. These innovations have propelled insights into neural development and disease, identifying cellular heterogeneity and molecular alterations in conditions like Alzheimer's and major depression. Despite challenges in cost, speed, and data analysis, spatial omics technologies continue to evolve, promising deeper insights into the molecular mechanisms of the brain and neurodegenerative diseases.

Keywords Spatial omics · Transcriptome · Epigenome · Neuroscience · Neurologic diseases

P. Guo · Y. Deng (✉)
Department of Pathology and Laboratory Medicine, Perelman School of Medicine, University of Pennsylvania, Philadelphia, PA, USA

Epigenetics Institute, Perelman School of Medicine, University of Pennsylvania, Philadelphia, PA, USA
e-mail: yanxiang.deng@pennmedicine.upenn.edu

A. C.-H. Yu et al. (eds.), *Systems Neuroscience*, Advances in Neurobiology 41,
https://doi.org/10.1007/978-3-031-69188-1_6

Abbreviations

ISH	In situ hybridization
ISS	In situ sequencing
FISH	fluorescence in situ hybridization
smFISH	single-molecule FISH
osmFISH	cyclic-ouroboros smFISH
split-FISH	split-probe strategy with multiplexed FISH
seqFISH	sequential fluorescence in situ hybridization
MERFISH	Multiplexed Error-Robust Fluorescence In Situ Hybridization
RCA	rolling-circle amplification
HybISS	hybridization-based in situ sequencing
SBH	sequence-by-hybridization
STARmap	spatially resolved transcript amplicon readout mapping
NGS	Next generation sequencing
Boloramis	barcoded oligonucleotides ligated on RNA amplified for multiplexed and parallel in-situ analyses
Baristaseq	Barcode in situ targeted sequencing
Exseq	Expansion Sequencing
HDST	High-definition spatial transcriptomics
SPLiT-seq	Split-pool ligation-based transcriptome sequencing
Stereo-seq	spatial temporal enhanced resolution omics sequencing
Pixel-seq	polony-indexed library-sequencing
ATAC-seq	Assay for Transposase-Accessible Chromatin using sequencing
CUT&Tag	cleavage under targets and tagmentation
MulTI-Tag	multiple target identification by tagmentation
NTT-seq	nanobody-tethered transposition followed by sequencing
Nano-CT	Nanobody-CUT&Tag
scM&T-seq	single-cell combined methylome and transcriptome sequencing
sc-GEM	single-cell analysis of genotype, expression, and methylation
CITE-seq	cellular indexing of transcriptomes and epitopes by sequencing
SNARE-seq	single-nucleus chromatin accessibility and mRNA expression sequencing
SHARE-seq	simultaneous high-throughput ATAC and RNA expression with sequencing
DBiT-seq	deterministic barcoding in tissue for spatial omics sequencing
LCM	laser capture microdissection
ROI	region of interest
ST	spatial transcriptomics
RCTD	robust cell type decomposition
PBN	parabrachial nucleus
AD	Alzheimer's disease
DAM	disease microglia
MDD	major depressive disorder
ALS	amyotrophic lateral sclerosis

1 Introduction

In the human brain, billions of neurons are interconnected by trillions of synapses forming the central nervous system (Colón-Ramos, 2009). The formation and maintenance of neuronal activities and behavior require complex, coordinated activities of diverse neuronal and non-neuronal cells. Within those cells, a vast array of molecular, anatomical, morphological, physiological, and plastic properties determine cell identity and network dynamics. Neurons, serving as the foundational components of brain circuits, exhibit intricate and diverse cellular compositions that intricately underlie their sophisticated functionalities. Enhancing comprehension of neurons and network organization and dynamics involves the identification of gene transcription, epigenetic conditions, and protein expression, all of which contribute to cell identity.

Conventional techniques, such as in situ hybridization (ISH) and immunohistochemistry, analyze neuron connections and classify cells in the nervous system by mapping RNA and proteins within tissue samples (Gray et al., 2004; Lein et al., 2007). However, these methods have significant limitations in terms of resolution, scale, robustness, comprehensiveness, and throughput when analyzing neuron types. We are learning more about the cell types of brain samples due to recent technological advances, especially in transcriptomics and proteomics. These methods have rapidly evolved, making it possible to characterize cells and tissues in development and disease in a comprehensive manner. They also enable researchers to enhance their insights into various biological processes and disease mechanisms. For example, bulk RNA sequencing allows the simultaneous analysis of gene expression within a sample. The development of single-cell genomic sequencing allows researchers to identify and characterize the remarkable full transcriptome of individual cells, empowering a wide range of discoveries. Nevertheless, these single-cell technologies only characterized dissociated cells derived from intact tissues of interest, so the molecular spatial context was lost in the process. This challenge can be overcome with new technologies that combine multiplexed fluorescence, DNA, RNA, and isotope labels, enabling the measurement of spatial information with spatial context maintained.

In recent years, there has been an explosion in methods for spatial omics, which vary in the capabilities to capture diverse information across different length scales (Table 1). The high degree of multiplexing was first demonstrated by proof of principle, and now advanced methods reveal the spatial organization of gene expression and transcriptional types in brain tissue (La Manno et al., 2016; Shah et al., 2016). A range of the latest spatial omics methods, including genomics, epigenomics, transcriptomics, and proteomics, provides multiplexing information for studying neural diseases molecularly in a spatially resolved manner. The use of spatial omics measurements enables the identification of cellular heterogeneity and complexity of tissue architectures, in addition to inspecting dynamic molecular alterations under both physiological and pathological, during development and disease processes. Moreover, the number of commercially available technologies will also increase with the widespread adoption of these technologies. This section aims to summarize the current state of these advanced technologies and describe the challenges associated with spatial omics.

Table 1 A table summarizing the advantages of in-situ imaging-based and NGS-based spatial omics technologies

technologies	type	sample	target	spatial resolution	detection efficiency
SeqFISH	mFISH	tissue	RNA	submicrometer	10000 genes
MERFISH	mFISH	tissue	RNA	submicrometer	10000 genes
osmFISH	mFISH	tissue	RNA	submicrometer	48 genes
split-FISH	mFISH	tissue	RNA	submicrometer	317 genes
HybISS	NGS	tissue	RNA	submicrometer	124 genes
STARmap	NGS	tissue	RNA	submicrometer	1020 genes
FISSEQ	NGS	tissue	RNA	nanoscale resolution	8742 genes
Exseq	NGS	tissue	RNA	1 μm	transcriptome
Seq-Scope	NGS	tissue	RNA	0.5~0.8 μm	transcriptome
Stereo-seq	NGS	tissue	RNA	0.6 μm	transcriptome
Pixel-seq	NGS	tissue	RNA	1 μm	transcriptome
spatial transcriptomes (ST)	NGS	tissue	RNA	55 μm	transcriptome
DBiT-seq	NGS	tissue	RNA, protein	10 μm	transcriptome, 20 proteins
Epigenomic MERFISH	mFISH	tissue	histone modification	submicrometer	epigenome
spatial-CUT&Tag	NGS	tissue	histone modification	20 μm	epigenome
spatial-ATAC-seq	NGS	tissue	DNA	20 μm	genome
spatial ATAC	NGS	tissue	DNA	55 μm	genome
spatial-CUT&Tag-RNA-seq	NGS	tissue	histone modification, RNA	20 μm	genome, transcriptome
spatial-ATAC-RNA-seq	NGS	tissue	DNA, RNA	20 μm	epigenome, transcriptome

1.1 *Spatial Omics Technologies*

1.1.1 FISH-Based Approaches

Spatial omics techniques and data-analysis methods that utilize spatial gene expression data to resolve tissue architecture have boomed in the last decade. The 1960s started to use in situ hybridization (ISH) to detect the presence and position of DNA or RNA in cells and tissues. Since in situ hybridization was first used to detect genes in a spatial context (Gall and Pardue, 1969; John et al., 1969), various ISH forms have been used to visualize gene expression. In 1980, a method named fluorescence in situ hybridization (FISH) was adapted to direct detection using fluorochrome-labeled RNA (Bauman et al., 1980). Even though advanced FISH methods have been widely used in cancer diagnosis and have also advanced our understanding of cancer research, these methods still have limitations. For example, they detect only a limited number of targets at a time, and researchers must choose the targets of interest beforehand. A more advanced FISH method is single-molecule fluorescence in situ hybridization (smFISH) (Femino et al., 1998), which allows the recording of positional information for individual mRNA molecules within a cell. However, the number of spectrally resolvable fluorophores constrains smFISH's ability to detect more RNA molecules in a single experiment. Several variations of smFISH have been developed as a means to overcome these challenges, such as osmFISH

(Codeluppi et al., 2018), split-FISH (Goh et al., 2020), seqFISH (Eng et al., 2019), and MERFISH (Xia et al., 2019). For example, the technique of seqFISH enables the detection of hundreds of genes in single cells and allows a deeper understanding of the structure of the mouse brain (Shah et al., 2016). Using a spatial histone modification-profiling approach (Epigenomic MERFISH), another study generated an atlas of hundreds of promoters and enhancers in embryonic and adult mouse brains (Lu et al., 2022). Epigenomic MERFISH involves in situ tagmentation and transcription followed by similar imaging steps to those of MERFISH. Though these tools can resolve cellular heterogeneity spatially, they are still limited by high experimental costs, long imaging acquisition time, and complex data analysis requirements.

1.1.2 In Situ Sequencing-Based (ISS) Approaches

ISS method is based on the combination of padlock probe, rolling circle amplification, and next-generation sequencing (Ke et al., 2013). In contrast to FISH-based techniques, ISS technologies yield spatial transcriptome information by sequencing (Fig. 1), which can be divided into two categories: gene barcodes (targeted) and short cDNA fragments (untargeted). Targeted spatial transcriptome focuses on specific genes or gene regions, which is suitable for investigating specific cellular processes. The next iteration of the ISS method is hybridization-based in situ sequencing (HybISS), which uses a sequence-by-hybridization (SBH) chemistry approach to define the location of mRNA molecules within the tissue (Gyllborg et al., 2020). In another example, spatially resolved transcript amplicon readout mapping (STARmap) has used a different padlock-probe strategy, in which each targeted RNA is labeled with pairs of DNA probes followed by enzymatic amplification (Wang et al., 2018). Using a combination of hydrogel-tissue chemistry, targeted signal amplification, and in situ sequencing, STARmap has been verified to simultaneously identify hundreds to around one thousand genes in cell culture and mouse brain samples. Another targeted method Baristaseq, which used a combination of gap-filling polymerase and padlock probe to create a DNA circle, has been used to map long-range neuronal projections and trace lineage (Chen et al., 2018).

In each of the targeted methods aforementioned, a short barcode is synthesized or filled out on the padlock probe. However, the untargeted methods generate the optical barcode using the RNA sequence itself. Untargeted spatial transcriptome enables a comprehensive profiling of the entire transcriptome, which allows the discovery of novel gene interactions, cell clusters, and tissue heterogeneity. This approach was pioneered by fluorescence in situ sequencing (FISSEQ) and its later iteration, ExSeq (Alon et al., 2021; Lee et al., 2014). FISSEQ depends on in situ reverse transcription and RCA-amplified cDNAs, which are cross-linked to the cell matrix and ligated by CircLigase. The fluorescent barcode is hybridized to cDNA amplicons and then sequenced within a biological sample. ExSeq, which combines expansion microscopy with RNA anchoring to an expandable hydrogel,

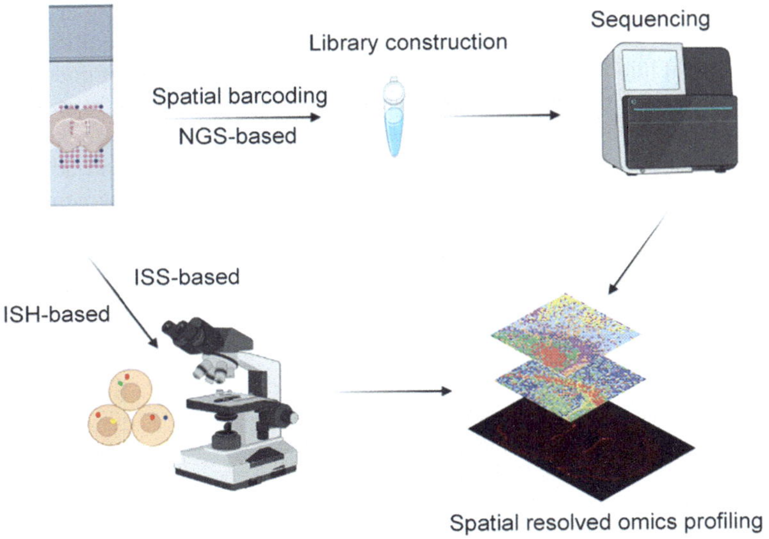

Fig. 1 Schematic workflow of spatial omics. Summary of spatial omics technologies: In situ hybridization (ISH)-based, in situ sequencing (ISS)-based, and next-generation sequencing (NGS)-based methods. (The diagram was drawn with Adobe Photoshop software)

has demonstrated that it can detect tens to hundreds of genes in mouse tissues and human breast cancer samples (Alon et al., 2021).

1.1.3 Next-Generation Sequencing-Based Approaches

With the rapid development of NGS-based techniques in the last decade, it has become easier to understand biological processes in greater detail. These techniques are based on single-cell RNA sequencing technology with the addition of a spatial barcode before next-generation sequencing to encode positional information (Fig. 1). Despite the rapid growth of NGS-based techniques in recent years, their key feature lies within their resolution, throughput, and multiplexing. Ståhl et al. generated the first NGS-based spatial transcriptomic technology that enables the sequencing of whole transcriptomes in mouse tissues and human breast cancer (Ståhl et al., 2016). Compared with ISH- and most ISS-based methods, spatial transcriptome is unbiased to capture all polyadenylated transcripts, depending on their binding to spatially barcoded microarray slides. This method can provide quantitative gene expression data for tissue sections with a relatively low resolution in hundreds of micrometers. The commercialized technique, Visium (Tirado-Lee, 2020), has improved the resolution by using spatially barcoded mRNA-binding oligonucleotides grouped in spots (55 μm spot diameter with 100 μm center-to-center distance). On the other hand, another NGS-based method Slide-Seq, which is based on unique barcoded hydrogel beads, has achieved a higher resolution of 10 μm (Rodriques et al., 2019). A similar method, termed high-definition spatial transcriptomics (HDST), provides the transcriptome-wide coverage in many systems at a resolution of 2 μm, which is an actual single-cell resolution (Vickovic et al., 2019). While advancements in the capability to achieve comprehensive transcriptomics have addressed numerous issues, the constraints of these methods persist due to their dependence on solid beads rather than oligonucleotide arrays. Bead-based methods are limited by their mRNA capture efficiency in comparison to oligonucleotide array methods.

A method called spatial temporal enhanced resolution omics sequencing (Stereo-Seq), which combines DNA nanoball-patterned arrays and in situ RNA capture, is capable of achieving nanoscale resolution in mice embryos (Chen et al., 2022). As another method of sequencing-based analysis, Seq-Scope amplifies barcoded oligonucleotides using solid-phase amplification to obtain subcellular resolution (0.5–0.8 μm) (Cho et al., 2021). All of these methods aforementioned come with significant cost and speed limitation. However, another polony-based method, PIXEL-seq, could produce high-resolution oligonucleotide arrays and enable repeatable enzymatic replication of barcode-patterned gels (Fu et al., 2022). PIXEL-seq is an innovative approach since it has the capability of performing spatial transcriptomics at a resolution of 1.22 μm with impressive sensitivity. The parabrachial nucleus of the mouse has been used to demonstrate the efficiency of this method, and it is anticipated that this method will be adopted by many researchers.

Although spatial transcriptomics has enabled whole genome gene expression profiling in tissue, approaches for profiling and capturing epigenetic information at the tissue scale are lacking. Single-cell epigenomic assays are capable of resolving cell types and stating cell characterization from complex tissues, but they lose spatial information in the process. Spatially resolved chromatin accessibility profiling of tissue sections using next-generation sequencing (spatial-ATAC-seq) was recently developed to reveal region-specific epigenetic signatures in human and mouse brains (Deng et al., 2022b). A microfluidic deterministic barcoding method combined with embedded Tn5 transposition chemistry is used to identify genomic regions that can be accessed. Llorens-Bobadilla et al. also developed the spatial ATAC (Buenrostro et al., 2013), which combines ATAC-seq with tagmented DNA capture on a solid surface containing barcoded oligonucleotides (Llorens-Bobadilla et al., 2023). This method has been used to investigate the regulatory programs underlying spatial gene expression both during mouse organogenesis and lineage differentiation as well as in human pathology.

An intricate functional and spatial connection between different cell types and states creates the complexity of a neuron network. A comprehensive analysis of chromatin states will provide crucial insights into systematically characterizing cellular phenotypes in neuroscience research. Current methods for measuring chromatin states are based on next-generation sequencing at a single-cell resolution. Recent technological breakthroughs, such as Spatial-CUT&Tag (Deng et al., 2022a), have improved our understanding of cellular heterogeneity. Spatial-

CUT&Tag combines in situ CUT&Tag, microfluidic deterministic barcoding, and next-generation sequencing, providing the ability to profile histone modifications at the tissue scale. As a result of this method, histone modifications determine the epigenetic state of the cortical layer development and the spatial patterning of various cell types. However, more systematic strategies for profiling multiplexed information about genomics modalities are needed. In heterogeneous tissues, more innovative approaches are needed to fully characterize chromatin states at single-cell resolution, such as directly profiling multiple modalities in the same sample. Recently, one approach, termed multiple target identification by tagmentation (MulTI-Tag), enables the detection of up to three histone modifications in the same cells (Meers et al., 2022). To realize high accurate for on-target peaks, Meers et al. improved MulTI-Tag with several modifications: (1) load pA-Tn5 onto first antibody-conjugated adapters; (2) tagment target chromatins in sequence; (3) add a second antibody followed by pA-Tn5 loaded with adapter. Two similar methods also have been developed that enable the measurement of multiple modalities of histone modification and chromatin state via the transposition of nanobody-tethers followed by sequencing: nanobody-tethered transposition followed by sequencing NTT-seq (Stuart et al., 2022) and nano-CUT&Tag (nano-CT) (Bartosovic & Castelo-Branco, 2023). Such methods simultaneously profile three histone modifications by using nanobodies in place of protein A/G; however, MulTI-Tag takes a different strategy, which involves complex experimental workflows. The use of these methods has been demonstrated at the cellular level, but we believe the technology development will advance quickly and allow the detection of multiplex modalities at the tissue scale in the near future.

Multimodal profiling of genome, transcriptome, epigenome, and proteomes has been immensely valuable for providing biological insights into tissue organization. Multimodal assays have already been validated for more refined characterization of cell types and cell states. This is exemplified by many assays: (1) scM&T and sc-GEM for profiling of single-cell genome-wide methylome and transcriptome sequencing (Angermueller et al., 2016; Cheow et al., 2016); (2) CITE-seq for profiling of transcriptomes and epitopes (Stoeckius et al., 2017); (3) SNARE-seq and SHARE-seq for profiling of single-nucleus chromatin accessibility and transcriptomes (Chen et al., 2019; Ma et al., 2020); (4) DBiT-Seq for co-mapping transcriptomes and proteins at tissue scale (Liu et al., 2020); (5) spatial-ATAC-RNA-seq or spatial-CUT&Tag-RNA-seq for profiling of chromatin accessibility or histone modification and gene expression on the same tissue section at a resolution near single cell (Fan et al., 2022).

In all NGS-based approaches, the spatial barcodes are embedded in chips/glass slides or delivered to the tissue surface through microfluidic channels, and the barcoded RNAs and genomic DNAs are collected and processed for sequencing. Spatial omics technologies are improving resolution and sensitivity, allowing multiplexing dimensions of different modalities to provide deeper insights.

2 Insights into the Development and Disease

Considering that the nervous system is highly complex, the exact mechanisms, spatial regulations, and functions of the neuron and non-neuron cells involved in neuron diseases remain relatively unknown. The recent technological advances based on next-generation sequencing have provided unparalleled resolution and depth to neuroscience research, reshaping our understanding of neuron diseases.

Within the mouse brain, a variety of cell types coexists, such as neurons, glial cells, and endothelial cells. The first attempt to use single-cell capture to analyze transcriptomes in the developing mouse cortex (302 cells from brain tissues) was published in 2014 (Pollen et al., 2014), in which Pollen et al. developed an efficient method for unbiased cell-type classification, such as multiple progenitor and neuronal subtypes. By using large-scale single-cell RNA sequencing, another study analyzed the transcriptomes of 3,000 single cells derived from mouse somatosensory cortex and hippocampal CA1 region (Zeisel et al., 2015). A total of 47 molecularly distinct subclasses have been recovered, including all known major cell types in the cortex. SPLiT-seq (Split-pool ligation-based transcriptome sequencing) allows the profiling of single cells and has been employed for the analysis of 156,040 single-nucleus transcriptomes from mouse brains and spinal cord (Rosenberg et al., 2018). This study identified more than 100 cell types and provided a snapshot of gene expression patterns of the mouse central nervous system during early postnatal development. Single-cell sequencing also has been used to study the diversity of neural circuits, decipher the role of non-neuronal cells, and detect the molecular changes that occur in disease. Although this sequencing tool is powerful, it is not without its challenges. One challenge is that generating single-cell suspensions from brain tissues is difficult because neurons and non-neurons are tightly bound and difficult to isolate. Another challenge is that some NGS-based methods depend on fresh samples, which are inherently limited. In contrast to cancer and pathologies, obtaining fresh intact tissues for neurodegenerative diseases is extremely difficult. The other major challenge is that single-cell sequencing requires isolated cells, resulting in a loss of spatial context.

It is important to note that spatial omics is an emerging field, and spatial methods are becoming increasingly important to neuroscience. Technological advancements will continue to drive this field forward, providing new insights into neural circuits, disease mechanisms, and brain function in the future.

2.1 Spatial Omics of Neural Development

Early methods, such as LCM-seq (Nichterwitz et al., 2016) and Geo-seq (Chen et al., 2017), combine single-cell sequencing and laser capture microdissection (LCM), which only focus on the targeted acquisition of region of interest (ROI). An LCM-seq study has been conducted on neurons isolated from mouse tissues, which provides biological insight into the neuronal population and elucidates gene expression in distinct cellular populations in human and mouse. Hybridization-based techniques also have been applied to tissues, including the mouse and human brains. For example, in probabilistic cell typing by in situ sequencing (pciSeq), Qian et al. mapped the inhibitory neurons and revealed their spatial arrangement in the mouse hippocampal area CA1 (Fig. 2a) (Qian et al., 2020). MERFISH has been used to profile 4000 gene transcriptions, which produced molecularly defined and spatially resolved cell atlas of the human cortex (Fang et al., 2022). The main drawback of hybridization-based techniques is the requirement to decide on the targeted genes in advance, which increases labor intensity as targets increase. Spatial methods also provide a new interpretation of cell density and cell fate determination. In another study that used single-cell transcriptome profiling and MERFISH to examine microglia subtype identity and fate, the scientists found that the subtypes of the local pyramidal neurons determine microglia fate and state identity (Stogsdill et al., 2022).

During the past decade, spatial transcriptomics methods have advanced rapidly, which can overcome the limitations of biased gene selection and gene coverage. By using spatial transcriptomics (ST) (Ståhl et al., 2016), Ortiz et al. generated a systematic reconstruction of the neuroanatomical mouse brain and established a molecular atlas to define the spatial structure of brain regions (Ortiz et al., 2020). While a limitation of spatial transcriptomics methods is that individual measurements may contain contributions from multiple cells (Cable et al., 2022), a computational method, termed robust cell type decomposition (RCTD), was developed, which decomposes spatial and single-cell profiles of tissues. Therefore, this method is well-suited for defining cellular identity and organization in biological tissues. Another leap forward was based on the improvement of spatial resolution to 1–10 μm, such as Slide-seq (Rodriques et al., 2019), DBiT-Seq (Liu et al., 2020), Seq-Scope (Cho et al., 2021), Pixel-seq (Fu et al., 2022), and Stereo-seq (Chen et al., 2022). These methods increase sensitivity and enable partial access to subcellular data. For example, Pixel-seq was applied to create a cell atlas of the mouse parabrachial nucleus (PBN), and the author also analyzed changes in neuropathic pain-regulated spatial transcriptomes and cell-cell communication after nerve ligation (Fu et al., 2022).

Fully understanding cell identity in the central nervous system (CNS) requires profiling epigenetic information at the tissue scale. Spatial chromatin modification profiling (Deng et al., 2022a) and spatial-ATAC-seq (Deng et al., 2022b) have been verified in mouse development by the same lab. In the mouse brain, Spatial-CUT&Tag revealed epigenetic control over mouse cortex layers and spatial patterning of cell types. For example, Spatial-CUT&Tag data of H3K27ac modifications could divide radial glia into three clusters: Radial glia_C1, Radial glia_C2, and Radial glia_C3 (Fig. 2b). Moreover, spatial-ATAC-seq identified gene regulators, epigenetic landscapes, and tissue-regional epigenetic profiles during central nervous system development (Fig. 2c). These diverse spatial omics methods have crucial implications for facilitating cell types and cell identity in neuroscience research.

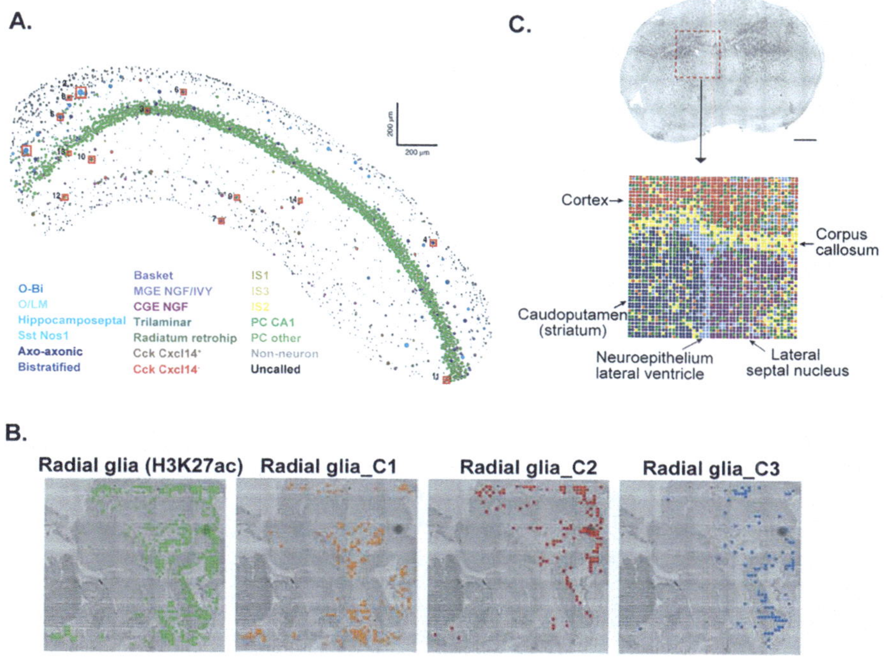

Fig. 2 Neuronal clusters exhibit regional specificity. (**a**) Spatial mapping of cell types across mouse hippocampus CA1. Cells are represented by pie charts of area proportional to the number of reads assigned to the cell. Reproduce from Qian (Qian et al., 2020). (**b**) Spatial mapping of E11 mouse embryo by spatial CUT&Tag. H3K27ac profiling of radial glial enabled the identification of three subpopulations. Reproduce from Deng (Deng et al., 2022a). (**c**) Spatial chromatin accessibility mapping of P21 mouse brain with a 20 μm pixel size. Unsupervised clustering analysis of each cluster in the mouse brain. Reproduce from Deng (Deng et al., 2022b). (The diagram was drawn with Adobe Photoshop software)

2.2 *Spatial Omics of Neural Disease*

As spatial omics methods provide an unbiased manner to analyze the multiple dimensions of the spatial organization, they have been applied to the study of neuronal diseases and neurodegenerative diseases, such as Alzheimer's disease (AD), major depression, and autism. A transformative work on Alzheimer's disease was published in 2017 (Keren-Shaul et al., 2017), which took advantage of scRNA-seq to identify a unique signature that contained a novel microglia type, termed disease microglia (DAM). This study demonstrated the potential of scRNA-seq to unveil new cell types associated with AD, which is conserved in mice and humans. In another exciting study, single-nucleus sequencing has been applied to generate single-cell transcriptomes profiling major depressive disorder (MDD), which is a complex and heterogeneous disease (Nagy et al., 2020). In this study,

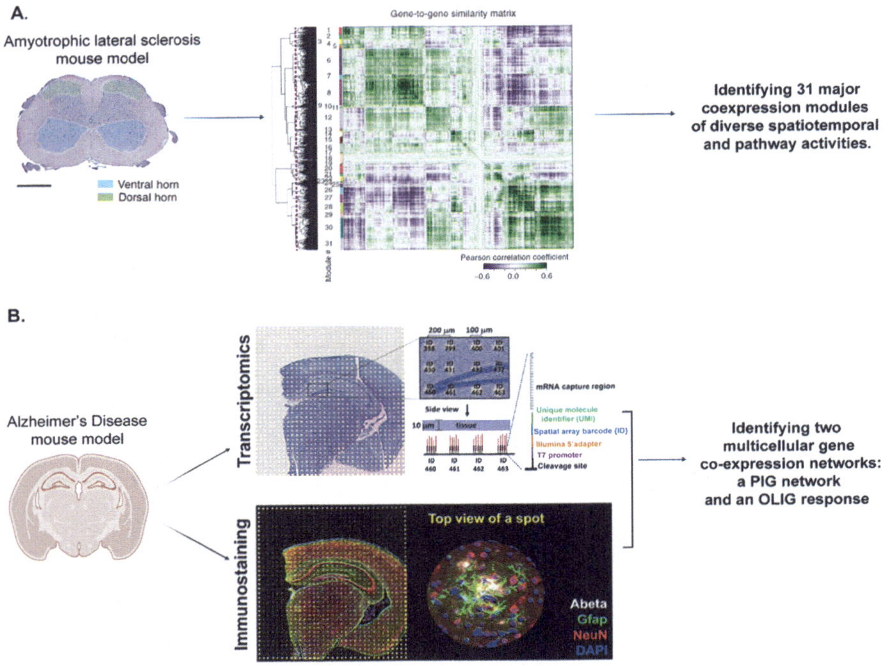

Fig. 3 Illustrative examples of spatial profiling of gene expression during disease progression. (**a**) Biclustering of the mouse spatial gene expression measurements reveals spatially and temporally coexpressed genes in amyotrophic lateral sclerosis. The dashed vertical purple line in the dendrogram denotes the break. Identifiers of coexpression modules are listed. Reproduced from Maniatis (Maniatis et al., 2019). (**b**) Spatially resolved transcriptomics profiles in Alzheimer's disease mouse model. Sequential coronal sections of mouse brains were collected, and the adjacent sections were used for spatial transcriptomics and immunostaining. Reproduced from Chen (Chen et al., 2020). (The diagram was drawn with Adobe Photoshop software)

snRNA-seq was shown to be effective at dissecting subtle phenotypes in brain tissue with improved resolution. Even though bulk gene expression studies have produced comprehensive changes in autism patients, these data are unable to provide specific information about the types of cells affected. A recent study has used single-nucleus RNA sequencing in cortical tissue from autism patients to show that dysregulation of specific groups of genes in projection neurons correlates with autism and molecular changes in cortical circuits are linked to autism (Velmeshev et al., 2019). Spatial transcriptomics (ST) has been leveraged to map amyotrophic lateral sclerosis (ALS), in which 31 distinct, disease-relevant gene modules were identified for ALS mice across key time points (Fig. 3a) (Maniatis et al., 2019). In another study, spatial transcriptomics (ST) and in situ sequencing (ISS) were combined to identify two multicellular gene co-expression networks in Alzheimer's disease (Fig. 3b) (Chen et al., 2020). By using Visium platform, Maynard et al. mapped the spatial gene expression patterns in the human dorsolateral prefrontal cortex, providing numerous insights into cell populations' spatial and molecular definitions (Maynard et al., 2021). Gregory et al. introduced the first description of spatial transcriptomics (100 μm) in postmortem cortical samples from ALS patients, which identified genes contributing to regional ALS vulnerabilities and provided new therapeutic and diagnostic possibilities (Gregory et al., 2020). The spatial transcriptomics technique was applied to a mouse model of Alzheimer's disease to identify genes affecting stress responses and transcription in both the hippocampus and olfactory bulb (Navarro et al., 2020). Among them, there are genes that are spatially differentially expressed, which may contribute to the better understanding of AD pathology.

Taken together, these examples illustrate the opportunities of directly profiling gene expression data at the single-cell level. However, spatial transcriptomics provides a more powerful tool for understanding the cellular context of the brain, enabling further studies of the specific types of cells in healthy and diseased brains.

3 Conclusion

Single-cell sequencing has driven breakthroughs in biology and neuroscience research as well. Over the past decade, many cutting-edge spatial omics technologies have led to tremendous discoveries in neuroscience and the molecular mechanisms underlying neurodegenerative diseases. This section outlines the current state-of-the-art genomic approaches and their applications in revealing insights into the underlying neuroscience. Using future innovations, epigenome, transcriptome, and protein expression may be analyzed in the same cell, making it possible to analyze the gene regulatory network at the tissue scale comprehensively.

Spatial transcriptomics has made significant advances in our understanding of neural circuitry and neurodegenerative disease, such as amyotrophic lateral sclerosis (Maniatis et al., 2019) and Alzheimer's disease (Chen et al., 2020). However, spatial transcriptomics only provides one layer of spatial information at the tissue scale. Multiple layers of spatial information, such as transcriptome, epigenome, and proteome, are required to build a more comprehensive picture in neuroscience. There has been recent progress in co-profiling two layers of information at a cellular level in a spatially resolved manner. For example, spatial-ATAC-RNA-seq and spatial-CUT&Tag-RNA-seq have been applied for co-profiling of genome-wide chromatin accessibility or histone modification in conjunction with whole transcriptomes at mouse and human hippocampus tissue section (Fan et al., 2022). These approaches will facilitate the study of neurodegenerative diseases, which will lead to a greater understanding of the field. Previously, DBiT-seq has been demonstrated to co-profiling of transcriptome and a panel of proteins (Liu et al., 2020). Future innovations will enable epigenome, transcriptome, and proteome profiles to be co-profiled on the same sample, which will allow a 3D tissue atlas to be reconstructed to visualize spatial changes. On all these fronts, the combination of multiplex omics data poses another challenge in data analysis.

Integrating spatial information across multiple modalities needs new technologies and computational tools. Taken together, researchers are revolutionizing biology research with spatial omics, opening new frontiers in neuroscience, and revealing unprecedented information about neural disease. We envision that spatial omics methods will transform our knowledge about dynamic changes and complex tissue architectures during development and disease.

References

Alon, S., Goodwin, D. R., Sinha, A., Wassie, A. T., Chen, F., Daugharthy, E. R., Bando, Y., Kajita, A., Xue, A. G., & Marrett, K. (2021). Expansion sequencing: Spatially precise in situ transcriptomics in intact biological systems. *Science, 371*, eaax2656.

Angermueller, C., Clark, S. J., Lee, H. J., Macaulay, I. C., Teng, M. J., Hu, T. X., Krueger, F., Smallwood, S. A., Ponting, C. P., & Voet, T. (2016). Parallel single-cell sequencing links transcriptional and epigenetic heterogeneity. *Nature Methods, 13*, 229–232.

Bartosovic, M., & Castelo-Branco, G. (2023). Multimodal chromatin profiling using nanobody-based single-cell CUT&Tag. *Nature Biotechnology, 41*, 794–805.

Bauman, J., Wiegant, J., Borst, P., & Van Duijn, P. (1980). A new method for fluorescence microscopical localization of specific DNA sequences by in situ hybridization of fluorochrome-labelled RNA. *Experimental Cell Research, 128*, 485–490.

Buenrostro, J. D., Giresi, P. G., Zaba, L. C., Chang, H. Y., & Greenleaf, W. (2013). Transposition of native chromatin for fast and sensitive epigenomic profiling of open chromatin, DNA-binding proteins and nucleosome position. *Nature Methods, 10*, 1213–1218.

Cable, D. M., Murray, E., Zou, L. S., Goeva, A., Macosko, E. Z., Chen, F., & Irizarry, R. A. (2022). Robust decomposition of cell type mixtures in spatial transcriptomics. *Nature Biotechnology, 40*, 517–526. https://doi.org/10.1038/s41587-021-00830-w

Chen, A., Liao, S., Cheng, M., Ma, K., Wu, L., Lai, Y., Qiu, X., Yang, J., Xu, J., & Hao, S. (2022). Spatiotemporal transcriptomic atlas of mouse organogenesis using DNA nanoball-patterned arrays. *Cell, 185*, 1777–1792. e1721.

Chen, J., Suo, S., Tam, P. P. L., Han, J.-D. J., Peng, G., & Jing, N. (2017). Spatial transcriptomic analysis of cryosectioned tissue samples with Geo-seq. *Nature Protocols, 12*, 566–580. https://doi.org/10.1038/nprot.2017.003

Chen, S., Lake, B. B., & Zhang, K. (2019). High-throughput sequencing of the transcriptome and chromatin accessibility in the same cell. *Nature Biotechnology, 37*, 1452–1457.

Chen, W.-T., Lu, A., Craessaerts, K., Pavie, B., Frigerio, C. S., Corthout, N., Qian, X., Laláková, J., Kühnemund, M., & Voytyuk, I. (2020). Spatial transcriptomics and in situ sequencing to study Alzheimer's disease. *Cell, 182*, 976-991. e919.

Chen, X., Sun, Y.-C., Church, G. M., Lee, J. H., & Zador, A. M. (2018). Efficient in situ barcode sequencing using padlock probe-based BaristaSeq. *Nucleic Acids Research, 46*, e22–e22.

Cheow, L. F., Courtois, E. T., Tan, Y., Viswanathan, R., Xing, Q., Tan, R. Z., Tan, D. S., Robson, P., Loh, Y.-H., & Quake, S. R. (2016). Single-cell multimodal profiling reveals cellular epigenetic heterogeneity. *Nat Methods, 13*, 833–836.

Cho, C.-S., Xi, J., Si, Y., Park, S.-R., Hsu, J.-E., Kim, M., Jun, G., Kang, H. M., & Lee, J. H. (2021). Microscopic examination of spatial transcriptome using Seq-Scope. *Cell, 184*, 3559–3572. e3522.

Codeluppi, S., Borm, L. E., Zeisel, A., La Manno, G., van Lunteren, J. A., Svensson, C. I., & Linnarsson, S. (2018). Spatial organization of the somatosensory cortex revealed by osmFISH. *Nature Methods, 15*, 932–935.

Colón-Ramos, D. A. (2009). Synapse formation in developing neural circuits. *Curr Top Dev Biol, 87*, 53–79.

Deng, Y., Bartosovic, M., Kukanja, P., Zhang, D., Liu, Y., Su, G., Enninful, A., Bai, Z., Castelo-Branco, G., & Fan, R. (2022a). Spatial-CUT&Tag: spatially resolved chromatin modification profiling at the cellular level. *Science, 375*, 681–686.

Deng, Y., Bartosovic, M., Ma, S., Zhang, D., Kukanja, P., Xiao, Y., Su, G., Liu, Y., Qin, X., & Rosoklija, G. B. (2022b). Spatial profiling of chromatin accessibility in mouse and human tissues. *Nature, 609*, 375–383.

Eng, C.-H. L., Lawson, M., Zhu, Q., Dries, R., Koulena, N., Takei, Y., Yun, J., Cronin, C., Karp, C., & Yuan, G.-C. (2019). Transcriptome-scale super-resolved imaging in tissues by RNA seqFISH+. *Nature, 568*, 235–239.

Fan, R., Zhang, D., Deng, Y., Kukanja, P., Bartosovic, M., Su, G., Bao, S., Liu, Y., Xiao, Y., & Ma, S. (2022). Spatially resolved epigenome-transcriptome co-profiling of mammalian tissues at the cellular level.

Fang, R., Xia, C., Close, J. L., Zhang, M., He, J., Huang, Z., Halpern, A. R., Long, B., Miller, J. A., Lein, E. S., & Zhuang, X. (2022). Conservation and divergence of cortical cell organization in human and mouse revealed by MERFISH. *Science, 377*, 56–62. https://doi.org/10.1126/science.abm1741

Femino, A. M., Fay, F. S., Fogarty, K., & Singer, R. H. (1998). Visualization of single RNA transcripts in situ. *Science, 280*, 585–590.

Fu, X., Sun, L., Dong, R., Chen, J. Y., Silakit, R., Condon, L. F., Lin, Y., Lin, S., Palmiter, R. D., & Gu, L. (2022). Polony gels enable amplifiable DNA stamping and spatial transcriptomics of chronic pain. *Cell, 185*, 4621–4633. e4617.

Gall, J. G., & Pardue, M. L. (1969). Formation and detection of RNA-DNA hybrid molecules in cytological preparations. *Proceedings of the National Academy of Sciences of the United States of America, 63*, 378–383.

Goh, J. J. L., Chou, N., Seow, W. Y., Ha, N., Cheng, C. P. P., Chang, Y.-C., Zhao, Z. W., & Chen, K. H. (2020). Highly specific multiplexed RNA imaging in tissues with split-FISH. *Nature Methods, 17*, 689–693.

Gray, P. A., Fu, H., Luo, P., Zhao, Q., Yu, J., Ferrari, A., Tenzen, T., Yuk, D. I., Tsung, E. F., Cai, Z., et al. (2004). Mouse brain organization revealed through direct genome-scale TF expression analysis. *Science (New York, N.Y.), 306*, 2255–2257. https://doi.org/10.1126/science.1104935

Gregory, J., McDade, K., Livesey, M., Croy, I., Marion de Proce, S., Aitman, T., Chandran, S., & Smith, C. (2020). Spatial transcriptomics identifies spatially dysregulated expression of GRM3 and USP47 in amyotrophic lateral sclerosis. *Neuropathology and Applied Neurobiology, 46*, 441–457.

Gyllborg, D., Langseth, C. M., Qian, X., Choi, E., Salas, S. M., Hilscher, M. M., Lein, E. S., & Nilsson, M. (2020). Hybridization-based in situ sequencing (HybISS) for spatially resolved transcriptomics in human and mouse brain tissue. *Nucleic Acids Research, 48*, e112–e112.

John, H., Birnstiel, M., & Jones, K. (1969). RNA-DNA hybrids at the cytological level. *Nature, 223*, 582–587.

Ke, R., Mignardi, M., Pacureanu, A., Svedlund, J., Botling, J., Wählby, C., & Nilsson, M. (2013). In situ sequencing for RNA analysis in preserved tissue and cells. *Nature Methods, 10*, 857–860.

Keren-Shaul, H., Spinrad, A., Weiner, A., Matcovitch-Natan, O., Dvir-Szternfeld, R., Ulland, T. K., David, E., Baruch, K., Lara-Astaiso, D., Toth, B., et al. (2017). A unique microglia type associated with restricting development of Alzheimer's disease. *Cell, 169*, 1276–1290.e1217. https://doi.org/10.1016/j.cell.2017.05.018

La Manno, G., Gyllborg, D., Codeluppi, S., Nishimura, K., Salto, C., Zeisel, A., Borm, L. E., Stott, S. R., Toledo, E. M., & Villaescusa, J. C. (2016). Molecular diversity of midbrain development in mouse, human, and stem cells. *Cell, 167*, 566-580. e519.

Lee, J. H., Daugharthy, E. R., Scheiman, J., Kalhor, R., Yang, J. L., Ferrante, T. C., Terry, R., Jeanty, S. S., Li, C., & Amamoto, R. (2014). Highly multiplexed subcellular RNA sequencing in situ. *Science, 343*, 1360–1363.

Lein, E. S., Hawrylycz, M. J., Ao, N., Ayres, M., Bensinger, A., Bernard, A., Boe, A. F., Boguski, M. S., Brockway, K. S., Byrnes, E. J., et al. (2007). Genome-wide atlas of gene expression in the adult mouse brain. *Nature, 445*, 168–176. https://doi.org/10.1038/nature05453

Liu, Y., Yang, M., Deng, Y., Su, G., Enninful, A., Guo, C. C., Tebaldi, T., Zhang, D., Kim, D., & Bai, Z. (2020). High-spatial-resolution multi-omics sequencing via deterministic barcoding in tissue. *Cell, 183*, 1665–1681. e1618.

Llorens-Bobadilla, E., Zamboni, M., Marklund, M., Bhalla, N., Chen, X., Hartman, J., Frisén, J., and Ståhl, P.L (2023). Solid-phase capture and profiling of open chromatin by spatial ATAC. Nature Biotechnology 1–4.

Lu, T., Ang, C. E., & Zhuang, X. (2022). Spatially resolved epigenomic profiling of single cells in complex tissues. *Cell, 185*, 4448–4464.e4417. https://doi.org/10.1016/j.cell.2022.09.035

Ma, S., Zhang, B., LaFave, L. M., Earl, A. S., Chiang, Z., Hu, Y., Ding, J., Brack, A., Kartha, V. K., & Tay, T. (2020). Chromatin potential identified by shared single-cell profiling of RNA and chromatin. *Cell, 183*, 1103–1116. e1120.

Maniatis, S., Äijö, T., Vickovic, S., Braine, C., Kang, K., Mollbrink, A., Fagegaltier, D., Andrusivová, Ž., Saarenpää, S., Saiz-Castro, G., et al. (2019). Spatiotemporal dynamics of molecular pathology in amyotrophic lateral sclerosis. *Science, 364*, 89–93. https://doi.org/10.1126/science.aav9776

Maynard, K. R., Collado-Torres, L., Weber, L. M., Uytingco, C., Barry, B. K., Williams, S. R., Catallini, J. L., Tran, M. N., Besich, Z., Tippani, M., et al. (2021). Transcriptome-scale spatial gene expression in the human dorsolateral prefrontal cortex. *Nature Neuroscience, 24*, 425–436. https://doi.org/10.1038/s41593-020-00787-0

Meers, M. P., Llagas, G., Janssens, D. H., Codomo, C. A., & Henikoff, S. (2022). Multifactorial profiling of epigenetic landscapes at single-cell resolution using MulTI-Tag. *Nature Biotechnology, 41*, 708. https://doi.org/10.1038/s41587-022-01522-9

Nagy, C., Maitra, M., Tanti, A., Suderman, M., Théroux, J.-F., Davoli, M. A., Perlman, K., Yerko, V., Wang, Y. C., Tripathy, S. J., et al. (2020). Single-nucleus transcriptomics of the prefrontal cortex in major depressive disorder implicates oligodendrocyte precursor cells and excitatory neurons. *Nature Neuroscience, 23*, 771–781. https://doi.org/10.1038/s41593-020-0621-y

Navarro, J. F., Croteau, D. L., Jurek, A., Andrusivova, Z., Yang, B., Wang, Y., Ogedegbe, B., Riaz, T., Støen, M., & Desler, C. (2020). Spatial transcriptomics reveals genes associated with dysregulated mitochondrial functions and stress signaling in Alzheimer disease. *Iscience, 23*, 101556.

Nichterwitz, S., Chen, G., Aguila Benitez, J., Yilmaz, M., Storvall, H., Cao, M., Sandberg, R., Deng, Q., & Hedlund, E. (2016). Laser capture microscopy coupled with Smart-seq2 for precise spatial transcriptomic profiling. *Nature Communications, 7*, 12139. https://doi.org/10.1038/ncomms12139

Ortiz, C., Navarro, J. F., Jurek, A., Märtin, A., Lundeberg, J., & Meletis, K. (2020). Molecular atlas of the adult mouse brain. *Science Advances, 6*, eabb3446.

Pollen, A. A., Nowakowski, T. J., Shuga, J., Wang, X., Leyrat, A. A., Lui, J. H., Li, N., Szpankowski, L., Fowler, B., & Chen, P. (2014). Low-coverage single-cell mRNA sequencing reveals cellular heterogeneity and activated signaling pathways in developing cerebral cortex. *Nature Biotechnology, 32*, 1053–1058.

Qian, X., Harris, K. D., Hauling, T., Nicoloutsopoulos, D., Muñoz-Manchado, A. B., Skene, N., Hjerling-Leffler, J., & Nilsson, M. (2020). Probabilistic cell typing enables fine mapping of closely related cell types in situ. *Nature Methods, 17*, 101–106. https://doi.org/10.1038/s41592-019-0631-4

Rodriques, S. G., Stickels, R. R., Goeva, A., Martin, C. A., Murray, E., Vanderburg, C. R., Welch, J., Chen, L. M., Chen, F., & Macosko, E. Z. (2019). Slide-seq: A scalable technology for measuring genome-wide expression at high spatial resolution. *Science, 363*, 1463–1467.

Rosenberg, A. B., Roco, C. M., Muscat, R. A., Kuchina, A., Sample, P., Yao, Z., Graybuck, L. T., Peeler, D. J., Mukherjee, S., & Chen, W. (2018). Single-cell profiling of the developing mouse brain and spinal cord with split-pool barcoding. *Science, 360*, 176–182.

Shah, S., Lubeck, E., Zhou, W., & Cai, L. (2016). In situ transcription profiling of single cells reveals spatial organization of cells in the mouse hippocampus. *Neuron, 92*, 342–357.

Ståhl, P. L., Salmén, F., Vickovic, S., Lundmark, A., Navarro, J. F., Magnusson, J., Giacomello, S., Asp, M., Westholm, J. O., & Huss, M. (2016). Visualization and analysis of gene expression in tissue sections by spatial transcriptomics. *Science, 353*, 78–82.

Stoeckius, M., Hafemeister, C., Stephenson, W., Houck-Loomis, B., Chattopadhyay, P. K., Swerdlow, H., Satija, R., & Smibert, P. (2017). Simultaneous epitope and transcriptome measurement in single cells. *Nature Methods, 14*, 865–868.

Stogsdill, J. A., Kim, K., Binan, L., Farhi, S. L., Levin, J. Z., & Arlotta, P. (2022). Pyramidal neuron subtype diversity governs microglia states in the neocortex. *Nature, 608*, 750–756. https://doi.org/10.1038/s41586-022-05056-7

Stuart, T., Hao, S., Zhang, B., Mekerishvili, L., Landau, D.A., Maniatis, S., Satija, R., & Raimondi, I (2022). Nanobody-tethered transposition enables multifactorial chromatin profiling at single-cell resolution. Nature Biotechnology 1–7.

Tirado-Lee, L. (2020). Introducing Visium gateway: Your entry to the world of spatial gene expression.

Velmeshev, D., Schirmer, L., Jung, D., Haeussler, M., Perez, Y., Mayer, S., Bhaduri, A., Goyal, N., Rowitch, D. H., & Kriegstein, A. R. (2019). Single-cell genomics identifies cell type–specific molecular changes in autism. *Science, 364*, 685–689.

Vickovic, S., Eraslan, G., Salmén, F., Klughammer, J., Stenbeck, L., Schapiro, D., Äijö, T., Bonneau, R., Bergenstråhle, L., & Navarro, J. F. (2019). High-definition spatial transcriptomics for in situ tissue profiling. *Nat Methods, 16*, 987–990.

Wang, X., Allen, W. E., Wright, M. A., Sylwestrak, E. L., Samusik, N., Vesuna, S., Evans, K., Liu, C., Ramakrishnan, C., & Liu, J. (2018). Three-dimensional intact-tissue sequencing of single-cell transcriptional states. *Science, 361*, eaat5691.

Xia, C., Fan, J., Emanuel, G., Hao, J., & Zhuang, X. (2019). Spatial transcriptome profiling by MERFISH reveals subcellular RNA compartmentalization and cell cycle-dependent gene expression. *Proceedings of the National Academy of Sciences of the United States of America, 116*, 19490–19499.

Zeisel, A., Muñoz-Manchado, A. B., Codeluppi, S., Lönnerberg, P., La Manno, G., Juréus, A., Marques, S., Munguba, H., He, L., & Betsholtz, C. (2015). Cell types in the mouse cortex and hippocampus revealed by single-cell RNA-seq. *Science, 347*, 1138–1142.

Part III
Application and Clinical Practice

Alzheimer's Disease from Modeling to Mechanism Research

Xiaoyan Sun and Weiqi Zhang

Abstract As our population continues to age, the search for effective therapeutic strategies to combat neurodegenerative diseases, particularly Alzheimer's disease (AD), has become more pressing than ever. For over a decade, researchers have focused on the amyloid cascade hypothesis in their pursuit of new drugs for AD. However, with numerous drugs targeting this hypothesis failing in clinical trials, it is clear that AD's pathogenesis is complex, and each individual may display significant heterogeneity. Consequently, treatment has shifted to focus on multiple targets and early AD detection. Furthermore, there is an urgent need to develop new models that address the shortcomings of current rodent models, which have species differences. The organoid model, a newly developed model, appears to be the future direction, but it must overcome some system immaturity problems.

Keywords Alzheimer's disease · Amyloid cascade hypothesis · Animal model · Organoid

X. Sun
CAS Key Laboratory of Genomic and Precision Medicine, Beijing Institute of Genomics, Beijing, China

China National Center for Bioinformation, Beijing, China

University of Chinese Academy of Sciences, Beijing, China

W. Zhang (✉)
CAS Key Laboratory of Genomic and Precision Medicine, Beijing Institute of Genomics, Beijing, China

China National Center for Bioinformation, Beijing, China

University of Chinese Academy of Sciences, Beijing, China

Institute for Stem Cell and Regeneration, Chinese Academy of Sciences, Beijing, China
e-mail: zhangwq@big.ac.cn

A. C.-H. Yu et al. (eds.), *Systems Neuroscience*, Advances in Neurobiology 41,
https://doi.org/10.1007/978-3-031-69188-1_7

1 Introduction

With the development of society and the advancement of medical and health conditions, life expectancy has increased, and world demographics are changing, with a rapid increase in the proportion of the elderly. Around 60 million people worldwide suffer from dementia, which refers to memory and cognitive deterioration (Collaborators, 2022). Most dementia (~70%) is caused by Alzheimer's disease (AD), a slowly progressive neurogenic disease characterized by the accumulation of neurological plaques and neurofibrillary tangles in the brain and gradually leading to neuronal and synaptic loss, brain atrophy, and death neuronal (Scheltens et al., 2021). There are about 50 million AD patients worldwide, which is expected to double every 5 years. By 2050, AD patients will exceed 150 million (Scheltens et al., 2021), making it the third most disabling and fatal disease after cardiovascular diseases and malignancies. Despite extensive basic and clinical research, no milestone has been achieved in the prevention and treatment of AD. The current treatments can only improve symptoms, not prevent or cure AD. AD-related research is urgent.

Alzheimer's disease was initially identified and described in 1907, followed in the 1980s and 1990s by many critical studies and the isolation of amyloid (Glenner & Wong, 1984a, b). Through genetic linkage analysis and cloning, the gene of amyloid precursor protein (APP) was identified (Kang et al., 1987; St George-Hyslop et al., 1987), and the first missense mutation of APP was detected in the AD family (Chartier-Harlin et al., 1991; Hardy, 2017; Hardy & Allsop, 1991). The scientific community is gradually recognizing the importance of amyloid in AD. In 1992, Hardy et al. proposed the amyloid cascade hypothesis (Hardy & Higgins, 1992), which suggests that the aggregation of amyloid-β (Aβ) triggers a series of cascade reactions, eventually leading to the pathology and symptoms of AD. These cascades include aggregation of hyperphosphorylated Tau (p-Tau) to form neurofibrillary tangles (NFTs), as well as inflammation and oxidative stress, leading to neuronal dysfunction and death (Selkoe & Hardy, 2016). Alongside these findings, the clinical diagnosis of AD has also developed. The continuum of AD is gradually determined (Ferrari & Sorbi, 2021), as it is not static dementia characterized by memory loss and impaired self-care ability, but a continuous progression from normal cognition and mild cognitive impairment (MCI) to dementia (Ferrari & Sorbi, 2021). In normal cognitive individuals, amyloid deposition happens first, followed by changes such as NFT accumulation, synaptic dysfunction, and neuronal damage, until the burden of damage is severe enough to cause behavioral cognitive impairment, initially as in MCI, and then further worsens until a diagnosis of dementia is made (Jack et al., 2010).

The study of the preclinical stages of AD has gained momentum among scientists due to the discovery of AD neuropathology and the concept of a continuum. The diagnosis of AD has gradually advanced from a clinical description to a neuropathological aspect, resulting in the development of biomarkers to assist with diagnosis (Ferrari & Sorbi, 2021). In 2018, NIA-AA (National Institute on Aging-Alzheimer's Association) recommended the A/T/N system as the diagnostic standard for future clinical trials, a crucial breakthrough representing the standardization of early clinical diagnosis and intervention of AD (Jack et al., 2018). The A/T/N system involves biomarkers such as amyloidosis (A), tau pathology (T), and neurodegeneration (N) where A can be identified by cortical amyloid PET (Amy-PET) and cerebrospinal fluid Aβ levels (Aβ CSF), T can be detected by p-Tau in CSF and Tau-PET, N represents biomarkers of neurodegeneration or neuronal damage, including brain atrophy in magnetic resonance imaging (MRI), hypometabolism fluorodeoxyglucose (FDG)-PET in bilateral temporoparietal lobes, and total Tau in CSF (Jack et al., 2018). The A/T/N system emphasizes that positive Aβ is a necessary condition for the pathological diagnosis of AD. Regardless of whether the patient has normal cognitive function, MCI, or dementia, as long as A is positive, the patient is included in the Alzheimer's disease continuum. Nevertheless, the A/T/N system has not been extensively utilized in clinical diagnosis and treatment due to limitations in the equipment (Ferrari & Sorbi, 2021).

After nearly a century of research and development, the essential features of AD have been identified. However, the complex nature of the disease means that it cannot be attributed to a single hypothesis (Ferrari & Sorbi, 2021; Pimplikar, 2009; Scheltens et al., 2021). AD exhibits a range of heterogeneity across its causes, genetics, clinical features, and pathology, which poses a significant challenge for the research (Ferrari & Sorbi, 2021). More importantly, the current AD research models are mainly rodents, and the success rate of drug discovery is very low due to species differences (Drummond & Wisniewski, 2017). Multiple clinical studies have been declared unsuccessful. The changes in the global aging population structure have further intensified the urgency of research on neurodegenerative diseases such as AD. It is vital to consolidate previous research and develop innovative models and technologies.

This chapter briefly reviews the pathogenesis of AD, such as the cholinergic hypothesis, the amyloid cascade hypothesis, and the neuroinflammation hypothesis, as well as the corresponding drug research, emphasizing the multifactorial pathogenesis of AD. Furthermore, we discuss the strengths and weaknesses of frequently used AD research models and evaluate their reliability. We believe that the continued enhancement and refinement of brain organoids will be an essential research paradigm in the future and point out the challenges and areas that lie ahead.

2 Multifactorial AD Neuropathology

Numerous hypotheses have been put forth over the years to explain the pathogenesis of AD, such as the cholinergic hypothesis, the amyloid cascade hypothesis, and the neuroinflammation hypothesis. However, the precise pathogenesis of AD is complex and still not fully understood. It is gradually recognized that AD is a complex neurodegenerative disease involving multiple regulatory factors and networks. Consequently, treating AD necessitates a comprehensive approach that simultaneously addresses multiple factors.

2.1 Cholinergic Hypothesis

One of the earliest theories regarding the development of AD is the cholinergic theory. Acetylcholine (ACh) is an important excitatory neurotransmitter secreted by cholinergic neurons, which is involved in signal transduction of advanced behaviors such as learning and memory. Choline acetyltransferase (ChAT) catalyzes the synthesis of acetylcholine from choline, acetyl-coenzyme A (acetyl-CoA), and ATP in presynaptic neurons. ACh is then released by vesicular follicles into the synaptic cleft, where it is quickly degraded by acetylcholinesterase (AChE) on postsynaptic neurons to acetate and choline (Chen et al., 2022). Clinical studies have shown that individuals with AD often have reduced cholinergic neurons, severe Ach deficiency, and a significant decrease in ChAT activity, resulting in impaired cholinergic signaling (Bowen et al., 1976). At present, the drugs used to treat AD mainly consist of cholinesterase inhibitors, such as donepezil, galantamine, huperzine A, and other drugs aimed at increasing the levels of acetylcholine and improving cognition and cellular function (Chen et al., 2022). However, these drugs only alleviate cognitive impairment in AD and do not provide a cure (Chen et al., 2022) (Fig. 1).

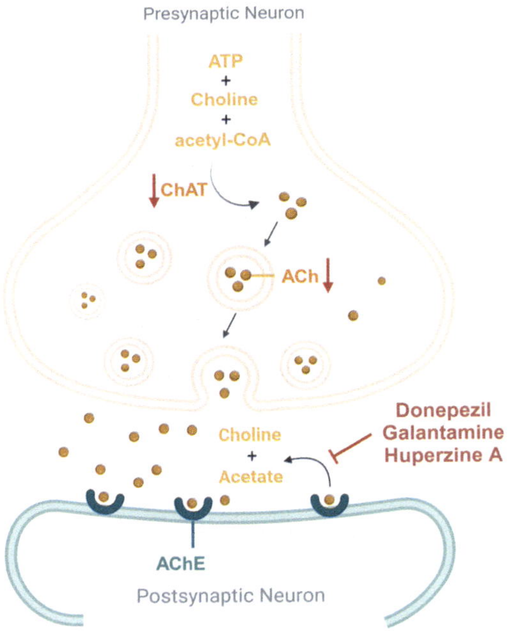

Fig. 1 The cholinergic hypothesis suggests that the significant reduction of cholinergic neurons results in lower ChAT activity, which synthesizes acetylcholine, and severe depletion of cholinergic neurotransmitters, leading to a decline in cognitive function. However, inhibiting the degradation of acetylcholine can help alleviate some of these symptoms of cognitive decline

2.2 *Amyloid Cascade Hypothesis*

In recent years, the widely accepted and yet highly controversial hypothesis is the amyloid cascade hypothesis, first proposed by John Hardy and David Allsop in 1992, which suggests that the chronic imbalance between production and clearance of Aβ peptides leads to the formation of neurotoxic Aβ deposits and neurofibrillary tangle, ultimately results in widespread neuronal/synaptic dysfunction and cell death (Hardy & Higgins, 1992; Selkoe & Hardy, 2016). While the deterministic nature of this model implies that Aβ is the primary molecular cause of AD and causes a cascade of downstream events, Frisoni et al. have recently proposed a probabilistic model of AD that could deepen our understanding of the hypothesis (Frisoni et al., 2022).

The cleaving of APP by β-secretase (β-site APP cleaving enzyme, BACE) and γ-secretase produces Aβ. The α-secretase cleaves APP to release the soluble extracellular domain (sAPPα) and a membrane-bound COOH terminal fragment (C83), which is subsequently cleaved by γ-secretase to release p3 and the APP intracellular domain (AICD). In the amyloidogenic pathway, APP is cleaved by β-secretase to generate soluble APPβ (sAPPβ) and an intracellular fragment (C99). C99 is further cleaved by γ-secretase to release AICD and Aβ, mainly Aβ40 and Aβ42. Aβ gradually accumulates and forms amyloid plaques. Aβ42 aggregates into plaques more readily than Aβ40 and is a major component of amyloid plaques (Ferrari & Sorbi, 2021). Tau is a microtubule-associated protein that attaches to microtubules, maintains microtubule integrity and function, and is involved in axonal transport. Hyperphosphorylation of tau impairs its ability to bind to microtubules, leading to microtubule disassembly, and it also accumulates, eventually forming NFTs in the cytoplasm of neurons (Brion, 1998). The accumulation of Aβ and tau can lead to a series of neuroinflammations, such as oxidative stress, activation of matrix metalloproteinases, activation of astrocytes and microglia, and release of inflammatory cytokines, leading to a vicious cycle (Wu et al., 2022).

The amyloid cascade hypothesis is supported by certain gene mutations related to the production and clearance of Aβ that can lead to Alzheimer's disease (AD), such as amyloid precursor protein (APP), presenilin (PSEN 1, PSEN 2), and apolipoprotein E (APOE). Mutations that cause Aβ accumulation largely contribute to the emergence of AD symptoms, such as cognitive impairment (Bateman et al., 2011). However, there are also protective mutations that can reduce Aβ secretion or increase Aβ clearance, potentially reducing the risk of AD (Jonsson et al., 2012). Studies have shown that the hyperphosphorylation of Tau can lead to the overproduction of Aβ and NFTs, ultimately leading to neurodegeneration. It is worth noting that AD patients with tau pathology alone are almost absent, while those with tau pathology almost always have amyloid plaques, suggesting that Aβ is a prerequisite for tau pathology (Jack et al., 2019). Clinical trials have also found that treatment with anti-Aβ drugs can decrease the content of p-Tau/Tau in the patient's cerebrospinal fluid, indicating that Aβ plays a crucial role in the onset of AD and is an upstream factor of Tau (Salloway et al., 2021).

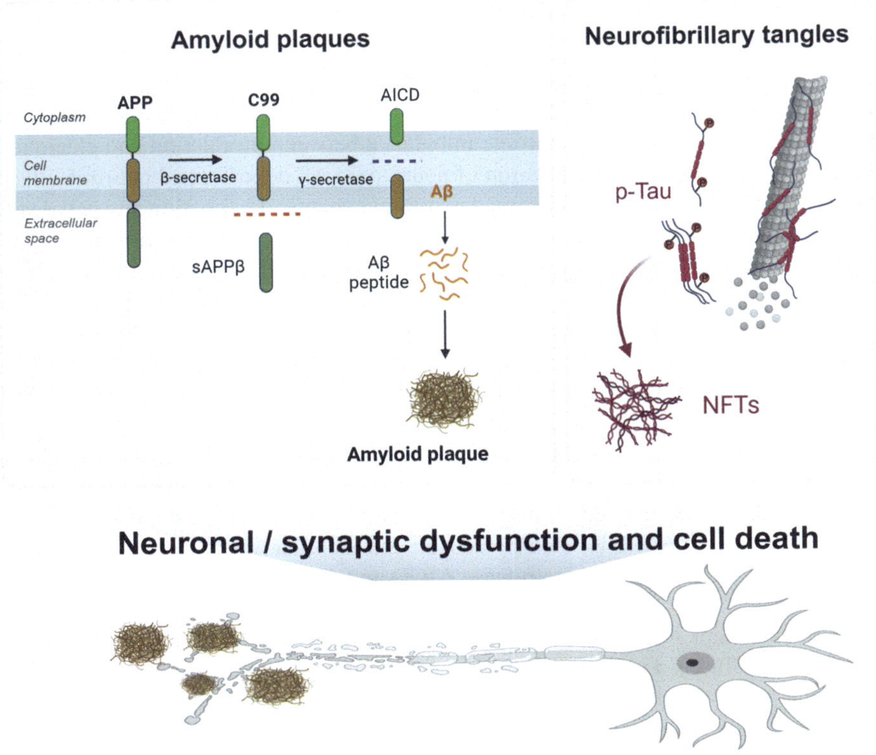

Fig. 2 When the APP protein is improperly cleaved, it forms Aβ, which then becomes insoluble amyloid plaques. This excess amyloid protein leads to a range of harmful processes in the brain, including the formation of neurofibrillary tangles, oxidative stress, and neuroinflammation. Over time, this can result in synaptic dysfunction and the death of neurons

According to the amyloid cascade hypothesis, Aβ is believed to be the earliest molecular trigger of the disease, leading to a series of downstream events (Jack et al., 2010, 2019). However, it is important to note that disruption of Aβ homeostasis alone may not be enough to cause a complete AD pathology. Studies on the transgenic mouse model of familial Alzheimer's disease (FAD) have shown that even with massive deposition of amyloid plaques, NFTs were not formed (Esquerda-Canals et al., 2017; Sasaguri et al., 2022). Several anti-Aβ drugs have been tested in clinical trials, such as AN-1792, solanezumab, bapineuzumab, semagacestat, and tarenflurbil, but failed to show efficacy in the final clinical stage (Cummings et al., 2019) (Fig. 2).

2.3 *Neuroinflammation*

Neuroinflammation is an immune response associated with glial cell activation in the central nervous system. Studies in recent years have found that neuroinflammation has become one of the major neuropathologies of AD. Inflammatory markers present in the AD brain, as well as AD risk genes associated with innate immune function, support the idea that neuroinflammation plays a significant role in AD pathogenesis. For example, TREM 2 (triggering receptor expressed on myeloid cells 2), which is highly expressed in microglia, was identified as an essential risk factor for AD (Whalley, 2021). Microglia and astrocytes are two types of cells that are crucial in the process of neuroinflammation (Leng & Edison, 2021; Mrak, 2012). In the early stages of AD, activated microglia gather near amyloid plaques, uptake, and clear extracellular plaques. However, when amyloid plaque accumulates excessively, microglia lose their ability to clear it away, leading to persistent neuroinflammation and further exacerbating neuronal damage and amyloid deposition, as pro-inflammatory cytokines like IL-1, TNF-α, and IL-6 are produced (Kinney et al., 2018). Similarly, astrocytes also can remove amyloid plaques. Studies have found that in animal models of AD, reactive astrocytes can release excess GABA (γ-aminobutyric acid) and glutamate, resulting in synaptic loss and memory impairment (Chang et al., 2013; Jo et al., 2014). The interaction between astrocytes and microglia influences the pathological process of AD. For example, astrocyte-derived interleukin-3 (IL-3) can shift the function of microglia from killing neurons to clearing amyloid protein deposition and NFT (McAlpine et al., 2021).

Recent studies suggest that neuroinflammation not only arises from Aβ activation but also plays a crucial role in the early progression of amyloidogenesis, independently of amyloid plaques (Onyango et al., 2021; von Bernhardi et al., 2015). Hur et al. found that inflammatory cytokines can induce the expression of interferon-induced transmembrane protein 3 (IFITM 3) in neurons and astrocytes, which binds γ-secretase, boosts γ-secretase activity, and in turn, increases Aβ production, highlighting the ability of neuroinflammation to promote amyloid plaque formation (Hur et al., 2020).

3 Modeling AD Neuropathology

Experimental models are a critical tool in gaining a better understanding of AD pathogenesis and testing new therapies. Over the past few decades, numerous models have made invaluable contributions to our knowledge of AD pathology and drug development, including human postmortem brain samples, animal models, cell lines, and more recently developed organoid models (Drummond & Wisniewski, 2017; Heuer et al., 2012; Papaspyropoulos et al., 2020; Penney et al., 2020; Yagi et al., 2011). In this section, we summarize current Alzheimer's disease models and discuss their reliability and limitations in simulating AD pathology.

3.1 *Animal Models*

3.1.1 Non-human Primates

Due to species advantages, one of the animal models that can most completely reproduce AD pathology is the non-human primate (NHP). NHPs naturally exhibit AD-like disease with aging, characterized by amyloid plaque formation and cognitive impairment (Rosen et al., 2011). However, a recent study found that AD-like NHP may be caused by aging rather than AD pathology, resulting in mild cognitive impairment in NHP, instead of the widespread cognitive impairment seen in human AD (Heuer et al., 2012). Unlike human AD, NHPs with AD-like diseases do not have NFTs (Lemere et al., 2004; Perez et al., 2013). However, developing spontaneous AD-like diseases in NHPs can take over 10 years. Therefore, several NHP models that accelerate AD pathology by injecting Aβ oligomers or introducing misfolded Tau were developed (Beckman et al., 2021; Yue et al., 2021). Batch injection of synthetic Aβ oligomers into the brain parenchyma has induced early pathological features of AD in cynomolgus monkeys, such as amyloid plaques accumulation, tau tangles, and neuroinflammation (Yue et al., 2021). Despite their usefulness, the high feeding costs and long time spans required for NHP still limit the application.

3.1.2 Rodent Models

To date, the vast majority of AD models are transgenic mice expressing mutated human genes that exhibit amyloid plaques (achieved by expressing human APP alone or co-expressed with human PSEN 1) and neurofibrillary tangles (via expression of human microtubule-associated protein tau (MAPT)), as well as a few rat models. The original transgenic mouse model expressed wild-type human APP, but this type of mouse did not show typical AD-associated pathology even with increased Aβ (Games et al., 1995; Hartman et al., 2005). Subsequently, multiple transgenic mouse models expressing APP and PSEN1 with FAD mutations have emerged, recapitulating AD features such as amyloid plaques, synaptic loss, and cognitive impairment, including PDAPP (Hartman et al., 2005), Tg2576 (Hsiao et al., 1996), APP23 (Sturchler-Pierrat et al., 1997), and APPPS1 (Radde et al., 2006), 5xFAD (Oakley et al., 2006).

Although transgenic mice with one or more FAD mutations can form amyloid plaques that are typical of AD, they are unable to form NFTs, which is another key feature of AD, and only partially exhibited hyperphosphorylation of Tau. To replicate both plaques and NFTs, a variety of transgenic mice expressing mutant forms of APP and MAPT have been created (Oddo et al., 2003). Among these, the 3xTg mouse model is widely used and is currently considered the most comprehensive mouse model that can recreate AD pathology (Oddo et al., 2003). Recently, AD knock-in mice have been developed to address the defect of APP overexpression present in the various transgenic mice aforementioned (Saito et al., 2014). Such overexpressed APP protein can be cleaved into Aβ and some non-physiological protein fragments, disrupting the normal physiological functions of mice, leading to various phenotypes that are unrelated to the AD (Drummond & Wisniewski, 2017; Sasaguri et al., 2022). Knock-in mice produce excess Aβ but do not express APP protein and, notably, cannot produce NFT (Li et al., 2014; Saito et al., 2014).

In addition to the widely utilized FAD model, numerous mouse models of sporadic Alzheimer's disease (SAD) have been developed. Although pathogenic mutations for SAD have not yet been identified, a large number of genetic loci that increase or decrease the risk of SAD have been reported, with APOE being by far the most important (Nakai et al., 2021). Several mice carrying the APOE gene modification have been developed, including APOE knock-in (APOE-KI), APOE knock-out (APOE-KO), and APOE-targeted replacement (APOE-TR) mice, which have also been bred with APP transgenic mice to help further study the relationship between APOE and amyloid (Esquerda-Canals et al., 2017).

Although transgenic rodents are commonly used as models for AD, they have their limitations and cannot fully replicate AD pathology. Firstly, depending on genetic background, promoters, and mutation sites, various transgenic mice show significant heterogeneity in plaque characteristics, synaptic damage, and cognitive impairment, making it difficult to compare results (Drummond & Wisniewski, 2017; Sasaguri et al., 2022). Secondly, due to species differences, rodent models cannot fully mimic AD pathologies, such as non-physiological overexpression of APP and Tau, lack of NFTs, absence of extensive neuronal damage, and cognitive impairment that occurs at different times compared with humans (Drummond & Wisniewski, 2017; Sasaguri et al., 2022). Moreover, most of these are FAD models, but clinically, FAD only accounts for about 5%, mainly SAD. These limitations may contribute to the high failure rate of clinical trials of drugs.

3.2 Cellular Models of AD

3.2.1 In Vitro Cell Cultures

Cell models have been widely utilized in AD research to explore pathology and evaluate new drugs, including primary neurons (mainly rat primary hippocampal cells), neuroblastoma, and various neurons derived from induced pluripotent stem cells (iPSCs). Thanks to advancements in iPSC technology and neural cell differentiation systems, there are now more cell models available than ever before (Penney et al., 2020). Nearly a decade ago, researchers started using patient cells with FAD and SAD to create iPSCs, which were then used to generate neurons displaying AD features, such as increased Aβ42/Aβ40 ratios and phosphorylated tau. However, these neurons were unable to form extra-amyloid plaques and NFTs (Israel et al., 2012; Yagi et al., 2011). Since then, researchers have conducted extensive studies using iPSC-derived neurons from AD, including glutamatergic neurons, GABAergic neurons, cholinergic neurons, and dopaminergic neurons. These neurons all replicate AD cellular phenotypes to different degrees, such as elevated levels of Aβ and phosphorylated Tau, enlarged endosomes, mitochondrial dysfunction, activation of oxidative stress, increased DNA damage, and sensitivity to Aβ toxicity (Chang et al., 2019; Ochalek et al., 2017; Ortiz-Virumbrales et al., 2017; Penney et al., 2020; Sproul et al., 2014; Vazin et al., 2014; Woodruff et al., 2013).

In addition, several studies have discovered a substantial amount of new genes associated with SAD through genome-wide association studies (GWAS), with a majority of them being expressed mainly in non-neuronal cells, such as microglia and astrocytes, in the brain (Jonsson et al., 2013; Lambert et al., 2013; Whalley, 2021). This indicates that non-neuronal cells have a crucial part to play in the pathogenesis of AD. IPSC-derived astrocytes carrying FAD or SAD mutations show a series of AD-related changes, such as morphological changes, increased Aβ42 release, oxidative stress, cholesterol accumulation, and changes in cytokine release (Jones et al., 2017; Oksanen et al., 2017). They also have a diminished neuroprotective effect, when co-cultured with neurons. (Jones et al., 2017). Little research has been done on newly developed iPSC-derived microglia, but an essential role for microglia-mediated neuroinflammation in AD has been identified (Brownjohn et al., 2018; Garcia-Reitboeck et al., 2018; Xu et al., 2019).

These in vitro differentiated cells have several advantages, such as human genetic background, ease of manipulation, and low research costs, but they also have significant limitations (Penney et al., 2020). The major issue is that these cellular models do not accurately represent the complex environment in the human brain, including the complex interactions between neurons and between neuronal and non-neuronal cells, which play an important role in the progression of AD (Cenini et al., 2021; Papaspyropoulos et al., 2020). This limitation is being partially addressed by developing 3D cell culture models.

3.2.2 Brain Organoids

Brain organoids are aggregates of specific cell types derived from progenitor cells, with the primary cell types and physiological architecture of the human brain (Trujillo & Muotri, 2018). Since 2013, when Lancaster et al. first developed self-organizing brain organoids (Lancaster et al., 2013), brain organoid technology has continued to mature and improve. In recent years, regional organoids have been created, such as forebrain organoids (Sloan et al., 2018), midbrain organoids (Jo et al., 2016), cerebellum organoids (Muguruma et al., 2015), the striatal organoid (Miura et al., 2020), and the choroid plexus organoid (Pellegrini et al.,

2020). Several studies have demonstrated that brain organoids cultured in vitro can simulate in vivo development with great accuracy, including neurogenesis, neuronal migration, cortical layering, and neural circuit establishment, which is a considerable advantage compared to 2D cells (Qian et al., 2019; Trujillo & Muotri, 2018). Patient-derived brain organoids created from iPSC have been utilized to study neurodevelopmental diseases, such as microcephaly (Krenn et al., 2021; Lancaster et al., 2013), Zika virus infections (Garcez et al., 2016), and psychiatric disorders (Paulsen et al., 2022), resulting in a deeper understanding of complex human diseases. However, there is still a lack of research on late-onset diseases, like neurodegenerative diseases in brain organoids, despite some preliminary studies being published (Di Lullo & Kriegstein, 2017; Grenier et al., 2020; Lee et al., 2017).

In 2014, Choi et al. first cultivated human neural progenitor cells expressing PSEN 1 and FAD mutant APP in 3D, established a brain organoid model of FAD, and simultaneously observed the presence of amyloid plaques and NFTs. They also found that glycogen synthase kinase 3 (GSK 3) regulates Aβ-mediated Tau phosphorylation (Choi et al., 2014). Subsequently, several studies have established iPSC-derived brain organoids from FAD and SAD patients (Gonzalez et al., 2018; Papaspyropoulos et al., 2020; Raja et al., 2016). The brain organoids from FAD patients with multiple APP repeats or PSEN1 spontaneously produced amyloid deposition and hyperphosphorylated Tau in an age-dependent manner, which can be significantly reduced by β- and γ-secretase inhibitors (Raja et al., 2016). The APP gene is located on chromosome 21, and patients with trisomy may experience certain pathologies linked to AD. Recently, Gonzalez et al. developed brain organoids derived from FAD and DS (Down syndrome) patients with endogenous protein expression levels. As time passes, Aβ spontaneously accumulates in brain organoids, eventually forming amyloid plaques and NFTs, as well as some degree of apoptosis (Gonzalez et al., 2018). In addition, by co-cultivating neurons, astrocytes, and microglia on a 3D microfluidic platform, Park et al. established a FAD brain organoid system miming neurodegeneration and neuroinflammation, hastening the pathogenic cascade of AD (Park et al., 2018).

SAD models, which are difficult to establish using animal models, have recently been successfully established using brain organoids, mainly harboring APOE mutations. Brain organoids carrying the APOE4 mutation exhibit elevated Aβ42 secretion in neurons, impaired cholesterol accumulation and Aβ uptake in astrocytes, and altered morphology of microglia-like cells, suggesting that APOE4-AD pathology is mainly influenced by impaired clearance of Aβ through astrocyte- and microglia-mediated processes, which can be attenuated by converting APOE4 to APOE3 (Lin et al., 2018). APOE4 also exacerbates tau pathology in brain organoids, and APOE may bind p-Tau through different molecular pathways (Zhao et al., 2020). Moreover, exposing brain organoids to serum, simulating blood-brain barrier (BBB) leakage, has successfully modeled SAD, exhibiting increased Aβ and p-Tau, synapse loss, and neural network damage (Chen et al., 2021).

Brain organoids have demonstrated the ability to accurately replicate typical AD pathologies, such as Aβ deposition, NFTs, and neurodegeneration, but apparent shortcomings need to be addressed (Bi et al., 2021; Gerakis & Hetz, 2019). Firstly, brain organoid technology is immature, exhibiting low synaptic activity, lack of cellular diversity, and an incomplete immune system. Microglia and blood vessels not only play an important role in the maturation of brain organoids but also are vital factors in AD disease progression. Second, SAD, which accounts for the majority of AD, primarily develops in the context of aging, but the current brain organoids are in the fetal stage, which may impact the results obtained. Co-culture with other types of cells and cross-application of bioengineering such as microfluidic systems, bioprinting technology, and ECM regulation may help address these issues in the future.

4 Conclusion

Over the past few decades, significant scientific research has been conducted on AD, leading to a deeper understanding of its pathogenesis, clinical diagnosis, and drug development. Central pathogenic mechanisms of AD include amyloid plaques, hyperphosphorylated Tau, and associated neurodegeneration. Additionally, the blood-brain barrier, neuroinflammation, the brain's immune system, and the gastrointestinal microbiota have also been identified as having a profound impact. As a result of these discoveries, clinical diagnosis has improved, and early intervention has become crucial for preventing irreversible neurotoxicity caused by amyloid plaques and NFTs. However, due to the lack of early samples from AD patients and shortcomings in current research models, cellular and molecular changes in the early stages of AD remain unclear. Moving forward, reliable research models will be necessary to gain a deeper understanding of AD pathogenesis, develop early diagnostic indicators, and create more effective interventions to enable early diagnosis and multi-targeting of AD.

References

Bateman, R. J., Aisen, P. S., De Strooper, B., Fox, N. C., Lemere, C. A., Ringman, J. M., Salloway, S., Sperling, R. A., Windisch, M., & Xiong, C. (2011). Autosomal-dominant Alzheimer's disease: A review and proposal for the prevention of Alzheimer's disease. *Alzheimer's Research & Therapy, 3*, 1. https://doi.org/10.1186/alzrt59

Beckman, D., Chakrabarty, P., Ott, S., Dao, A., Zhou, E., Janssen, W. G., Donis-Cox, K., Muller, S., Kordower, J. H., & Morrison, J. H. (2021). A novel tau-based rhesus monkey model of Alzheimer's pathogenesis. *Alzheimers Dement, 17*, 933–945. https://doi.org/10.1002/alz.12318

Bi, F. C., Yang, X. H., Cheng, X. Y., Deng, W. B., Guo, X. L., Yang, H., Wang, Y., Li, J., & Yao, Y. (2021). Optimization of cerebral organoids: A more qualified model for Alzheimer's disease research. *Translational Neurodegeneration, 10*, 27. https://doi.org/10.1186/s40035-021-00252-3

Bowen, D. M., Smith, C. B., White, P., & Davison, A. N. (1976). Neurotransmitter-related enzymes and indices of hypoxia in senile dementia and other abiotrophies. *Brain, 99*, 459–496. https://doi.org/10.1093/brain/99.3.459

Brion, J. P. (1998). Neurofibrillary tangles and Alzheimer's disease. *European Neurology, 40*, 130–140. https://doi.org/10.1159/000007969

Brownjohn, P. W., Smith, J., Solanki, R., Lohmann, E., Houlden, H., Hardy, J., Dietmann, S., & Livesey, F. J. (2018). Functional studies of missense TREM2 mutations in human stem cell-derived microglia. *Stem Cell Reports, 10*, 1294–1307. https://doi.org/10.1016/j.stemcr.2018.03.003

Cenini, G., Hebisch, M., Iefremova, V., Flitsch, L. J., Breitkreuz, Y., Tanzi, R. E., Kim, D. Y., Peitz, M., & Brustle, O. (2021). Dissecting Alzheimer's disease pathogenesis in human 2D and 3D models. *Molecular and Cellular Neurosciences, 110*, 103568. https://doi.org/10.1016/j.mcn.2020.103568

Chang, J., Liu, F., Lee, M., Wu, B., Ting, K., Zara, J. N., Soo, C., Al Hezaimi, K., Zou, W., Chen, X., et al. (2013). NF-kappaB inhibits osteogenic differentiation of mesenchymal stem cells by promoting beta-catenin degradation. *Proceedings of the National Academy of Sciences of the United States of America, 110*, 9469–9474. https://doi.org/10.1073/pnas.1300532110

Chang, K. H., Lee-Chen, G. J., Huang, C. C., Lin, J. L., Chen, Y. J., Wei, P. C., Lo, Y. S., Yao, C. F., Kuo, M. W., & Chen, C. M. (2019). Modeling Alzheimer's disease by induced pluripotent stem cells carrying APP D678H mutation. *Molecular Neurobiology, 56*, 3972–3983. https://doi.org/10.1007/s12035-018-1336-x

Chartier-Harlin, M. C., Crawford, F., Houlden, H., Warren, A., Hughes, D., Fidani, L., Goate, A., Rossor, M., Roques, P., Hardy, J., et al. (1991). Early-onset Alzheimer's disease caused by mutations at codon 717 of the beta-amyloid precursor protein gene. *Nature, 353*, 844–846. https://doi.org/10.1038/353844a0

Chen, X., Sun, G., Tian, E., Zhang, M., Davtyan, H., Beach, T. G., Reiman, E. M., Blurton-Jones, M., Holtzman, D. M., & Shi, Y. (2021). Modeling sporadic Alzheimer's disease in human brain organoids under serum exposure. *Advanced Science (Weinheim), 8*, e2101462. https://doi.org/10.1002/advs.202101462

Chen, Z. R., Huang, J. B., Yang, S. L., & Hong, F. F. (2022). Role of cholinergic signaling in Alzheimer's disease. *Molecules, 27*. https://doi.org/10.3390/molecules27061816

Choi, S. H., Kim, Y. H., Hebisch, M., Sliwinski, C., Lee, S., D'Avanzo, C., Chen, H., Hooli, B., Asselin, C., Muffat, J., et al. (2014). A three-dimensional human neural cell culture model of Alzheimer's disease. *Nature, 515*, 274–278. https://doi.org/10.1038/nature13800

Collaborators, G. B. D. D. F. (2022). Estimation of the global prevalence of dementia in 2019 and forecasted prevalence in 2050: An analysis for the Global Burden of Disease Study 2019. *The Lancet Public Health, 7*, e105–e125. https://doi.org/10.1016/S2468-2667(21)00249-8

Cummings, J., Lee, G., Ritter, A., Sabbagh, M., & Zhong, K. (2019). Alzheimer's disease drug development pipeline: 2019. *Alzheimer's & Dementia (New York), 5*, 272–293. https://doi.org/10.1016/j.trci.2019.05.008

Di Lullo, E., & Kriegstein, A. R. (2017). The use of brain organoids to investigate neural development and disease. *Nature Reviews. Neuroscience, 18*, 573–584. https://doi.org/10.1038/nrn.2017.107

Drummond, E., & Wisniewski, T. (2017). Alzheimer's disease: Experimental models and reality. *Acta Neuropathologica, 133*, 155–175. https://doi.org/10.1007/s00401-016-1662-x

Esquerda-Canals, G., Montoliu-Gaya, L., Guell-Bosch, J., & Villegas, S. (2017). Mouse models of Alzheimer's disease. *Journal of Alzheimer's Disease, 57*, 1171–1183. https://doi.org/10.3233/JAD-170045

Ferrari, C., & Sorbi, S. (2021). The complexity of Alzheimer's disease: An evolving puzzle. *Physiological Reviews, 101*, 1047–1081. https://doi.org/10.1152/physrev.00015.2020

Frisoni, G. B., Altomare, D., Thal, D. R., Ribaldi, F., van der Kant, R., Ossenkoppele, R., Blennow, K., Cummings, J., van Duijn, C., Nilsson, P. M., et al. (2022). The probabilistic model of Alzheimer disease: The amyloid hypothesis revised. *Nature Reviews. Neuroscience, 23*, 53–66. https://doi.org/10.1038/s41583-021-00533-w

Games, D., Adams, D., Alessandrini, R., Barbour, R., Berthelette, P., Blackwell, C., Carr, T., Clemens, J., Donaldson, T., Gillespie, F., et al. (1995). Alzheimer-type neuropathology in transgenic mice overexpressing V717F beta-amyloid precursor protein. *Nature, 373*, 523–527. https://doi.org/10.1038/373523a0

Garcez, P. P., Loiola, E. C., Madeiro da Costa, R., Higa, L. M., Trindade, P., Delvecchio, R., Nascimento, J. M., Brindeiro, R., Tanuri, A., & Rehen, S. K. (2016). Zika virus impairs growth in human neurospheres and brain organoids. *Science, 352*, 816–818. https://doi.org/10.1126/science.aaf6116

Garcia-Reitboeck, P., Phillips, A., Piers, T. M., Villegas-Llerena, C., Butler, M., Mallach, A., Rodrigues, C., Arber, C. E., Heslegrave, A., Zetterberg, H., et al. (2018). Human induced pluripotent stem cell-derived microglia-like cells harboring TREM2 missense mutations show specific deficits in phagocytosis. *Cell Reports, 24*, 2300–2311. https://doi.org/10.1016/j.celrep.2018.07.094

Gerakis, Y., & Hetz, C. (2019). Brain organoids: A next step for humanized Alzheimer's disease models? *Molecular Psychiatry, 24*, 474–478. https://doi.org/10.1038/s41380-018-0343-7

Glenner, G. G., & Wong, C. W. (1984a). Alzheimer's disease and Down's syndrome: Sharing of a unique cerebrovascular amyloid fibril protein. *Biochemical and Biophysical Research Communications, 122*, 1131–1135. https://doi.org/10.1016/0006-291x(84)91209-9

Glenner, G. G., & Wong, C. W. (1984b). Alzheimer's disease: Initial report of the purification and characterization of a novel cerebrovascular amyloid protein. *Biochemical and Biophysical Research Communications, 120*, 885–890. https://doi.org/10.1016/s0006-291x(84)80190-4

Gonzalez, C., Armijo, E., Bravo-Alegria, J., Becerra-Calixto, A., Mays, C. E., & Soto, C. (2018). Modeling amyloid beta and tau pathology in human cerebral organoids. *Molecular Psychiatry, 23*, 2363–2374. https://doi.org/10.1038/s41380-018-0229-8

Grenier, K., Kao, J., & Diamandis, P. (2020). Three-dimensional modeling of human neurodegeneration: Brain organoids coming of age. *Molecular Psychiatry, 25*, 254–274. https://doi.org/10.1038/s41380-019-0500-7

Hardy, J. (2017). The discovery of Alzheimer-causing mutations in the APP gene and the formulation of the "amyloid cascade hypothesis". *The FEBS Journal, 284*, 1040–1044. https://doi.org/10.1111/febs.14004

Hardy, J., & Allsop, D. (1991). Amyloid deposition as the central event in the aetiology of Alzheimer's disease. *Trends in Pharmacological Sciences, 12*, 383–388. https://doi.org/10.1016/0165-6147(91)90609-v

Hardy, J. A., & Higgins, G. A. (1992). Alzheimer's disease: The amyloid cascade hypothesis. *Science, 256*, 184–185. https://doi.org/10.1126/science.1566067

Hartman, R. E., Izumi, Y., Bales, K. R., Paul, S. M., Wozniak, D. F., & Holtzman, D. M. (2005). Treatment with an amyloid-beta antibody ameliorates plaque load, learning deficits, and hippocampal long-term potentiation in a mouse model of Alzheimer's disease. *The Journal of Neuroscience, 25*, 6213–6220. https://doi.org/10.1523/JNEUROSCI.0664-05.2005

Heuer, E., Rosen, R. F., Cintron, A., & Walker, L. C. (2012). Nonhuman primate models of Alzheimer-like cerebral proteopathy. *Current Pharmaceutical Design, 18*, 1159–1169. https://doi.org/10.2174/138161212799315885

Hsiao, K., Chapman, P., Nilsen, S., Eckman, C., Harigaya, Y., Younkin, S., Yang, F., & Cole, G. (1996). Correlative memory deficits, Abeta elevation, and amyloid plaques in transgenic mice. *Science, 274*, 99–102. https://doi.org/10.1126/science.274.5284.99

Hur, J. Y., Frost, G. R., Wu, X., Crump, C., Pan, S. J., Wong, E., Barros, M., Li, T., Nie, P., Zhai, Y., et al. (2020). The innate immunity protein IFITM3 modulates gamma-secretase in Alzheimer's disease. *Nature, 586*, 735–740. https://doi.org/10.1038/s41586-020-2681-2

Israel, M. A., Yuan, S. H., Bardy, C., Reyna, S. M., Mu, Y., Herrera, C., Hefferan, M. P., Van Gorp, S., Nazor, K. L., Boscolo, F. S., et al. (2012). Probing sporadic and familial Alzheimer's disease using induced pluripotent stem cells. *Nature, 482*, 216–220. https://doi.org/10.1038/nature10821

Jack, C. R., Jr., Knopman, D. S., Jagust, W. J., Shaw, L. M., Aisen, P. S., Weiner, M. W., Petersen, R. C., & Trojanowski, J. Q. (2010). Hypothetical model of dynamic biomarkers of the Alzheimer's pathological cascade. *Lancet Neurology, 9*, 119–128. https://doi.org/10.1016/S1474-4422(09)70299-6

Jack, C. R., Jr., Bennett, D. A., Blennow, K., Carrillo, M. C., Dunn, B., Haeberlein, S. B., Holtzman, D. M., Jagust, W., Jessen, F., Karlawish, J., et al. (2018). NIA-AA research framework: Toward a biological definition of Alzheimer's disease. *Alzheimer's & Dementia, 14*, 535–562. https://doi.org/10.1016/j.jalz.2018.02.018

Jack, C. R., Wiste, H. J., Botha, H., Weigand, S. D., Therneau, T. M., Knopman, D. S., Graff-Radford, J., Jones, D. T., Ferman, T. J., Boeve, B. F., et al. (2019). The bivariate distribution of amyloid-beta and tau: Relationship with established neurocognitive clinical syndromes. *Brain, 142*, 3230–3242. https://doi.org/10.1093/brain/awz268

Jo, S., Yarishkin, O., Hwang, Y. J., Chun, Y. E., Park, M., Woo, D. H., Bae, J. Y., Kim, T., Lee, J., Chun, H., et al. (2014). GABA from reactive astrocytes impairs memory in mouse models of Alzheimer's disease. *Nature Medicine, 20*, 886–896. https://doi.org/10.1038/nm.3639

Jo, J., Xiao, Y., Sun, A. X., Cukuroglu, E., Tran, H. D., Goke, J., Tan, Z. Y., Saw, T. Y., Tan, C. P., Lokman, H., et al. (2016). Midbrain-like organoids from human pluripotent stem cells contain functional dopaminergic and neuromelanin-producing neurons. *Cell Stem Cell, 19*, 248–257. https://doi.org/10.1016/j.stem.2016.07.005

Jones, V. C., Atkinson-Dell, R., Verkhratsky, A., & Mohamet, L. (2017). Aberrant iPSC-derived human astrocytes in Alzheimer's disease. *Cell Death & Disease, 8*, e2696. https://doi.org/10.1038/cddis.2017.89

Jonsson, T., Atwal, J. K., Steinberg, S., Snaedal, J., Jonsson, P. V., Bjornsson, S., Stefansson, H., Sulem, P., Gudbjartsson, D., Maloney, J., et al. (2012). A mutation in APP protects against Alzheimer's disease and age-related cognitive decline. *Nature, 488*, 96–99. https://doi.org/10.1038/nature11283

Jonsson, T., Stefansson, H., Steinberg, S., Jonsdottir, I., Jonsson, P. V., Snaedal, J., Bjornsson, S., Huttenlocher, J., Levey, A. I., Lah, J. J., et al. (2013). Variant of TREM2 associated with the risk of Alzheimer's disease. *The New England Journal of Medicine, 368*, 107–116. https://doi.org/10.1056/NEJMoa1211103

Kang, J., Lemaire, H. G., Unterbeck, A., Salbaum, J. M., Masters, C. L., Grzeschik, K. H., Multhaup, G., Beyreuther, K., & Muller-Hill, B. (1987). The precursor of Alzheimer's disease amyloid A4 protein resembles a cell-surface receptor. *Nature, 325*, 733–736. https://doi.org/10.1038/325733a0

Kinney, J. W., Bemiller, S. M., Murtishaw, A. S., Leisgang, A. M., Salazar, A. M., & Lamb, B. T. (2018). Inflammation as a central mechanism in Alzheimer's disease. *Alzheimer's & Dementia (New York), 4*, 575–590. https://doi.org/10.1016/j.trci.2018.06.014

Krenn, V., Bosone, C., Burkard, T. R., Spanier, J., Kalinke, U., Calistri, A., Salata, C., Rilo Christoff, R., Pestana Garcez, P., Mirazimi, A., & Knoblich, J. A. (2021). Organoid modeling of Zika and herpes simplex virus 1 infections reveals virus-specific responses leading to microcephaly. *Cell Stem Cell, 28*(1362–1379), e1367. https://doi.org/10.1016/j.stem.2021.03.004

Lambert, J. C., Ibrahim-Verbaas, C. A., Harold, D., Naj, A. C., Sims, R., Bellenguez, C., DeStafano, A. L., Bis, J. C., Beecham, G. W., Grenier-Boley, B., et al. (2013). Meta-analysis of 74,046 individuals identifies 11 new susceptibility loci for Alzheimer's disease. *Nature Genetics, 45*, 1452–1458. https://doi.org/10.1038/ng.2802

Lancaster, M. A., Renner, M., Martin, C. A., Wenzel, D., Bicknell, L. S., Hurles, M. E., Homfray, T., Penninger, J. M., Jackson, A. P., & Knoblich, J. A. (2013). Cerebral organoids model human brain development and microcephaly. *Nature, 501*, 373–379. https://doi.org/10.1038/nature12517

Lee, C. T., Bendriem, R. M., Wu, W. W., & Shen, R. F. (2017). 3D brain organoids derived from pluripotent stem cells: Promising experimental models for brain development and neurodegenerative disorders. *Journal of Biomedical Science, 24*, 59. https://doi.org/10.1186/s12929-017-0362-8

Lemere, C. A., Beierschmitt, A., Iglesias, M., Spooner, E. T., Bloom, J. K., Leverone, J. F., Zheng, J. B., Seabrook, T. J., Louard, D., Li, D., et al. (2004). Alzheimer's disease abeta vaccine reduces central nervous system abeta levels in a non-human primate, the Caribbean vervet. *The American Journal of Pathology, 165*, 283–297. https://doi.org/10.1016/s0002-9440(10)63296-8

Leng, F., & Edison, P. (2021). Neuroinflammation and microglial activation in Alzheimer disease: Where do we go from here? *Nature Reviews Neurology, 17*, 157–172. https://doi.org/10.1038/s41582-020-00435-y

Li, H., Guo, Q., Inoue, T., Polito, V. A., Tabuchi, K., Hammer, R. E., Pautler, R. G., Taffet, G. E., & Zheng, H. (2014). Vascular and parenchymal amyloid pathology in an Alzheimer disease knock-in mouse model: Interplay with cerebral blood flow. *Molecular Neurodegeneration, 9*, 28. https://doi.org/10.1186/1750-1326-9-28

Lin, Y. T., Seo, J., Gao, F., Feldman, H. M., Wen, H. L., Penney, J., Cam, H. P., Gjoneska, E., Raja, W. K., Cheng, J., et al. (2018). APOE4 causes widespread molecular and cellular alterations associated with Alzheimer's disease phenotypes in human iPSC-derived brain cell types. *Neuron, 98*(1141–1154), e1147. https://doi.org/10.1016/j.neuron.2018.05.008

McAlpine, C. S., Park, J., Griciuc, A., Kim, E., Choi, S. H., Iwamoto, Y., Kiss, M. G., Christie, K. A., Vinegoni, C., Poller, W. C., et al. (2021). Astrocytic interleukin-3 programs microglia and limits Alzheimer's disease. *Nature, 595*, 701–706. https://doi.org/10.1038/s41586-021-03734-6

Miura, Y., Li, M. Y., Birey, F., Ikeda, K., Revah, O., Thete, M. V., Park, J. Y., Puno, A., Lee, S. H., Porteus, M. H., & Pasca, S. P. (2020). Generation of human striatal organoids and cortico-striatal assembloids from human pluripotent stem cells. *Nature Biotechnology, 38*, 1421–1430. https://doi.org/10.1038/s41587-020-00763-w

Mrak, R. E. (2012). Microglia in Alzheimer brain: A neuropathological perspective. *International Journal of Alzheimer's Disease, 2012*, 165021. https://doi.org/10.1155/2012/165021

Muguruma, K., Nishiyama, A., Kawakami, H., Hashimoto, K., & Sasai, Y. (2015). Self-organization of polarized cerebellar tissue in 3D culture of human pluripotent stem cells. *Cell Reports, 10*, 537–550. https://doi.org/10.1016/j.celrep.2014.12.051

Nakai, T., Yamada, K., & Mizoguchi, H. (2021). Alzheimer's disease animal models: Elucidation of biomarkers and therapeutic approaches for cognitive impairment. *International Journal of Molecular Sciences, 22*. https://doi.org/10.3390/ijms22115549

Oakley, H., Cole, S. L., Logan, S., Maus, E., Shao, P., Craft, J., Guillozet-Bongaarts, A., Ohno, M., Disterhoft, J., Van Eldik, L., et al. (2006). Intraneuronal beta-amyloid aggregates, neurodegeneration, and neuron loss in transgenic mice with five familial Alzheimer's disease mutations: Potential factors in amyloid plaque formation. *The Journal of Neuroscience, 26*, 10129–10140. https://doi.org/10.1523/JNEUROSCI.1202-06.2006

Ochalek, A., Mihalik, B., Avci, H. X., Chandrasekaran, A., Teglasi, A., Bock, I., Giudice, M. L., Tancos, Z., Molnar, K., Laszlo, L., et al. (2017). Neurons derived from sporadic Alzheimer's disease iPSCs reveal elevated TAU hyperphosphorylation, increased amyloid levels, and GSK3B activation. *Alzheimer's Research & Therapy, 9*, 90. https://doi.org/10.1186/s13195-017-0317-z

Oddo, S., Caccamo, A., Shepherd, J. D., Murphy, M. P., Golde, T. E., Kayed, R., Metherate, R., Mattson, M. P., Akbari, Y., & LaFerla, F. M. (2003). Triple-transgenic model of Alzheimer's disease with plaques and tangles: Intracellular Abeta and synaptic dysfunction. *Neuron, 39*, 409–421. https://doi.org/10.1016/s0896-6273(03)00434-3

Oksanen, M., Petersen, A. J., Naumenko, N., Puttonen, K., Lehtonen, S., Gubert Olive, M., Shakirzyanova, A., Leskela, S., Sarajarvi, T., Viitanen, M., et al. (2017). PSEN1 mutant iPSC-derived model reveals severe astrocyte pathology in Alzheimer's disease. *Stem Cell Reports, 9*, 1885–1897. https://doi.org/10.1016/j.stemcr.2017.10.016

Onyango, I. G., Jauregui, G. V., Carna, M., Bennett, J. P., Jr., & Stokin, G. B. (2021). Neuroinflammation in Alzheimer's disease. *Biomedicine, 9*. https://doi.org/10.3390/biomedicines9050524

Ortiz-Virumbrales, M., Moreno, C. L., Kruglikov, I., Marazuela, P., Sproul, A., Jacob, S., Zimmer, M., Paull, D., Zhang, B., Schadt, E. E., et al. (2017). CRISPR/Cas9-correctable mutation-related molecular and physiological phenotypes in iPSC-derived Alzheimer's PSEN2 (N141I) neurons. *Acta Neuropathologica Communications, 5*, 77. https://doi.org/10.1186/s40478-017-0475-z

Papaspyropoulos, A., Tsolaki, M., Foroglou, N., & Pantazaki, A. A. (2020). Modeling and targeting Alzheimer's disease with organoids. *Frontiers in Pharmacology, 11*, 396. https://doi.org/10.3389/fphar.2020.00396

Park, J., Wetzel, I., Marriott, I., Dreau, D., D'Avanzo, C., Kim, D. Y., Tanzi, R. E., & Cho, H. (2018). A 3D human triculture system modeling neurodegeneration and neuroinflammation in Alzheimer's disease. *Nature Neuroscience, 21*, 941–951. https://doi.org/10.1038/s41593-018-0175-4

Paulsen, B., Velasco, S., Kedaigle, A. J., Pigoni, M., Quadrato, G., Deo, A. J., Adiconis, X., Uzquiano, A., Sartore, R., Yang, S. M., et al. (2022). Autism genes converge on asynchronous development of shared neuron classes. *Nature, 602*, 268–273. https://doi.org/10.1038/s41586-021-04358-6

Pellegrini, L., Bonfio, C., Chadwick, J., Begum, F., Skehel, M., & Lancaster, M. A. (2020). Human CNS barrier-forming organoids with cerebrospinal fluid production. *Science, 369*. https://doi.org/10.1126/science.aaz5626

Penney, J., Ralvenius, W. T., & Tsai, L. H. (2020). Modeling Alzheimer's disease with iPSC-derived brain cells. *Molecular Psychiatry, 25*, 148–167. https://doi.org/10.1038/s41380-019-0468-3

Perez, S. E., Raghanti, M. A., Hof, P. R., Kramer, L., Ikonomovic, M. D., Lacor, P. N., Erwin, J. M., Sherwood, C. C., & Mufson, E. J. (2013). Alzheimer's disease pathology in the neocortex and hippocampus of the western lowland gorilla (Gorilla gorilla gorilla). *The Journal of Comparative Neurology, 521*, 4318–4338. https://doi.org/10.1002/cne.23428

Pimplikar, S. W. (2009). Reassessing the amyloid cascade hypothesis of Alzheimer's disease. *The International Journal of Biochemistry & Cell Biology, 41*, 1261–1268. https://doi.org/10.1016/j.biocel.2008.12.015

Qian, X., Song, H., & Ming, G. L. (2019). Brain organoids: Advances, applications and challenges. *Development, 146*. https://doi.org/10.1242/dev.166074

Radde, R., Bolmont, T., Kaeser, S. A., Coomaraswamy, J., Lindau, D., Stoltze, L., Calhoun, M. E., Jaggi, F., Wolburg, H., Gengler, S., et al. (2006). Abeta42-driven cerebral amyloidosis in transgenic mice reveals early and robust pathology. *EMBO Reports, 7*, 940–946. https://doi.org/10.1038/sj.embor.7400784

Raja, W. K., Mungenast, A. E., Lin, Y. T., Ko, T., Abdurrob, F., Seo, J., & Tsai, L. H. (2016). Self-organizing 3D human neural tissue derived from induced pluripotent stem cells recapitulate Alzheimer's disease phenotypes. *PLoS One, 11*, e0161969. https://doi.org/10.1371/journal.pone.0161969

Rosen, R. F., Walker, L. C., & Levine, H., 3rd. (2011). PIB binding in aged primate brain: Enrichment of high-affinity sites in humans with Alzheimer's disease. *Neurobiology of Aging, 32*, 223–234. https://doi.org/10.1016/j.neurobiolaging.2009.02.011

Saito, T., Matsuba, Y., Mihira, N., Takano, J., Nilsson, P., Itohara, S., Iwata, N., & Saido, T. C. (2014). Single app knock-in mouse models of Alzheimer's disease. *Nature Neuroscience, 17*, 661–663. https://doi.org/10.1038/nn.3697

Salloway, S., Farlow, M., McDade, E., Clifford, D. B., Wang, G., Llibre-Guerra, J. J., Hitchcock, J. M., Mills, S. L., Santacruz, A. M., Aschenbrenner, A. J., et al. (2021). A trial of gantenerumab or solanezumab in dominantly inherited Alzheimer's disease. *Nature Medicine, 27*, 1187–1196. https://doi.org/10.1038/s41591-021-01369-8

Sasaguri, H., Hashimoto, S., Watamura, N., Sato, K., Takamura, R., Nagata, K., Tsubuki, S., Ohshima, T., Yoshiki, A., Sato, K., et al. (2022). Recent advances in the modeling of Alzheimer's disease. *Frontiers in Neuroscience, 16*, 807473. https://doi.org/10.3389/fnins.2022.807473

Scheltens, P., De Strooper, B., Kivipelto, M., Holstege, H., Chetelat, G., Teunissen, C. E., Cummings, J., & van der Flier, W. M. (2021). Alzheimer's disease. *Lancet, 397*, 1577–1590. https://doi.org/10.1016/S0140-6736(20)32205-4

Selkoe, D. J., & Hardy, J. (2016). The amyloid hypothesis of Alzheimer's disease at 25 years. *EMBO Molecular Medicine, 8*, 595–608. https://doi.org/10.15252/emmm.201606210

Sloan, S. A., Andersen, J., Pasca, A. M., Birey, F., & Pasca, S. P. (2018). Generation and assembly of human brain region-specific three-dimensional cultures. *Nature Protocols, 13*, 2062–2085. https://doi.org/10.1038/s41596-018-0032-7

Sproul, A. A., Jacob, S., Pre, D., Kim, S. H., Nestor, M. W., Navarro-Sobrino, M., Santa-Maria, I., Zimmer, M., Aubry, S., Steele, J. W., et al. (2014). Characterization and molecular profiling of PSEN1 familial Alzheimer's disease iPSC-derived neural progenitors. *PLoS One, 9*, e84547. https://doi.org/10.1371/journal.pone.0084547

St George-Hyslop, P. H., Tanzi, R. E., Polinsky, R. J., Haines, J. L., Nee, L., Watkins, P. C., Myers, R. H., Feldman, R. G., Pollen, D., Drachman, D., et al. (1987). The genetic defect causing familial Alzheimer's disease maps on chromosome 21. *Science, 235*, 885–890. https://doi.org/10.1126/science.2880399

Sturchler-Pierrat, C., Abramowski, D., Duke, M., Wiederhold, K. H., Mistl, C., Rothacher, S., Ledermann, B., Burki, K., Frey, P., Paganetti, P. A., et al. (1997). Two amyloid precursor protein transgenic mouse models with Alzheimer disease-like pathology. *Proceedings of the National Academy of Sciences of the United States of America, 94*, 13287–13292. https://doi.org/10.1073/pnas.94.24.13287

Trujillo, C. A., & Muotri, A. R. (2018). Brain organoids and the study of neurodevelopment. *Trends in Molecular Medicine, 24*, 982–990. https://doi.org/10.1016/j.molmed.2018.09.005

Vazin, T., Ball, K. A., Lu, H., Park, H., Ataeijannati, Y., Head-Gordon, T., Poo, M. M., & Schaffer, D. V. (2014). Efficient derivation of cortical glutamatergic neurons from human pluripotent stem cells: A model system to study neurotoxicity in Alzheimer's disease. *Neurobiology of Disease, 62*, 62–72. https://doi.org/10.1016/j.nbd.2013.09.005

von Bernhardi, R., Eugenin-von Bernhardi, L., & Eugenin, J. (2015). Microglial cell dysregulation in brain aging and neurodegeneration. *Frontiers in Aging Neuroscience, 7*, 124. https://doi.org/10.3389/fnagi.2015.00124

Whalley, K. (2021). TREM2 stems the flow of AD pathology. *Nature Reviews Neuroscience, 22*, 260–261. https://doi.org/10.1038/s41583-021-00456-6

Woodruff, G., Young, J. E., Martinez, F. J., Buen, F., Gore, A., Kinaga, J., Li, Z., Yuan, S. H., Zhang, K., & Goldstein, L. S. (2013). The presenilin-1 DeltaE9 mutation results in reduced gamma-secretase activity, but not total loss of PS1 function, in isogenic human stem cells. *Cell Reports, 5*, 974–985. https://doi.org/10.1016/j.celrep.2013.10.018

Wu, T., Lin, D., Cheng, Y., Jiang, S., Riaz, M. W., Fu, N., Mou, C., Ye, M., & Zheng, Y. (2022). Amyloid Cascade hypothesis for the treatment of Alzheimer's disease: Progress and challenges. *Aging and Disease, 13*, 1745–1758. https://doi.org/10.14336/AD.2022.0412

Xu, M., Zhang, L., Liu, G., Jiang, N., Zhou, W., & Zhang, Y. (2019). Pathological changes in Alzheimer's disease analyzed using induced pluripotent stem cell-derived human microglia-like cells. *Journal of Alzheimer's Disease, 67*, 357–368. https://doi.org/10.3233/JAD-180722

Yagi, T., Ito, D., Okada, Y., Akamatsu, W., Nihei, Y., Yoshizaki, T., Yamanaka, S., Okano, H., & Suzuki, N. (2011). Modeling familial Alzheimer's disease with induced pluripotent stem cells. *Human Molecular Genetics, 20*, 4530–4539. https://doi.org/10.1093/hmg/ddr394

Yue, F., Feng, S., Lu, C., Zhang, T., Tao, G., Liu, J., Yue, C., & Jing, N. (2021). Synthetic amyloid-beta oligomers drive early pathological progression of Alzheimer's disease in nonhuman primates. *iScience, 24*, 103207. https://doi.org/10.1016/j.isci.2021.103207

Zhao, J., Fu, Y., Yamazaki, Y., Ren, Y., Davis, M. D., Liu, C. C., Lu, W., Wang, X., Chen, K., Cherukuri, Y., et al. (2020). APOE4 exacerbates synapse loss and neurodegeneration in Alzheimer's disease patient iPSC-derived cerebral organoids. *Nature Communications, 11*, 5540. https://doi.org/10.1038/s41467-020-19264-0

Inflammatory Demyelinating Diseases of the Central Nervous System

Minshu Li and Qiang Liu

Abstract Over the past decades, a large number of immunomodulatory or immunosuppressive treatments have been approved to treat central nervous system (CNS) demyelinating disorders such as multiple sclerosis (MS). Owing to the heterogeneity of patients with CNS demyelinating diseases, there is no clinical treatment that can adequately control all disease subtypes. Although significant progress has been made for relapsing–remitting MS, effective management of the progressive phase of MS has not yet been achieved. This is at least in part caused by our incomplete understanding of the mechanisms driving disease progression, despite our increasing knowledge regarding the underlying cellular and molecular mechanisms. Here, we summarized our current knowledge regarding the mechanisms of CNS demyelinating disorders and their animal models to identify open questions and challenges for existing concepts. We also discussed potential strategies for the future design of immune therapies to treat CNS demyelinating disorders.

Keywords CNS demyelinating disorders · Neuroinflammation · Autoimmunity · Multiple sclerosis · Immune therapy

Abbreviations

ADCC	antibody-mediated cell-dependent cytotoxicity
ADEM	acute disseminated encephalomyelitis
AHSCT	autologous hematopoietic stem cell transplantation
ALCAM	activated leukocyte cell adhesion molecule
APCs	antigen-presenting cells
APRIL	a proliferation-inducing ligand
AQP4	aquaporin-4

M. Li · Q. Liu (✉)
Department of Neurology, Tianjin Neurological Institute, Tianjin Medical University General Hospital, Tianjin, China
e-mail: qliu@tmu.edu.cn

A. C.-H. Yu et al. (eds.), *Systems Neuroscience*, Advances in Neurobiology 41,
https://doi.org/10.1007/978-3-031-69188-1_8

AQP4-Ab	aquaporin-4 antibodies
AZA	azathioprine
BAFF/BLyS	B-cell activating factor/B-lymphocyte stimulating factor
BBB	blood-brain barrier
BCMA	B cell maturation antigen
Bmem	memory B cell
BTK	Bruton's tyrosine kinase
CDC	complement-dependent cytotoxicity
CHI3L1	polysaccharidase 3-like protein 1
CIS	clinically isolated syndrome
CNS	central nervous system
CSF	cerebrospinal fluid
CXCL	CXC chemokine ligand
CXCR	CXC-chemokine receptor
DCs	dendritic cells
DHODH	dihydroorotate dehydrogenase
DICAM	dual immunoglobulin domain containing cell adhesion molecule
DMTs	disease-modifying therapies
EAE	experimental autoimmune encephalomyelitis
EBV	Epstein-Barr virus
G-CSF	granulocyte colony stimulating factor
GM-CSF	granulocyte-macrophage colony-stimulating factor
HCA2	hydroxycarboxylic acid receptor 2
HSCs	hematopoietic stem cells
ICAM-1	intercellular adhesion molecule 1
IFN-β	interferon-β
IGF-1	insulin-like growth factor-1
IL	interleukin
JAK	Janus kinase
JAML	junctional adhesion molecules like protein
KFLC	kappa free light chain
KIRs	killer cell immunoglobulin-like receptors
LFA-1	lymphocyte function-associated antigen 1
LT	lymphotoxin
mAbs	monoclonal antibodies
MAC	membrane attack complex
MBP	myelin basic protein
MCAM	melanoma cell adhesion molecule
MHC II	major histocompatibility complex II
MMF	mycophenolate mofetil
MOG	myelin oligodendrocyte glycoprotein
MOGAD	MOG antibody-associated disease
MS	multiple sclerosis
NFL	neurofilament light chains
NMO-IgG	neuromyelitis optica immunoglobulin G

NMOSD	neuromyelitis optica spectrum disorders
Nrf2	nuclear factor erythroid 2-related factor 2
NSC	neural stem cell
OCB	oligoclonal bands
OPC	oligodendrocyte precursor cells
PPMS	primary progressive multiple sclerosis
PRMS	progressive-relapsing multiple sclerosis
RIPK1	receptor-interacting serine/threonine-protein kinase 1
RORγt	retinoic acid-related orphan receptor γt
RRMS	relapsing-remitting multiple sclerosis
RTX	rituximab
S1PR	sphingosine-1-phosphate receptor
sNFL	serum neurofilament light chains
SPMS	secondary progressive multiple sclerosis
TGF-β	transforming growth factor beta
Th	T helper
TNF	tumor necrosis factor
Tregs	regulatory T cells
VCAM-1	vascular cell adhesion molecule 1
VEGF	vascular endothelial growth factors
VLA-4	very late antigen 4

1 Overview

Inflammatory demyelinating diseases of the central nervous system (CNS) encompass a group of clinical syndromes related to autoimmune diseases and are characterized by the loss of myelin in the CNS and the infiltration of inflammatory cells around small blood vessels. This type of disease includes multiple sclerosis (MS), neuromyelitis optica spectrum disorders (NMOSD), acute disseminated encephalomyelitis (ADEM), and so on. The course of disease often follows a relapsing stage, with a significant proportion of patients experiencing irreversible neurological impairment, leading to lifelong disability. Due to unclear etiology and complex pathogenesis, inflammatory demyelinating diseases of the CNS exhibit diverse clinical symptoms and heterogeneity of disease subtypes, and there is currently no effective cure for these diseases. Of particular interest is that the pathologies of these diseases are primarily driven by the classical immune response, highlighting the intricate interplay between the immune component and the central nervous system, and the disease spectrum has a high heterogeneity and complexity; it has been a hot field for neuroimmune scientists to conduct basic immunological research and clinical drug therapy exploration, and research achievements and theories in CNS-specific immunology will provide clues for other CNS diseases' investigation, such as cerebrovascular diseases, neurodegenerative diseases, and brain aging. In recent years, with the development of science and technology, advances have been made in treatment, immunological mechanisms, and clinical diagnostic approaches for inflammatory demyelinating diseases of the CNS. This chapter will take MS (driven by cellular immunity) and NMOSD (predominant by humoral immunity) as examples, focusing on reviewing the pathological mechanism, immunological basis, clinical diagnosis, and treatment research progress of these diseases, so as to pave the way for further advancements in research and clinical management.

2 Clinical Characteristics

MS is an immune-mediated, chronic, and disabling CNS inflammatory demyelinating disease. It commonly presents with an acute/subacute onset, characterized by dissemination in space and time. Dissemination in space refers to the involvement of multiple lesion sites, including the cerebrum, brainstem, cerebellum, and spinal cord, either simultaneously or successively. The clinical manifestations vary widely, with common symptoms including visual impairment, diplopia, sensory disturbances, motor dysfunction, ataxia, and bladder or bowel dysfunction. Dissemination in time refers to a relapsing-remitting course, extensive lesions involvement can lead to eventual disability or even death of MS patients. The prevalence of MS is about 142.81: 100,000 in Europe, 117.49:100,000 in America, 8.76:100,000 in Africa, and 8.62:100,000 in South East Asia (Walton et al., 2020). Classical MS typically affects young individuals, with an average age of onset between 20 and 40 years, while cases in patients younger than 10 or older than 50 years are less common (Graves et al., 2023). The disease is more prevalent in females, with a female-to-male ratio of 2:1 globally, even as high as 4:1 in some countries (Walton et al., 2020). Based on the clinical course, MS can be classified into four types: relapsing-remitting MS (RRMS), secondary progressive MS (SPMS), primary progressive MS (PPMS), and progressive-relapsing MS (PRMS).

Neuromyelitis optica (NMO), also known as Devic's syndrome, is a CNS inflammatory demyelinating disease characterized primarily by the involvement of the optic nerves and spinal cord. It is clinically characterized by severe optic neuritis and longitudinally extensive transverse myelitis (more than three vertebral segments). It often presents a progressive or remission and relapse trend. In 2007, Wingerchuk et al. proposed the concept of NMO spectrum disease (NMOSD), which is characterized by the presence of NMO IgG in the serum, including recurrent or bilateral optic neuritis and longitudinally extensive transverse myelitis (LETM) apart from the classical NMO (Wingerchuk et al., 2007). NMOSD can occur at any age but is more common in individuals aged 40–50 years, with a higher prevalence in females than males. The male-to-female ratio of recurrent cases was about 1:2.3–11 (Wingerchuk et al., 2015; Tian et al., 2020). NMOSD often presents with blurred vision, bilateral limb weakness, sensory loss, sensory abnormalities, band-like sensations, and urinary and bowel dysfunction. Clinical symptoms are generally more severe than those in MS, with a rapid disease progression and a high early relapse rate. Approximately 80% of NMOSD patients have a specific antibody marker in their serum, known as NMO immunoglobulin G (NMO-IgG) or aquaporin-4 antibodies (AQP4 Ab) (Papp et al., 2021). According to the diagnostic criteria for NMOSD established by the International Panel for NMO Diagnosis (IPND) in 2015, the detection of AQP4 antibodies has been included as a diagnostic criterion for NMOSD (Wingerchuk et al., 2015).

3 Pathology and Immunological Basis

Although MS and NMO belong to immune-mediated demyelinating diseases of the CNS, they are independent disease entities in terms of immune mechanism, pathological features, clinical treatment, and prognosis. MS is a demyelinating disease of the CNS dominated by cellular immunity mediated by $CD4^{+}$T cells and supplemented by humoral immunity involving B lymphocytes. It predominantly affects the subcortical, periventricular, subtentorial, and spinal cords. Microscopically pathological findings showed that monocytes, lymphocytes, and plasma cells had cuff infiltration around blood vessels in the acute stage, which form a "vascular sheath." Myelin was disintegrated and lost, and axons were relatively intact. Myelin regeneration occurred in the late stage of the lesion, the inflammatory changes were replaced by astrogliosis, constituting sclerotic plaques of a later period. NMO is a CNS demyelinated disease that is mainly mediated by humoral immunity against aquaporin-4 (AQP4). It mainly involves damage to the optic nerve and spinal cord. The pathological basis of NMO is different from MS, mainly autoimmune astrocyte injury, and secondary demyelination. Major pathological features include demyelination and hardened plaques of the white and gray matter of the spinal cord, local necrosis, cavity formation, loss of AQP4, acute axonal injury, and perivascular inflammatory cell infiltration including macrophages lymphocytes (mainly B lymphocytes), neutrophils and eosinophils. Antibodies and complement were seen as rosette deposits around blood vessels. In recent years, with the deepening of research, we further explicit pathological view of these diseases, elucidating the function of immune cell subsets and the regulatory network between different organs and tissues that contribute to the pathological injury; these advances will be elaborated in detail.

3.1 Pathology and Immunological Mechanism of MS

3.1.1 Pathological Features of MS

According to the presence of inflammatory activity, MS lesions are divided into active lesions, chronic active or smoldering lesions (mixed lesions), and inactive lesions. The typical initial pathological feature of MS is focal demyelinating lesions in the white matter, known as active lesions, in which macrophages/microglia are distributed throughout the lesion, accompanied by immune cell infiltration surrounding central veins. These lesions are characterized by blood-brain barrier (BBB) leakage, deposition of plasma proteins, infiltration of B cells, T cells (mainly $CD8^+$) and myeloid cells, demyelination, reactive gliosis, loss of oligodendrocytes, and varying degrees of axonal injury. Although axons and neurons are largely preserved in early MS, sustained disease progression leads to gradual loss of neuronal axons, resulting in brain atrophy accompanied by ventricular enlargement, which correlates with patient disability. The number and characteristics of microglia at the sites of these focal lesions can be used to determine the stage of MS lesions: in acute active lesions, occurring within the first few days, myelin protein is present within the myeloid cells (macrophages and microglia); in subsequent active phases, within a few weeks of onset, myeloid cells no longer contain myelin proteins and are filled with lipids (Kuhlmann et al., 2017). Astrocytes show morphological and functional changes in active lesions, cell body hypertrophy, and high expression of the glial fibrillary acidic protein (GFAP), pro-inflammatory cytokines, chemokines, and remyelination signaling molecules (Rawji et al., 2020).

As the progress of MS, the number of active lesions gradually decreases, and chronic active lesions dominate. In chronic active lesions, there is a hypocellular lesion center with very few microglia/macrophages and moderate T-cell infiltration. Adjacent to the central region, a more extensive active microglia/macrophage exhibiting foamy or branched phagocytic phenotypes limited and localized to the lesion border. In addition, demyelination, axonal injury, and BBB disruption, as well as immune cell infiltration, occur in such lesions (Frischer et al., 2015). In this type of lesion, iron accumulation was found in microglia and macrophages which were in a pro-inflammatory activated state at the lesion margins (Gillen et al., 2018) as well as in astrocytes (Popescu et al., 2017) in the inactive gliotic centers. Lesions containing iron accumulation are more destructive (Dal-Bianco et al., 2021), and repeated exposure to iron may impact the antioxidant pathways of astrocytes, leading to astrocyte loss and tissue rarefaction in regions adjacent to iron-positive astrogliosis zones (Popescu et al., 2017). Chronic active lesions are associated with disease progression in MS and are more common in elderly patients (Reich et al., 2018). This perhaps due to the defects of the CNS repair and/or dysregulation of immune response in elderly patients.

In patients with a prolonged disease course, the number of chronic inactive lesions increases, but the size of inactive lesions in the brains and spinal cords of progressive MS patients decreases with disease duration. Inactive lesions were shown with small numbers of glial and immune cells and areas of old demyelinating damage. Astrocytes in inactive lesions typically have smaller cell bodies and contain numerous elongated processes but with no transcriptional activity, forming multiple sclerotic glial scars in white matter lesions (Frischer et al., 2015). There are no significant changes in the composition of infiltrating T cells in the lesions, but there is an increased proportion of B cells and plasma cells (Frischer et al., 2009), and a decreased number of infiltrating cells is observed. Though, microglia and macrophages are almost absent but remain chronically activated state.

It is noteworthy that the CNS pathology in MS is not limited to white matter demyelinating lesions. Diffuse injury, atrophy, disruption of the BBB, perivascular astrocyte proliferation, and non-demyelinating inflammation have been detected in both white and gray matter, possibly exceeding the extent of white matter demyelination (Frischer et al., 2015). Furthermore, inflammatory immune cell infiltration and ectopic lymphoid follicles have been observed in the meninges of MS patients, which are associated with severe cortical pathology, neurodegeneration, and progressive disease course (Magliozzi et al., 2010). Therefore, multiple site lesions as well as complex immune response in CNS microenvironments contribute to the pathology of MS.

3.1.2 Immunological Mechanisms of MS

Accumulating research has demonstrated that MS is an autoimmune disease primarily mediated by $CD4^+$ Th1 cells. The peripheral priming of naïve T cells and reactivation in the specific environment of CNS are essential steps in experimental autoimmune encephalomyelitis (EAE) and MS. In recent years, with the introduction of new technologies such as single-cell sequencing, new mechanisms of autoreactive T cell activation, and interaction between different immune organs and immune cells have been gradually revealed, and a series of important scientific questions have been answered, for example: (1) How peripheral self-reactive T cells are activated and reach the CNS? (2) What mechanisms drive their reactivation within the CNS and contribute to tissue damage? (3) What role does the bone marrow, as a central immune organ, play a role in the origin and evolution of T-cell-mediated autoimmune responses in MS? (4) How do B cells, as crucial immune cells, contribute to the pathogenesis of MS? (5) What is the relationship between immunoregulation and remyelination? These advanced research results would enrich and develop our understanding of the immune mechanism of MS and provide an important clue for us to find potential drug targets (see Fig. 1).

Activation and Migration of Self-reactive T Lymphocytes

It is currently believed that the activation of peripheral myelin-specific reactive T cells serves as a triggering factor in the onset of MS. Two main hypotheses have been proposed to explain their activation: cross-immune reaction with foreign antigens and CNS antigens including myelin basic protein (MBP), MOG, and PLP, leaking into the cervical lymph nodes. Foreign antigens like viruses, bacteria, in tissues such as the skin, intestines, or lungs are captured by antigen-presenting cells (APCs), which subsequently migrate to draining lymph nodes and activate pathogen-reactive T cells that exhibit cross-reactivity with myelin-antigens in the CNS (Gran et al., 1999). For example, Epstein-Barr virus (EBV) infection can

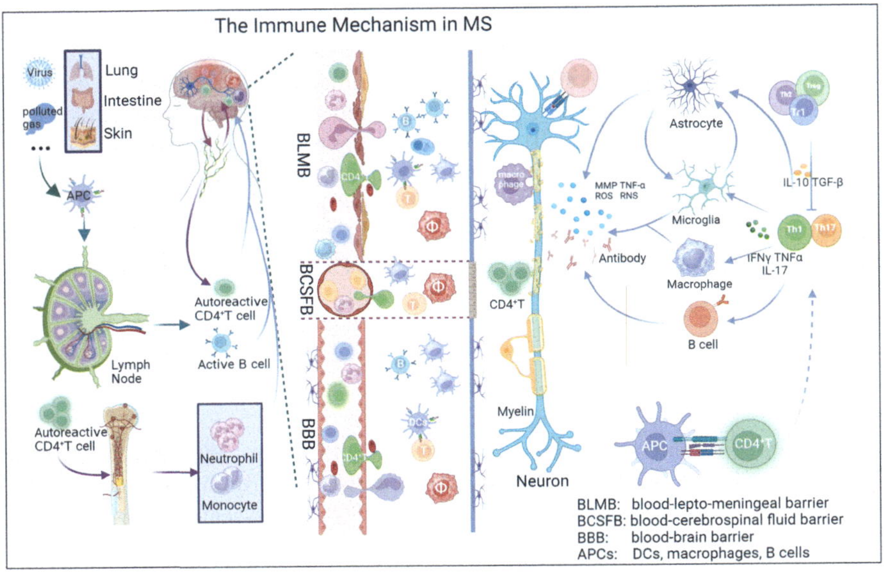

Fig. 1 The immunological mechanism of MS. Exogenous pathogens, such as viruses and polluted gas, can lead to the activation and migration of antigen-presenting cells (APCs) from peripheral tissues, including the lungs, intestines, and skin, to cervical lymph nodes. APCs activate autoreactive T cells by presenting cross-reactive antigens. CNS antigens including myelin basic protein (MBP), MOG, and PLP, may also leak to the periphery in the cervical lymph nodes to activate autoreactive T cells. Autoreactive T cells and other peripheral immune cells (e.g., B cells, neutrophils, monocytes) invade the CNS via BLMB, BCSFB, and BBB and interact with local APC subsets. Notably, B cells, plasma cells, T cells, and follicular dendritic cells form ectopic follicle-like aggregates in the meninges, potentially secreting various soluble inflammatory mediators that promote activation of autoreactive T cells. Autoreactive T cells are reactivated by APCs (such as microglia, DCs, and B cells), differentiate into pathogenic Th subsets (mainly Th1 and Th17), and secrete TNF-α, IFN-γ, IL-17, and other pro-inflammatory cytokines, promoting the pathological responses of astrocytes, microglia, macrophages, and B cells. Bidirectional communication between astrocytes and microglia further regulates their activities and CNS inflammation. Demyelination and axonal damage are caused by various mechanisms, including the secretion of MMPs, TNF-α, ROS, as well as soluble neurotoxic molecules by astrocytes, microglia, and macrophages; antibody secretion by B cells; cytotoxic effects of $CD8^+$ T cells; and phagocytosis and ADCC by macrophages. Tregs and other immune regulatory cells can suppress these pathogenic processes by secreting immunomodulatory cytokines such as IL-10 and TGF-β. In addition to CNS infiltration, self-reactive T cells can be home to the bone marrow, promoting a shift in hematopoiesis toward the myeloid lineage, leading to an increased influx of neutrophils and monocytes into the CNS, exacerbating the damage. This figure was drawn with a commercial tool (BioRender)

lead to molecular mimicry between foreign microbial antigens and self-antigens including MBP, GlialCAM (HEPACAM), and Anoctamin 2 (TMEM16B). Alternatively, it is possible that CNS soluble antigens present in the cerebrospinal fluid (CSF) activate T cells in cervical lymph nodes (Ransohoff & Engelhardt, 2012). These self-reactive T cells, upon activation and polarization into pro-inflammatory subtypes, subsequently migrate to the CNS via the bloodstream, where they are reactivated by dendritic cells (DCs) and macrophages, participating in antigen presentation processes within the dural sinus (Rustenhoven et al., 2021). However, there is also evidence suggesting that antigen presentation occurs in the choroid plexus (Rodríguez-Lorenzo et al., 2020) and leptomeninges (Merlini et al., 2022). Regardless of how self-reactive lymphocytes are activated, a key pathological feature of MS and other autoimmune diseases is epitope spreading, whereby immune responses initially targeting a single epitope expand to other epitopes within the same antigen or even spread to other antigenic peptides, further amplifying the pathogenic autoimmune response.

Following peripheral activation, self-reactive T cells need to migrate to the CNS, which involves three major barriers: the blood-leptomeningeal barrier, the blood-cerebrospinal fluid barrier, and the BBB. Recently, a novel layer of meninges called the subarachnoid lymphatic-like membrane has been identified (Møllgård et al., 2023). Additionally, there have been reports of immune cell migration from the skull bone marrow to the CNS (Herisson et al., 2018), although their specific roles in EAE and MS patients has not been fully elucidated. The mechanisms underlying immune cell migration into the CNS remain unclear but likely involve cellular adhesion molecules, chemokines, and their receptor-mediated cell-cell interactions. MS lesions display an increased expression of endothelial adhesion molecules such as intercellular adhesion molecule 1 (ICAM-1) and vascular cell adhesion molecule 1 (VCAM-1). Interactions between these adhesion molecules and their ligands ICAM-1/LFA-1 and VCAM-1/VLA-4 as well as recently identified molecules including DICAM, MCAM, ALCAM, and JAML promote immune cells migration (Charabati et al., 2022; Lécuyer et al., 2017). For instance, natalizumab, a monoclonal antibody directed against the alpha chain of the VLA-4 integrin, hinders leucocytes trafficking across the BBB in MS, and pharmaceutically neutralizing DICAM can reduce murine and human Th17 cell migrating to the CNS and alleviate disease symptoms (Charabati et al., 2022). Therefore, for the treatment of MS, we may not only consider how to inhibit peripheral auto-reactive T cell activation but also intervene these cells enter into the CNS.

Reactivation, Polarization of T Lymphocytes, and CNS Immune Damage

The immune response of the CNS is unique. The antigen presentation process in the CNS is significantly different compared with the periphery. Under pathological conditions, the resident cells in the brain, such as microglia and astrocytes, can initiate T and B lymphocytes in the same way as dendritic cells and macrophages in the peripheral. In addition, the structural, cellular, and soluble factors that characterize the CNS create a microenvironment for peripherally homing immune cells that is distinct from the peripheral portion. Upon entering the CNS, immune cells will interact with a range of factors in the brain that they are not exposed to in the periphery, including glial cells.

Studies demonstrated that self-reactive immune cells in the CNS are reactivated by APCs including dendritic cells, macrophages, and microglia. Clonal expansion of $CD4^{+}$ T helper (Th) subsets, particularly T-bet-controlled IFN-γ^{+}Th1 cells, and RORγt-controlled IL-17^{+}Th17 cells, plays a pivotal role in the pathology of MS and EAE (Steinman, 2014). Single-cell analysis has revealed that the polarization of $CD4^{+}$ T cells into distinct Th subsets is driven by their local microenvironment and lineage-specific transcription factor expression (Schafflick et al., 2020). For example, the transcription factor aryl hydrocarbon receptor induces Th17 cells to convert into Treg cells in the presence of TGF-β (Gagliani et al., 2015) and IL-23 drives intestinal nonpathogenic Th17 cells to convert into pathogenic $CXCR6^{+}$ Th17 cells (Schnell et al., 2021), suggesting that these Th subsets may be interconverted in the CNS. Th1 and Th17 cells secrete pro-inflammatory cytokines such as IFN-γ, TNF-α, IL-17, and GM-CSF, which activate the macrophages, dendritic cells, astrocytes, and microglia, further aggravating CNS damage. In this context, astrocytes and microglia also secrete chemokines (e.g., CCL2, CXCL2, CCL5), cytokines (e.g., IL6, TNF, IL-1β, IFN-γ), and neurotoxic molecules (C1q, glutamate, RNS, and ROS) to promote CNS inflammation and brain injury (Giovannoni & Quintana, 2020; Domingues et al., 2016; Ajami et al., 2018; Acharjee et al., 2021; Liddelow et al., 2017; Wheeler et al., 2019). Therefore, remodeling the immune response by regulating the brain microenvironment could alleviate tissue damage.

It needs to be mentioned that although $CD4^{+}$ T cells play an important role in the pathological mechanism of MS, $CD8^{+}$ T cells are the main T cell population in brain lesions of patients with MS, exhibiting a significant increase in numbers within active or chronic active lesions and meningeal infiltrates. $CD8^{+}$ T cells in lesions display a tissue-resident memory phenotype and also exhibit activation markers (Fransen et al., 2020) as well as expression of the pro-inflammatory cytokine IL-17. These $CD8^{+}$ T cells have been detected as clonally expanded populations in the CSF of MS patients (Fransen et al., 2020) and are believed to be putative driving factors for compartmentalized autoimmune CNS damage (Frieser et al., 2022). How $CD8^{+}$T cells are activated and what key molecules contribute to this process? What is the potential role of the intervention $CD8^{+}$ T cells for MS treatment? These questions need to be further investigated.

Aberrant Bone Marrow Hematopoiesis Promotes MS Progression

Bone marrow, as a central immune organ, is the primary site of hematopoiesis that harbors hematopoietic stem and progenitor cells (HSPCs) to generate blood cells and to maintain immune homeostasis in the periphery and CNS (Herisson et al., 2018; Cugurra et al., 2021; Mazzitelli et al., 2022). Recent studies have revealed that bone marrow hematopoiesis may drive the progression of MS. Pathogenic T cells producing granulocyte-macrophage colony-stimulating factor (GM-CSF) were found in both MS patients and EAE models, triggering the migration of $Ly6C^{hi}$ classical monocytes from the bone marrow and enhancing their pathogenic activity, highlighting the role of $CD4^{+}$ T cell-monocyte interaction in CNS pathology

(Rasouli et al., 2020; Croxford et al., 2015). Shi et al., utilized single-cell sequencing, lineage analysis, and flow cytometry, to find that the activation of hematopoietic stem cells (HSCs) in bone marrow hematopoietic system develops toward a biased myeloid lineage commitment, resulting in a significant increase in the monocytes and neutrophils in MS patients. They further demonstrated that autoreactive T cells are able to migrate into the bone marrow through the CXCL12-CXCR4 axis and tilt hematopoietic stem/progenitor cell differentiation toward the myeloid system, resulting in bone marrow exporting myeloid cells to aggravate CNS inflammatory injury (Shi et al., 2022a). This indicates the influence of self-reactive T cells on bone marrow hematopoiesis and also suggests that aberrant bone marrow hematopoiesis drives the progression of MS and the restriction of bone marrow myelopoiesis as a potential therapy for MS.

Role of B Cells in MS

MS is primarily considered a T-cell-driven CNS demyelinating disease. However, the role of B cell-mediated humoral immunity in MS pathogenesis has gradually been valued since the discovery of elevated CSF-IgG levels in MS patients (Riddoch & Thompson, 1970). With the expanding therapeutic efficacy of CD20-targeting B cell-depleting therapy in RRMS and PPMS patients, attention has shifted toward understanding the diverse roles of different B cell subtypes in MS. These roles encompass not only antibody secretion but also cytokine production, antigen presentation, and ectopic follicle-like aggregate formation. The accumulation of B cells in meningeal ectopic follicles is associated with cortical pathology in MS and likely reflects the role of B-cell-derived granulocyte-macrophage colony-stimulating factor (GM-CSF), TNF-α, IL-6, IL-12, IL-15, and other soluble factors in promoting inflammation and neurotoxicity (Cencioni et al., 2021). Moreover, B cells play a pivotal role as APCs, inducing the proliferation of Th1 and Th17 cells that infiltrate the CNS in MS and promote disease progression.

Notably, memory B cell (Bmem) has emerged as a key B cell phenotype of interest in MS research due to antigen affinity maturation, long lifespan, and rapid response to stimulation. Once activated, Bmem can follow two pathways: (1) rapid differentiation into antibody-secreting cells (ASCs) and (2) reentry into secondary germinal centers for further affinity maturation and isotype switching. The differentiation of Bmem into ASCs facilitates the rapid and abundant production of high-affinity antibodies to supplement the antibody production by terminally differentiated plasma cells. Apart from their rapid differentiation into ASCs, Bmem also functions as effective APCs by expressing major histocompatibility complex II (MHC II), allowing them to effectively recognize antigens and process antigen presentation to activate other immune cells, including T cells. Lastly, Bmem can produce various cytokines, including TNF, GM-CSF, IL-6, lymphotoxin (LT), and IL-10 (DiSano et al., 2021). In MS, B cells reside in multiple compartments within the CNS, including the CSF, brain parenchyma, and meninges. Bmem is enriched in the CSF compared to peripheral blood and constitutes the majority of

B cells. However, studies on Bmem in MS have primarily focused on peripheral blood and CSF. Furthermore, there are significant discrepancies in the definition of Bmem, with most studies defining it solely based on CD27 expression. Therefore, careful consideration of surface markers used to define Bmem is essential for each study. Future research should address several key unresolved questions to provide deeper insights into Bmem in MS, including transport mechanisms, site-specific roles within the CNS, functional relevance in the immunopathogenesis of MS, and associations with clinical outcomes.

Immunomodulation and Myelin Regeneration

A clinical hallmark of MS is that patients often experience complete or partial recovery following the onset of MS owing to multiple anti-inflammatory processes. Anti-inflammatory cytokines produced in the CNS, such as IL-27 and IFN-β, are believed to act on lymphocytes and APCs to suppress inflammation (Mascanfroni et al., 2013; Sospedra & Martin, 2005). Anti-inflammatory microglia and macrophages contribute to this process by clearing debris and dead cells through phagocytosis and proteolysis. Additionally, $CD4^+$ regulatory T cells (Tregs) have been extensively studied in MS and EAE, particularly $CD4^+CD25^+FoxP3^+$ regulatory T cells. In EAE, Tregs have been shown to inhibit the development of autoimmune neuroinflammation and promote remyelination (Dombrowski et al., 2017), while in patients with MS, Tregs exhibit impaired suppressive function (Venken et al., 2008). Another subset of regulatory $CD4^+$ Tregs, known as type 1 regulatory T cells (TR1s), suppress microglia and astrocytes responses via the production of IL-10. Accordingly, it has been proposed that the relapse-remission process in MS reflects game theory between effector T cells and Treg cells, specifically, the expansion of CNS antigen-specific effector T cells (relapse), followed by the expansion of Tregs specific to the same antigen, leading to the control of the initial pathogenic response (remission). Subsequent epitope spreading leads to the activation of effector T cells with different antigen specificity, evading regulation by antigen-specific Tregs and initiating a new cycle of disease progression (Steinman, 2014). $CD8^+$ regulatory T cells characterized by the expression of inhibitory killer cell immunoglobulin-like receptors (KIRs) may suppress pathogenic T cells (Li et al., 2022a), B cells are also capable of limiting CNS inflammation, promoting oligodendrocyte generation, and facilitating myelin regeneration (Cencioni et al., 2021), but less is known about the regulatory mechanisms underlying this process.

The involvement of immune cells in the process of remyelination has been recognized. Specifically, microglia and macrophages with immunoregulatory characteristics are believed to facilitate remyelination through diverse mechanisms, including phagocytosis and proteolytic degradation of myelin debris, secretion of regulatory factors such as insulin-like growth factor-1 (IGF-1) and activin A, as well as metabolic and nutritional support (Rawji et al., 2016). Some pathways implicated in inhibiting remyelination have been identified as potential therapeutic targets, including Notch signaling, LINGO-1, muscarinic acetylcholine receptor signaling, Wnt, hyaluronic acid, fibrinogen, and chondroitin sulfate proteoglycans, which could be as potential therapeutic targets (Lubetzki et al., 2020). Targeted strategies for remyelination, alone or in combination with other approaches such as anti-inflammation may present an avenue to treat MS.

3.2 *The Pathology and Immunological Mechanism of NMOSD*

3.2.1 NMO Pathology

NMO is a CNS autoimmune disease dominated by humoral immunity, in which AQP4 antibody (AQP4-Ab) is detectable in the sera of more than 80% of NMOSD patients. The target antigens of AQP4-Ab are AQP4 protein located in the astrocyte end-foot as well as the ependymal membrane. AQP4-Ab is predominantly IgG1, which can trigger antibody-mediated complement-dependent cytotoxicity (CDC) and antibody-dependent cell-mediated cytotoxicity (ADCC) (Wu et al., 2019). CNS lesions in AQP4-Ab-positive patients are characterized by the deposition of IgG and IgM in the center of the blood vessel, which is most prominent around the blood vessel (corresponding to the high expression of AQP4 in the glial boundary membrane), immune cell infiltration, astrocytes loss, and (or) secondary injury of oligodendrocytes and neurons depending on the stage of the disease and the severity of the attack.

Studies based on post-mortem brain tissue show that active lesions together with inactive destructive or demyelinated lesions exist in the CNS of NMO patients. In the active lesion, there is marked inflammation, mainly distributed around the cuff-like changes that appear around the blood vessels and within the lesions, and the infiltrating inflammatory cells mainly include T cells, B cells, and macrophages (Misu et al., 2013). The lesions also contained a large number of neutrophils and a small number of eosinophils (Lucchinetti et al., 2002). Based on the extent of complement deposition, granulocyte infiltration, acute astrocyte alterations, astrocyte loss, demyelination, and axonal injury, the lesions can be classified into six types (Misu et al., 2013; Takai et al., 2021):

1. active NMO lesions associated with granulocyte infiltration, and perivascular deposition of activated complement, leading to astrocytes damage. While myelin is relatively well preserved in this phase.
2. complete tissue destruction and necrosis of astrocytes.
3. gliosis in the white matter fibers of the spinal cord (similar to secondary Wallerian degeneration).
4. mainly in the surrounding areas of astrocyte lysis and progenitor recruitment lesion (loss of selective AQP4).
5. astrocytopathic lesions with degenerated astrocytic clasmatodendrosis, mainly in the early stages of the disease, without significant myelin abnormalities in this phase, mainly in the periphery of type 1 and type 6 lesions.

6. lesions with astrocyte dystrophy and demyelination that are associated with oligodendrocyte apoptosis. In addition to astrocyte destruction, this phase of the lesion is observed in other chronic lesions. The lack of activated complement in this phase and the absence of deposition of C9neo (a residual component of the membrane attack complex), but mostly accompanied by deposition of C3d (one of the components of complement degradation), suggest a possible complement-mediated pathological mechanism.

The demyelination of NMO is a secondary damage caused by AQP4 antibodies-mediated immune attack to astrocytes, so NMOSD is now regarded as an immune astrocytopathy and eventually causes irreversible pathological damage, such as fibrosis and hyalinization in the vessel wall and lumen formation in the necrotic area (Lucchinetti et al., 2002; Popescu & Lucchinetti, 2012). Some perivascular lesions show rosette-like changes due to the deposition of immunoglobulin and complement, etc. (Wu et al., 2019).

3.2.2 Immunological Mechanisms of NMOSD

NMOSD is a B-cell-mediated demyelinating disease of the CNS. In the periphery, AQP4-specific B cells differentiate into plasma cells and produce anti-AQP4 antibodies, which cross the blood-brain barrier into the CNS during the acute phase of the disease and are deposited primarily on perivascular astrocyte foot processes that are highly expressing AQP4, leading to AQP4 internalization and function loss (Jarius et al., 2020). Binding of AQP4 to AQP4-Ab further triggers CDC and ADCC. Additionally, complement activation enhances the recruitment of activated macrophages, neutrophils, eosinophils, and basophils, resulting in the production of more cytokines and oxygen radicals, further exacerbation of neuronal and oligodendrocyte destruction (see Fig. 2). In recent years, the role of tissue specificity of B cells in the CNS, the function of T cells, and the role of glial cells in the pathological mechanism of NMOSD have been gradually revealed.

B Cells in NMOSD

As AQP4 antibody discovery, the importance of B cells in the pathogenesis of NMOSD was recognized. B cells, as the main cells of humoral immunity, play a core role in the production of pathogenic antibody, antigen presentation, and immune regulation during the pathological development of NMO. The study found that $CD19^{int}CD27^{high}CD38^{high}CD180^{-}$, a subset of B cells, is selectively expanded in the peripheral blood of NMOSD patients, and these cells have the function of plasma cells that can secrete AQP4-Ab with IL-6 stimulation (Chihara et al., 2011). The B cells in the CSF of NMOSD patients have a high mutational signal in somatic B cells, suggesting that the process of antigen recognition may occur within the CNS (Carnero Contentti & Correale, 2021). Similarly, increased levels of B-cell

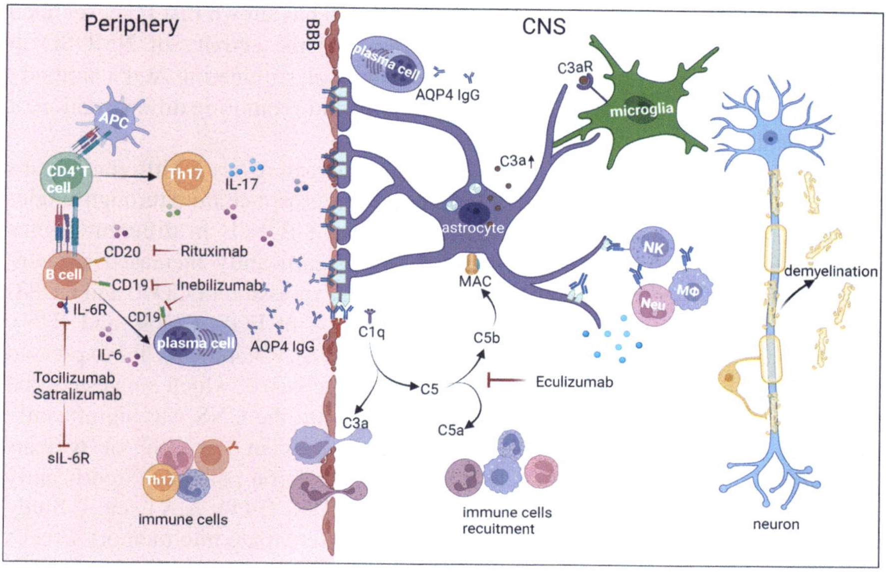

Fig. 2 The Pathophysiologic mechanisms and therapeutic targets in NMOSD. With the help of T cells, AQP4-specific B cells can differentiate into plasma cells that can produce anti-AQP4 antibody (AQP4-Ab) in the peripheral. AQP4-Ab can penetrate the blood-brain barrier and enter the central nervous system, which is mainly deposited on the foot processes of the astrocyte. Meanwhile, with interaction with B cells or dendritic cells and in the presence of IL-6, IL-23, and TGF-β stimulation, specific T cells differentiate into Th17 cells. The inflammatory environment further interrupts BBB integrity and facilitates immune cells and AQP4-Ab to enter into CNS. In the CNS, binding of AQP4-Ab with AQP4 not only leads to AQP4 internalization and function impairment but also induces CDC and ADCC. The product of complement cascade reaction, MAC, directly kills the astrocyte, and C3a and C5a not only increase the permeability of BBB but also act as chemokines to recruit peripheral neutrophils, basophils, eosinophils, mast cells, and macrophages to the CNS. In ADCC, neutrophils, macrophages, and NK cells bind to the Fc segment of AQP4-IgG and induce astrocyte damage via degranulation. In addition, activated astrocytes release more complement C3a, which binds to C3aR on the membrane of microglia to stimulate microglia to produce C1q. The mechanisms afore-described may produce cytotoxic effects on oligodendrocytes and neurons through bystander effects, leading to neural demyelination and neuron injury. Rituximab and Inebilizumab are used to deplete B cells by targeting CD20 and CD19, respectively. Tocilizumab, and Satralizumab: target IL-6R (including mIL-6R and sIL-6R) to reduce inflammatory response. Eculizumab: Inhibition of C5 cleavage to C5a and C5b. This figure was drawn with a commercial tool (BioRender)

activating factor (BAFF), proliferation-inducing ligand (APRIL), CXCL13, and IL-6 in the CSF of NMOSD patients may play an important role in the recruitment and maintenance of plasma cells producing AQP4 antibodies (Vaknin-Dembinsky et al., 2010; Kaneko et al., 2018). In addition, B cells with the function of antibody presentation may play an essential role in supporting follicular T cell development, which is involved in B cell differentiation and class switching (Molnarfi et al., 2013; Deenick et al., 2010). Cytokines produced by B cells including IL-6, TNF-α, and

GM- CSF promote the activity of NMOSD. Research has shown that IL-6, secreted by the pro-inflammatory memory B cells increases the activity of NMOSD in multiple aspects such as promoting plasma cell survival, stimulating AQP4 antibody production, exacerbating BBB integrity disruption, and promoting differentiation of pathological Th17 (Fujihara et al., 2020).

Recently, Zhang et al. (2021) analyzed B cell characteristics in different organs including the bone marrow, peripheral blood, and cerebrospinal fluid through single-cell sequencing and found that the characteristics of B cells in different organs were different. B cells in peripheral blood have significantly increased antigen-presenting ability (accompanied by increased signal expression of CD40 and CD83) and pro-inflammatory function (increased expression of FOS, CD69, and JUN). More importantly, the level of Chemokine receptor CXCR3 and CXCR4 expressed by peripheral blood B cells was also significantly increased, which suggested that the ability of pro-inflammatory B cells to migrate into the CNS was significantly enhanced. Compared with the periphery, the B cells in the CSF of patients with NMOSD have significantly activated type I interferon pathway (significantly increased type I interferon gene markers include IRF4, IFI44, MX1, etc.). Study shows that type I interferon can induce B cells to differentiate into memory B cells and antibody-secreting cells, thus help B cells produce pathogenic AQP4-IgG. This study suggests that B cells in NMOSD patients have tissue specificity.

T Cells in NMOSD

AQP4-Ab is an IgG1 subtype that requires T cell assistance in its production, indicating that AQP4-specific $CD4^{+}$T cells are involved in NMOSD pathology (Varrin-Doyer et al., 2012). Although there are few T cells in the lesions, it is reported that T cells may play a role in the imbalance between peripheral Immune tolerance and the production of AQP4 antibody (Papadopoulos & Verkman, 2012; Chan & Lee, 2021). Compared with healthy controls, peripheral blood T cells from NMOSD patients have a stronger response to AQP4 peptide p61-80 (Varrin-Doyer et al., 2012). Research found that AQP4 antigen stimulation can induce Th17 cell polarization and secrete Th17-related factors such as IL-6 and IL-21 (Linhares et al., 2013), which promotes endothelial activation and neutrophil transepithelial migration (Kebir et al., 2007). Increased IL-6 levels in CSF of NMOSD patients contribute to the survival of AQP4-specific Th17 cells and suppress FOXP3 regulatory T cells (Korn et al., 2008). In conclusion, these results suggest a potential role for AQP4-specific T cells in the initiation of both humoral and cellular immune responses.

Innate Immune Cells in NMOSD

The binding of AQP-4 Ab to AQP4 in the astrocyte triggers CDC and ADCC (Carnero Contentti & Correale, 2021). The pathological mechanism supporting complement involvement in NMOSD is reflected in the active deposition of immunoglobulins around blood vessels in demyelinating lesions, a vascular fibrosis and hyalinize can be seen in both active and inactive lesions (Lucchinetti et al., 2002). In this process, C1q binds to the Fc fragment of AQP4-Ab to form the membrane-penetrating complex (MAC) through the complement activation pathway (Zipfel & Skerka, 2009), causing astrocyte damage. In addition, C3a and C5a increase BBB permeability and chemotaxis of peripheral immune cells including neutrophils, basophils, eosinophils, mast cells, and macrophages (Vincent et al., 2008) into the CNS. In the ADCC, neutrophils, macrophages, and NK cells bind to the Fc segment of AQP4-Ab via the Fcγ receptor, triggering degranulation and causing cellular damage (Ratelade et al., 2012). Inflammatory molecules and toxic substances produced by these immune cells can also have toxic effects on adjacent cells, including oligodendrocytes and neurons (Tradtrantip et al., 2017; Duan et al., 2018).

Microglia, as a CNS resident immune cell, is also observed activation in the lesion (Carnero Contentti & Correale, 2021), which is mainly distributed in the high expression of AQP4 around the complement activation site (Lucchinetti et al., 2014). The potential role of microglia has been studied in animal model, it has reported that binding of AQP4-ab with AQP4 causes astrocytes produce complement fraction C3a (Howe et al., 2014), which binds to microglial C3aR to augment microglial activation and enhance microglial interaction with astrocytes (Chen et al., 2020). Microglia activation can also induce complement C1q production, which in turn promotes axonal injury and neurodegeneration (Dejanovic et al., 2018; Ten et al., 2010).

4 Clinical Treatment

Due to unclear etiology and complex pathogenesis, the clinical features of MS and NMO exhibit diverse symptoms and heterogeneity of disease subtypes, and there is no clinical treatment suitable for all disease subtypes. The treatment principles primarily aim to ameliorate symptoms during acute exacerbations and use disease-modifying therapies (DMTs) to mitigate disease progression and prevent relapses. With the revelation of the pathological characteristics and immunological mechanisms of such diseases, drugs for the treatment of MS and NMO have been developed from traditional immunosuppressive drugs to targeted therapy, and the safety and efficacy have been greatly improved. In recent years, monoclonal antibodies (mAbs) and new drugs have emerged in an endless stream, and it became the fastest-progressing area in the treatment of neurological diseases. This chapter aims to provide an overview of the clinical management of MS and NMOSD, with a particular emphasis on summarizing the therapeutic medications that have been marketed or are under clinical trials in recent years.

4.1 *Treatment of MS*

4.1.1 FDA-Approved DMTs

In the past two decades, disease modification therapy (DMT) has gradually become a long-term comprehensive management strategy for MS treatment. The widespread use of a series of disease-modifying drugs (DMD) represented by interferon-β (IFN-β), MS treatment has made great progress, DMDs can effectively reduce the frequency of MS relapse, reduce the degree of deterioration, and delay the natural course of the disease. Currently, there are more than 10 DMDs approved by the FDA for the clinical treatment of MS (see Fig. 3 and Table 1). The evolution of DMT for MS treatment has transitioned from interferons to agents like fingolimod, teriflunomide, CD20 mAb, and the ongoing development of Bruton's tyrosine kinase (BTK) inhibitors. Additionally, the mode of administration has shifted from injectable formulations to oral formulations, providing greater convenience for patients. The continuous expansion of therapeutic options heralds a new era in the field of MS treatment.

IFN-β is a type I interferon, which binds to type I interferon receptors, inhibits T cell proliferation and infiltration into the CNS, induces Treg generation, and inhibits B cell activation (Kieseier, 2011). IFNβ-1b was granted approval in the United States in 1993, establishing it as the earliest injectable therapeutic agent for MS and remaining a first-line treatment option. Subsequently, glatiramer acetate and IFN-β-1a were developed and also became first-line medications for MS treatment. Glatiramer acetate is a synthetic polypeptide resembling myelin basic protein (MBP). It was approved in 1996 in the United States. GA reduces MS disease activity and has displayed comparable therapeutic effect as high-dose interferon-β. The mechanism of GA action includes promoting expansion of GA-reactive T-helper 2 and regulatory T cells, as well as influence antigen-presenting cells (APCs)

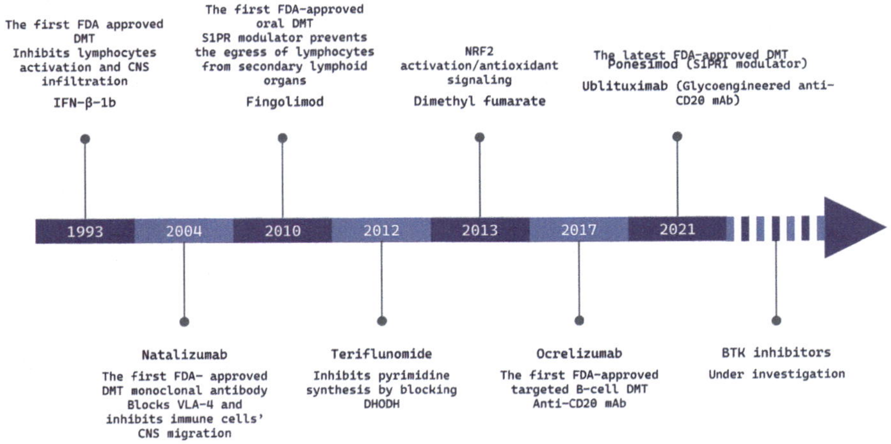

Fig. 3 The history of DMT drugs for MS

Table 1 FDA-approved MS disease-modifying drugs

Drug	Mechanism of action	Indication	Approval	Administration
IFN-β-1b IFN-β-1a	Binds to type I interferon receptors; Inhibits T cell division and CNS infiltration; induces Tregs and suppresses B cells activation	RRMS	1993 (IFN-β1b); 1996 (IFN-β1a) First line	IM or SC
Glatiramer acetate	A synthetic polypeptide resembling MBP; promotes expansion of GA-reactive T-helper 2 and regulatory T cells, as well as influence antigen-presenting cells (APCs)	RRMS, SPMS, CIS	1996, First line	IM or SC
Natalizumab	Anti-VLA-4 mAb, a potent inhibitor of leukocytes migration toward CNS	RRMS	2004, Second line	IV
Fingolimod	S1P receptor modulator ($S1P_1$ and $S1P_{3-5}$), prevents lymphocyte egress from lymphoid tissues in circulation, neuroprotective	RRMS	2010, First or second line	PO
Siponimod	S1P receptor modulator ($S1P_1$ and $S1P_5$), prevents lymphocyte egress from lymphoid tissues; neuroprotective	RRMS, SPMS	2019, Second line	PO
Ozanimod	S1P receptor modulator ($S1P_1$ and $S1P_5$), neuroprotective	RRMS	2020, First or second line	PO
Ponesimod	$S1P_1$ receptor modulator, prevents lymphocyte egress from lymphoid tissues; neuroprotective	RRMS	2021, First or second line	P.O
Teriflunomide	Inhibits pyrimidine synthesis by blocking DHODH	RRMS	2012, First line	PO
Dimethyl fumarate (DMF)	NRF2 pathway activation/antioxidant signaling; NF-κB pathway inhibition; Inhibition of neutrophil infiltration and microglia activation through activating HCA2	RRMS	2013, First line	PO
Diroximel fumarate (DRF)	A new drug from the fumarate; better GI tolerability. The mechanism of action of DRF is similar to that of DMF	RRMSSPMSCIS	2019, First line	PO
Alemtuzumab	Anti-CD52 mAb, depletion predominantly of CD52-bearing B and T cells	RRMS	2014, Second line	IV
Cladribine	Immunosuppressant, interferes with DNA synthesis	RRMS, SPMS	2019, Second line	PO
Mitoxantrone	Immunosuppressant, inhibits nucleic acid synthesis	RRMS, SPMS	2000, Third line	IV

NRF2 nuclear factor erythroid-2 related factor, *S1PR* sphingosine-1-phosphate receptor, *HCA2* hydroxycarboxylic acid receptor 2, *DHODH* dihydroorotate dehydrogenase, *RRMS* relapsing–remitting MS, *SPMS* secondary progressive MS, *CIS* clinically isolated syndrome, *GI* gastrointestinal, *IM* intramuscular injection, *SC* subcutaneous injection, *PO* per os, *IV* intravenous injection

(Schrempf & Ziemssen, 2007). Natalizumab, a humanized monoclonal antibody targeting the α4 integrin, was introduced in 2004, It can inhibit the migration of leucocytes to CNS. It has shown to be able to slow down the progression of symptom during a flare-up in patients with MS and reduce the rate of relapse in RRMS patients. However, it is worth emphasizing that due to the occurrence of rare adverse reactions such as progressive multifocal leukoencephalopathy, strict patient screening and close monitoring are required during application (Khoy et al., 2020).

In 2010, fingolimod was approved as the first oral DMT in 2010, offering a new treatment option for MS patients. Fingolimod and the later marketed siponimod, ozanimod, and ponesimod are all sphingosine-1-phosphate receptor (S1PR) modulators that inhibit the release of activated lymphocytes into the CNS (Kandjani et al., 2023). Following fingolimod, teriflunomide, and dimethyl fumarate were successively launched, expanding the range of oral DMTs. Teriflunomide is a reversible inhibitor of dihydroorotate dehydrogenase and an immunomodulator. It selectively inhibits the activity of dihydroorotate dehydrogenase, a key mitochondrial enzyme involved in de novo pyrimidine synthesis, thus preventing the production of new pyrimidines and inhibiting the proliferation of activated T and B cells (Hauser & Cree, 2020). Dimethyl fumarate and diroximel fumarate share similar mechanisms of action and pharmacological properties. They activate the nuclear factor erythroid 2-related factor 2 (Nrf2), which regulates immune responses and protects the CNS by suppressing oxidative stress and inflammation (Yadav et al., 2019). Diroximel fumarate has shown improved gastrointestinal tolerability compared to dimethyl fumarate. Alemtuzumab is another mAb drug after natalizumab, which targets CD52 and regulates and suppresses immune responses by eliminating CD52-positive immune cells (Syed, 2021). Furthermore, as B cells are now understood to be a central component of MS pathology, a series of mAbs targeting B cells such as ocrelizumab and ofatumumab have been launched, and Bruton's tyrosine kinase (BTK) inhibitors are also under development, which will be discussed in detail in the next section, "Targeted B Cell Therapy." Besides the aforementioned treatments, immunosuppressants such as mitoxantrone and cladribine have been FDA-approved for MS treatment, although they are not considered first-line therapies due to safety and tolerability concerns.

Table 1 FDA-approved MS disease-modifying drugs

NRF2 nuclear factor erythroid-2 related factor, *S1PR* sphingosine-1-phosphate receptor, *HCA2* hydroxycarboxylic acid receptor 2, *DHODH* dihydroorotate dehydrogenase, *RRMS* relapsing–remitt MS, *SPMS* secondary progressive MS, *CIS* clinically isolated syndrome, *GI* gastrointestinal, *IM* intramuscular injection, *SC* subcutaneous injection, *PO* per os, *IV* Intravenous injection

4.1.2 Targeted B Cell Therapy

As the discovery of abnormal antibodies produced in the CNS of MS patients, such as oligoclonal bands in the CSF and immunoglobulins (Ig) and complement detection in demyelinating lesions. The understanding of disease pathogenesis has evolved from the traditional T-cell-mediated perspective to encompass the intricate interplay between various immune cell populations, including B cells. Specifically, from a pathological perspective, B cell infiltration is more prominent in MS compared to other CNS inflammatory disorders. In early MS and individuals with active lesions, $CD20^+$ B cells predominantly reside around larger veins and exhibit pro-inflammatory functions. Pathological studies on the meninges in MS patients reveal B-cell-rich inflammatory aggregates like tertiary lymphoid follicles, which may facilitate B cell homing, survival, and functional activation, thereby sustaining chronic inflammation, demyelination, and neurodegeneration (Touil et al., 2018). From an immunological standpoint, there is an increased presence of B cells and plasma cells in the CSF of patients with MS, and their quantities are positively associated with intrathecal inflammation and Ig synthesis (Cepok et al., 2005). In addition, accumulating evidence suggests that antibody-independent functions of B cells play a critical role in mediating disease activity. As antigen-presenting cells, B cells demonstrate higher efficacy in antigen presentation and activation of T cells compared to non-B cell APCs, thereby inducing aberrant pro-inflammatory responses involving Th1, Th17, and myeloid cells, promoting cellular immune cascade. B cells also actively participate in the production of cytokines and chemokines, exerting both anti-inflammatory and pro-inflammatory effects, contributing to oligodendrocyte and neuronal damage (Baecher-Allan et al., 2018). Currently, selective B cell depletion strategies, notably employing anti-CD20 mAbs, have shown robust therapeutic efficacy, substantiating the critical role of B cells in driving the disease activity of MS.

Anti-CD20 Monoclonal Antibodies CD20 as a specific surface marker on human B lymphocytes, have function of regulating B cells' proliferation and differentiation. Unlike CD19, CD20 appears from pre-B cell to mature B cell stages but is not expressed on hematopoietic stem cells, progenitor B cells, or mature plasma cells, thus exhibiting higher selectivity for B cells and minimal impact on plasma cells and hematopoietic function. Anti-CD20 mAb achieves selective B cell depletion by targeting $CD20^+$ cells. A body of evidence indicates that anti-CD20 mAbs are associated with three mechanisms of action: antibody-dependent cell-mediated cytotoxicity (ADCC), complement-dependent cytotoxicity (CDC), and direct effects resulting from antibody binding to CD20 molecules, including inhibition of cell growth, alteration of the cell cycle, and induction of apoptosis (Margoni et al., 2022).

Rituximab, a human-mouse chimeric mAb, is the first generation of anti-CD20 mAb and the earliest clinical exploration of anti-CD20 mAbs for the treatment of MS in the world. Although it has not received FDA approval for MS treatment, studies have shown that it can reduce disease activity and relapse rates in MS patients (Salzer et al., 2016). Ocrelizumab, a humanized anti-CD20 mAb, was the first approved DMT that depletes B cells for the treatment of RRMS and PPMS in the United States in 2017. The antigenic epitope targeted by ocrelizumab overlaps with rituximab but differs in certain aspects. Moreover, its humanized nature diminishes immunogenicity and infusion-related reactions, thereby positioning it as a first-line treatment for MS (Lamb, 2022). Monoclonal antibodies have provided MS patients with a more precise treatment approach, but their safety and intravenous administration pose inconveniences. Notably, ofatumumab, introduced in 2020, possesses three characteristics: full humanization, small dose equivalence, and subcutaneous injection, which further reduces infusion-related reactions and makes it the only B-cell therapy currently available for patients to self-administer once a month (Hauser et al., 2020), providing more options for MS patients. However, as the humanization of anti-CD20 mAb increases, the specificity and affinity of antibody binding to antigens decrease. Therefore, ublituximab, a modified antibody targeting the Fc segment, has entered the market. Ublituximab is a novel glycoengineered anti-CD20 mAb that targets a unique epitope of the CD20 antigen on mature B lymphocytes. Ublituximab exhibited a statistically significant reduction in the annualized relapse rate (ARR) compared to teriflunomide in two global active-controlled Phase 3 clinical trials (Steinman et al., 2022). The information on anti-CD20 mAb for MS is summarized in Fig. 4.

Bruton Tyrosine Kinase (BTK) Inhibitors Although great progress has been made in the treatment of RRMS in the past 25 years, limited therapeutic options are

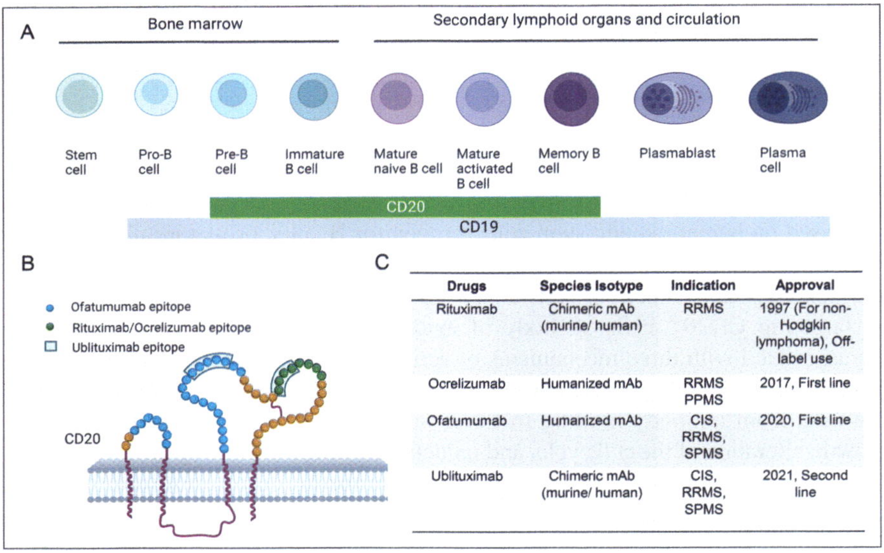

Drugs	Species Isotype	Indication	Approval
Rituximab	Chimeric mAb (murine/ human)	RRMS	1997 (For non-Hodgkin lymphoma), Off-label use
Ocrelizumab	Humanized mAb	RRMS PPMS	2017, First line
Ofatumumab	Humanized mAb	CIS, RRMS, SPMS	2020, First line
Ublituximab	Chimeric mAb (murine/ human)	CIS, RRMS, SPMS	2021, Second line

Fig. 4 Anti-CD20 monoclonal antibodies for MS treatment. (**a**) B cells originate from stem cells in the bone marrow and are defined according to the expression of specific cell surface antigens. CD20 is expressed in pre-B cells, mature and memory B cells, while CD19 appears from pro-B cells to plasma cells. Therefore, CD20 exhibits higher selectivity for B cells and preserves B-cell repopulation and humoral immune memory. (**b**) Schematic representation of the different CD20 epitopes recognized by the different anti-CD20 mAbs used in MS treatment. (**c**) main anti-CD20 mAbs for the treatment of various types of MS

available for patients with relapsing MS or PPMS. Autopsy pathological features showed that widespread activation of microglia and macrophages, astrogliosis, and persistence of lymphocyte activity in meningeal aggregates and dispersed in the CNS parenchyma. In the active lesion, activated microglia and macrophages are located at the lesion rim (Faissner et al., 2019; Yong & Yong, 2022; Choi et al., 2012). These results suggest that CNS-resident immune cells may contribute to driving the accumulation of disability. It may also explain why current therapies that only modulate the peripheral immune response have limited effects on PPMS. BTK is an intracellular signaling molecule that is produced in B cells and most hematopoietic lineages including monocytes, macrophages, microglia, mast cells, and neutrophils (Brunner et al., 2005; Torke et al., 2020; Garg et al., 2022). It plays an essential role in the maturation, survival, migration, and activation of B cells and microglia. Importantly, current BTK inhibitors possess the potential to effectively penetrate the BBB, ensuring simultaneous control of peripheral and CNS inflammation that contributes to the immunopathology of MS on both sides of BBB. These indicate that BTK inhibitor could be a highly promising therapeutic candidate for MS. Currently, six BTK inhibitors (including evobrutinib, tolebrutinib, fenebrutinib, remibrutinib, orelabrutinib, and BIIB091) that differ in selectivity, strength of inhibition, binding mechanisms and ability to modulate immune cells within the CNS are under investigation in clinical trials for MS, as shown in Table 2 (Li et al., 2022b).

4.1.3 Other Investigational Therapies for MS

In addition to the aforementioned drugs, there is ongoing research and development of new drugs for MS (see Table 3). SAR443820 is a CNS-penetrant small molecule inhibitor of receptor-interacting serine/threonine-protein kinase 1 (RIPK1), a key signaling protein involved in regulating neuroinflammation and cell death. Currently, SAR443820 is undergoing phase 2 clinical trials (Muehler et al., 2020). IMU-838 is an immunomodulator that selectively blocks dihydroorotate dehydrogenase (DHODH) in active T and B cells (Muehler et al., 2020). This mechanism potentially inhibits inflammatory cell activity while sparing other components of the immune system.

As DMT has evolved, the treatment goals for MS have expanded. Repairing damage and restoring function have become another major direction of drug research, currently focusing on reducing myelin damage and remyelination. IMCY-0141 is a synthetic peptide based on myelin oligodendrocyte glycoprotein (MOG), which stimulates the expansion of a population of antigen-specific cytolytic $CD4^+$ T cells that reset the body's autoimmune response. Other drugs such as antihistamines can promote oligodendrocyte precursor cells (OPC) differentiation to promote myelination (Green et al., 2017). Opicinumab antagonizes LINGO, a negative regulator of oligodendrocyte differentiation, and promotes myelin regeneration (Cadavid et al., 2017). High-dose biotin (MD1003) is also considered as a potential

Table 2 BTK inhibitors (under investigation)

Drugs	Phase	Condition	Result
Evobrutinib	Phase 2, Active, not recruiting, NCT02975349	RRMS	Patients with RMS who received 75 mg of evobrutinib had significantly fewer enhancing lesions during weeks 12 through 24 than those who received placebo. No significant difference in ARR or disability progression in trial groups (Montalban et al., 2019)
	Phase 3, active, not recruiting, NCT04338022; NCT04338061	RMS	Not published
Tolebrutinib	Phase2b, completed, NCT03889639; A long-term extension study (NCT03996291)	RMS	A dose-dependent reduction in new Gd-enhanced lesions; drug was well tolerated; reduction of acute inflammation (Reich et al., 2021)
	Phase3; active, not recruiting, [1]NCT04410978 [2]NCT04410991 [3]NCT04411641 [4]NCT04458051 (recruiting)	[1,2]RMS (RRMS and active SPMS) [3]Non-active SPMS [4]PPMS	Not published
Remibrutinib	Phase 3, recruiting NCT05156281; NCT05147220	RMS	Not published
Orelabrutinib	Phase 2, active, not recruiting, NCT04711148	RRMS	An interim analysis showed that orelabrutinib led to significant, dose-dependent reductions in lesion burden compared with a placebo
Fenebrutinib	Phase 3 [1]NCT04544449 (active, not recruiting) [2]NCT04586010 (active, not recruiting) [3]NCT04586023 (active, not recruiting)	[1]PPMS [2]RMS [3]RMS	Not published
BIIB091	Phase 2, (Recruiting) NCT05798520	RMS	Not published

RRMS relapsing–remitting MS, *RMS* relapsing multiple sclerosis, *PPMS* primary progressive MS, *ARR* annualized relapse rate

Table 3 Other investigational therapies for MS

Therapy	Mechanism of action	Clinical trial	Condition
SAR443820	Inhibiting inflammation and nerve cell death in the CNS by blocking RIPK1	Phase 2, recruiting, NCT05630547	RRMS, SPMS, PPMS
IMU-838	Blocking an enzyme called DHODH to suppress the metabolism of activated immune T-cells and B-cells	Phase 3, recruiting, NCT05134441	RMS (RRMS and active SPMS)
IMCY-0141	Synthetic peptide based on MOG	Phase 1/2, recruiting, NCT05417269	RRMS
ATA188	Allogeneic T-cell immunotherapy; targets EBV-infected B cells	Phase 1/2, active, not recruiting. NCT03283826	PPMS, SPMS, Positive EBV serology
AHSCT	Reset the immune system	Phase 3, recruiting, NCT04047628	RMS, RRMS, SPMS

RMS relapsing multiple sclerosis, *RRMS* relapsing–remitting MS, *PPMS* primary progressive MS, *SPMS* secondary progressive MS, *PMS* progressive MS, *RIPK1* receptor-interacting protein kinase 1, *DHODH* dihydroorotate dehydrogenase, *MOG* myelin oligodendrocyte glycoprotein, *EBV* Epstein-Barr virus, *HSC* hematopoietic stem cell, *NSC* neural stem cell, *AHSCT* autologous hematopoietic stem cell transplantation

remyelinating agent. While these experimental drugs have yet to demonstrate conclusive clinical significance, they offer promising avenues for advancing MS treatment.

In addition to pharmacological agents, cell therapies are undergoing clinical trials. ATA188 is an allogeneic T-cell immunotherapy that involves the engineering of T cells to specifically target Epstein-Barr virus (EBV)-infected B cells and plasma cells, thereby reducing their attack on the neural myelin sheath (Robinson & Steinman, 2022). Furthermore, stem cell transplantation has gained significant attention. Autologous hematopoietic stem cell transplantation (AHSCT) is the collection of blood stem cells from adult patients and usage of them to repopulate the bone marrow after patients undergo chemotherapy and immunosuppression to eradicate disease-causing self-reactive immune cells, aiming to "reset" MS patients' immune system to suppress it from attacking the CNS (Boffa et al., 2021). Neural stem cell transplantation, on the other hand, entails obtaining neural stem cells/predecessor cells from donated aborted fetuses (10–12 weeks of gestational age) and injecting them into the spinal cord of patients with progressive MS. This approach has demonstrated successful reduction of disease biomarkers and aims to promote remyelination for neuronal protection (Genchi et al., 2023).

4.2 Treatment of NMOSD

NMOSD is a rare demyelinating disease of the CNS. It causes multifocal CNS inflammation that primarily affects the optic nerves and spinal cord, and it typically results in attacks of visual loss and paralysis. Treatment of NMOSD includes acute phase treatment, and maintenance phase treatment (Preventive therapy), (Prevention and treatment of recurrence), In the acute phase, high-dose steroids (Nakamura et al., 2010; Gal et al., 2012; Collongues & de Seze, 2011; Okamoto et al., 2008) and plasma exchange (Chan & Lee, 2021; Li et al., 2021) are preferred. For resistant cases of persistent inflammation without improvement after both steroids and plasma exchange, intravenous cyclophosphamide or immunoglobulin (IVIg) therapies could be considered according to patients' condition (Roed et al., 2005; Elsone et al., 2014). The general principles of preventative therapy are focusing on balance effectiveness with short-and long-term side effects. There are many immunosuppressive drugs for maintenance phase treatment including azathioprine, rituximab, mycophenolate mofetil, methotrexate, etc. Among these drugs, azathioprine (AZA) and rituximab (RTX) are recommended as first-line therapy, and other immunosuppressive drugs such as methotrexate, mycophenolate mofetil (MMF) and mitoxantrone are recommended as second-line therapy. For refractory NMOSD, second-line therapies such as alkylating agent cyclophosphamide (Shi et al., 2022b; Wang et al., 2020) and topoisomerase II inhibitor mitoxantrone (Enriquez et al., 2019; Cabre et al., 2013) may have a good effect.

In recent years, with the exploration of the immunopathological mechanism of NMOSD, targeted therapy for pathogenic links such as complement, AQP4-IgG, and inflammatory factors is gradually deepening, and some drugs have received FDA approval and entered clinical applications, including inebilizumab and rituximab that target B cells, anti-interleukin-6 (IL-6) receptor antibodies (tocilizumab and satralizumab), and complement inhibitors (eculizumab), etc., which are summarized in Table 4. It is worth noting that some drugs used to treat MS, such as interferon β, fingolimod, natalizumab, may cause worsening of NMOSD (Palace et al., 2019; Sato et al., 2013).

With the emergence of new drugs, a large number of clinical trials exploring the treatment of NMO are underway, including targeting the complement system (C1-esterase inhibitor and anti-C5 mAb), depleting B cell or inhibiting B cell activity or proliferation mAb (second- and third-generation CD20) and small molecule drugs (BTK inhibitor, JAK inhibitor, BAFF inhibitor), blocking NMO-IgG activity (FcRN mAb), maintaining the integrity of the blood-brain barrier (humanized VEGF mAb), and autologous stem cell transplantation, etc., as shown in Table 5 for details. It is worth noting that although immune-targeted drugs have made great progress in preventing relapse of NMOSD, the clinical application of these medicines are still facing many challenges such as high prices. Importantly, the efficacy of new therapeutic medicines is still limited due to the lack of a large amount of clinical data to confirm their efficacy. In addition, immune-targeted drugs are mostly developed for AQP4 antibody-positive patients, while patients with AQP4-Ab-negative or MOG antibodies may have different clinical and immunological features compared

Table 4 Clinical application of DMTs in NMOSD

	Drugs	State	Mechanism	Indication
Targeting complement system	Eculizumab (Pittock et al., 2013, 2019, 2022)	2019, FDA approved	Inhibits C5 cleavage to the activated form (C5a and C5b)	Adult patients with AQP4-ab$^+$ NMOSD
Targeting B cells	Rituximab (Tahara et al., 2020; Kim et al., 2013)	Recommended by Chinese guideline (2021)	Anti- CD20 mAb	Adult patients with AQP4-Ab$^+$ NMOSD
	Inebilizumab (Cree et al., 2019)	2020, FDA approved	Anti-CD19 mAb	Adult patients with AQP4-Ab$^+$ NMOSD
Targeting the IL-6 receptor	Tocilizumab (Araki et al., 2014; Lotan et al., 2021; Zhang et al., 2020)	Recommended by Chinese guideline (2021)	First humanized anti-IL-6 receptor monoclonal antibody	Adult patients with AQP4-Ab$^+$ NMOSD
	Satralizumab (Traboulsee et al., 2020; Yamamura et al., 2019)	2020, FDA approved	Binds to membrane as well as to solubilized IL-6 receptors	AQP4-Ab$^+$ NMOSD in children and adults over 12 years of age

to those of AQP4-positive patients, whether these drugs are suitable for this subtype of patients need to be further explored.

5 Laboratory Tests for Inflammatory Demyelinating Diseases of CNS

The feature of patients with inflammatory demyelinating diseases of CNS are often complex and diverse in terms of clinical symptoms, imaging, disease progression, and responsiveness to drug therapy, so the differential diagnosis and personalized management of patients rely highly on laboratory testing. With the development of biotechnology and a deeper understanding of the pathophysiological processes of the disease, laboratory tests for inflammatory demyelinating diseases of CNS have evolved from the traditional routine analysis of chemical composition (sugar, chloride, total cell and protein analysis in the cerebrospinal fluid, etc.) to qualitative and quantitative analysis of immune-related molecules (IgG index, oligoclonal bands, immune cell subtype analysis, etc.) and specific neurological antibodies (Anti-AQP4 antibodies, anti-myelin antibodies), the tests are gradually diversifying and refining. Neurological antibody testing is an important indicator for early diagnosis of disease as well as fine classification of diseases. In recent years, the application of single-molecule assay (Simoa) to test neuronal proteins in the

Table 5 Drugs under investigation for NMOSD

	Drugs	Mechanism	phase /state	Condition	Results
Targeting complement system	Ravulizumab (Pittock et al., 2023)	Anti-C5 mAb	Phase3, completed [Clinical Trials.gov, NCT04201262]	Adult patients with AQP4-Ab$^+$ NMOSD	Significantly reduced relapse risk, a safety profile consistent with those of eculizumab and ravulizumab
	C1-esterase inhibitor (Levy & Mealy, 2014)	Human C1-inhibitor that suppress the complement pathways	Phase 1, completed, [Clinical Trials.gov, NCT01759602]	NMO/NMOSD presenting with acute transverse myelitis and optic neuritis	a safe add-on therapy for patients with NMOSD, reducing neurologic damage and improving outcomes
Targeting B cells	Ublituximab[a] (Mealy & Levy, 2019)	Specifically binds to the trans-membrane antigen CD20 and induce B cell lysis	Phase 1, completed, [Clinical Trials.gov, NCT02276963]	Adult NMOSD patients	A safe add-on therapy for NMOSD patients presenting with acute transverse myelitis and optic neuritis; a promising benefit on durability of remission when B cell depletion is achieved
	Divozilimab	Anti-CD20 mAb	Phase 3, recruiting, [Clinical Trials.gov NCT05730699]	Adult NMOSD patients	not published
	Ofatumumab	Second-generation anti-CD20 mAb with a fully humanization	Phase1/2, Recruiting, [Clinical Trials.gov, NCT05504694]	Adult patients with AQP4-Ab$^+$ NMOSD	not published
	MIL62	Third-generation anti-CD20 mAb with stronger ADCC and anti-tumor activities	Phase3, Recruiting, [Clinical Trials.gov, NCT05314010]	Adult NMOSD patients	not published

Targeting plasma cells	Bortezomib (Zhang et al., 2017)	An inhibitor of the 26 s proteasome that depletes long-lived plasma cells	Completed, [Clinical Trials.gov, NCT02893111]	Highly relapsing AQP4-Ab^{+} NMOSD	Reduces annual recurrence rates, and AQP4-Ab titer for highly active NMOSD
	Daratumumab	Anti-CD38 mAb that targets ACS	Phase2/3, Recruiting, [Clinical Trials.gov, NCT05403138]	Adult patients with AQP4-Ab^{+} NMOSD	not published
BTK^{b} inhibitor	Zanubrutinib	BTK^{b} inhibitor	Phase 2, Recruiting, [Clinical Trials.gov, NCT05356858]	Adult patients with AQP4-Ab^{+} NMOSD	not published
	Orelabrutinib	BTK^{b} inhibitor	Unknown status [Clinical Trials.gov, NCT05284175]	Patients with NMOSD	not published
Others	Batoclimab (Wang et al., 2023)	Anti-neonatal fc receptor (antiFcRn) mAb	Phase 1b, completed	Adult NMOSD patients	A safe add-on therapy to IVMP and tolerated in patients with NMOSD
	Bevacizumab (Mealy et al., 2015)	Humanized VEGF mAb that inhibits the interaction of VEGF to Flt1 and KDR receptors on the surface of endothelial cells	Phase 1b, completed, [Clinical Trials.gov, NCT01777412]	Adult NMOSD patients	A safe add-on therapy to high-dose corticosteroids for NMOSD presenting with an acute relapse

(continued)

Table 5 (continued)

	Drugs	Mechanism	phase /state	Condition	Results
	CAR T (Qin et al., 2023)	Chimeric antigen receptor (CAR) T-cell therapy that targets BCMA	Phase1, Completed [Clinical Trials.gov, NCT05828212]	Relapsed/refractory AQP4-Ab^{+} NMOSD	A manageable safety profile and therapeutic potentials for patients with relapsed/refractory AQP4-Ab^{+}NMOSD.
	Hematopoietic stem cell transplantation; (Burt et al., 2019)	Infusion of participant's own stem cells[b]	Phase1/2, Completed, [Clinical Trials.gov, NCT00787722]	Adult patients with AQP4-Ab^{+} NMOSD	Obtain a prolonged drug-free remission in 12 patients with NMOSD; 9/11 AQP4-IgG become negative
	Belimumab	A human IgG1λ antibody that blocks the bioactivity of soluble BAFF	Phase1/2, Completed, [Clinical Trials.gov, NCT05154734]	Adult NMOSD patients	not published
	Baricitinib	JAK1/JAK2 inhibitor, inhibit bone marrow regulation of B cell proliferation and differentiation	Phase1/2, Recruiting, [Clinical Trials.Gov, NCT05792462]	Adult NMOSD patients	not published

[a]Has been granted orphan drug status by the FDA for NMOSD as an experimental drug; ASCs, Antibody-secreting cells; BCMA, B-cell maturation antigen
[b]Hematopoietic Stem Cell Transplantation was performed after cyclophosphamide; rATG, rituximab, and methylprednisolone conditioning regimen
JAK Janus kinase, *BAFF* B-cell activating factor/B-lymphocyte stimulating factor (BLyS), *BTK* Bruton tyrosine

peripheral blood including NFL, GFAP, etc., further enriches and develops the auxiliary diagnosis system for neuroimmune diseases.

5.1 Cerebrospinal Fluid Examination

Due to CSF being directly linked to the brain and spinal cord, laboratory tests for CSF can better reflect the changes in CNS of inflammatory demyelinating disease, it can provide essential evidence for the diagnosis of MS or NMO and to rule out relevant differential diagnoses.

Cerebrospinal Fluid Routine analysis Patients with demyelinating diseases (e.g., MS, NMOSD) have mostly normal lumbar puncture pressure, the appearance of CSF is colorless and transparent, the number of mononuclear cells may be mildly or moderately elevated, the count of leukocytes in MS patients generally does not exceed $50*10^6$/L, but it exceeds $100*10^6$/L in 30% of NMOSD patients in the acute phase. The content of sugar and chloride is often normal, and about 25% of patients have a mild-to-moderate elevation of proteins, predominantly immunoglobulins (mostly <1 g/L).

Cerebrospinal Fluid Cytology Immunoreactive cells such as activated lymphocytes, plasma cells, and activated monocytes can be observed. In the acute phase, small lymphocytes are usually present, accompanied by activated lymphocytes and plasma cells, and occasionally multinucleated cells, which is a sign of disease activity. In the remission stage, activated monocytes and macrophages are commonly found, and the cytology between attacks could be completely normal.

IgG Intrathecal Synthesis The detection of intrathecal lgG synthesis is an important test for the clinical diagnosis of MS. Immunoglobulins are commonly increased in the CSF of MS patients, and the content of lgG is predominantly elevated. At present, the detection methods for IgG intrathecal synthesis include qualitative and quantitative detection.

lgG Index The lgG index is a quantitative indicator of intrathecal IgG synthesis, which is increased in 70–75% of MS patients. The calculation formula for the lgG index is followed as: [CSF-lgG/ S (serum) – lgG]/[CSF-Alb (albumin)/S-Alb]. The upper limit value is 0.7, exceeding which indicates an increase in IgG intrathecal synthesis. The premise for judging the intrathecal synthesis of IgG is that the ratio of CSF-Alb/S-Alb is normal, which indicates that the function of blood–brain barrier (BBB). When the CSF-Alb/S-Alb ratio is normal and the CSF-IgG/S-IgG ratio increases by more than four times during the course of the disease, intrathecal synthesis can be determined.

lgG Oligoclonal Bands (OCB) An important qualitative indicator of intrathecal synthesis of lgG, which can be detected in the CSF in 80–95% of MS patients (Dobson et al., 2013; Bourahoui et al., 2004). Intrathecal OCB refers to the formation of several narrow, discontinuous bands in the gamma globulin region by the isoelectric focusing electrophoresis method. Detection of oligoclonal lgG bands should be performed simultaneously with the CSF and serum. The presence of two or more bands in paired samples of cerebrospinal fluid and serum is considered positive for OCB. OCB examination is an important guide for MS diagnosis and treatment, and studies have shown that intrathecal synthesis of lgG precedes the appearance of clinical imaging lesions and is an important indicator for early differential diagnosis of MS to other CNS demyelinating diseases. In addition, OCB has the potential to monitor disease activity and evaluate prognosis, such as predicting the conversion of Clinically Isolated Syndrome (CIS) to MS (Stangel et al., 2013). In general, OCB-positive MS patients have a higher degree of disease disability than OCB-negative patients. In relapsing-remitting MS, OCB-positive patients also have a higher chance of transitioning to progressive MS, and the number of OCB bands early in the disease correlates with the long-term prognosis of MS patients (Farina et al., 2017). However, intrathecal IgG OCB synthesis is not MS-specific and can be seen in a variety of diseases, such as other inflammatory or non-inflammatory diseases (Stangel et al., 2013). Therefore, if the result of is positive for a patient, further analysis needs to be conducted in conjunction with clinical symptoms.

5.2 Antibodies Test

Aquaporin 4 (AQP4) Antibodies Approximately 80% of NMO/NMOSD patients have aquaporin-4 antibodies (AQP4-Ab) in serum (Jarius et al., 2020). AQP4-Ab is a highly specific antibody for the diagnosis of NMOSD. The diagnostic criteria for NMOSD proposed in 2015 were stratified based on AQP4 antibodies. AQP4-Ab$^+$ patients only need to have one of the six core features to be diagnosed, while AQP4-Ab$^+$ patients need to have two core features and strict MRI attachment conditions are proposed as support for diagnosis (Wingerchuk et al., 2015). Patients with AQP4-Ab$^+$NMOSD have a higher risk of relapse than those with AQP4-Ab$^-$NMOSD. It is noteworthy that although the tiers of AQP4-Ab$^+$ in serum do not always correlate with the activity of the disease, it can indicate the severity of the disease at the attack phase (Liu et al., 2021). Their titers may decrease with immunotherapy in the acute phase and can be used as an indicator of treatment efficacy.

Myelin Oligodendrocyte Glycoprotein (MOG) Antibodies In patients with AQP4-Ab$^-$ NMOSD, approximately 20–40% patients are positive for an antibody against MOG antibody (MOG-Ab) (Sato et al., 2014; Narayan et al., 2018; Carnero Contentti et al., 2021). However, subsequent studies found that MOG-IgG positivity was not limited to NMOSD, and less than one-third of serum MOG-IgG-positive patients met the diagnostic criteria for NMOSD (Dos Passos et al., 2018). In 2018, two international groups defined MOG antibody-associated disease (MOGAD) as an independent disease entity independent of MS and AQP4-Ab$^+$ NMOSD (Jarius et al., 2018; López-Chiriboga et al., 2018). MOG antibodies are mostly present in patients with optic neuritis and/or myelitis, but rarely in the classical MS (Mayer & Meinl, 2012). Compared to AQP4 seropositive patients, MOGAD more commonly affects young Caucasian males, with a relatively better long-term clinical outcome. High titers of MOG antibodies are more specifically present in the sera of pediatric acute disseminated encephalomyelitis (ADEM) patients. Serum titers are related to disease activity and treatment status. In the process of long-term follow-up, it was found that antibody titers in children with ADEM gradually decreased or even became negative, and decreased antibody titers often indicated a good clinical prognosis in ADEM patients, but in patients with NMOSD or MS, the titers mostly continued to fluctuate at the original level or increased again (Di Pauli et al., 2011; Pröbstel et al., 2011).

Myelin Basic Protein (MBP) Antibodies MBP is an intracellular myelin-associated protein. Studies have shown that the level of MBP antibody in relapsing MS patients and asymptomatic patients with gadolinium-enhancing lesions first increases and then rapidly decreases within a week of relapse, making MBP a novel biological marker of MS progression (Burman et al., 2014). However, this antibody can also be present in patients with other diseases and in healthy populations. Therefore, its specificity for the diagnosis of MS remains to be verified.

5.3 Nerve Injury Biomarkers

Neurofilament Light Chain (NFL) NFL is an axonal skeleton protein that is released into the CSF and peripheral blood after axonal damage and therefore can be detected in both CSF and serum. The study found that the content of the NFL in CSF or serum could reflect MS disease activity, progression, and efficacy of therapy. Studies have shown that the level of serum NFL (sNFL) in patients with recurrent MS is related to the size of MRI brain lesions and cerebral atrophy. Regular examination of sNFL helps to assess the clinical disease activity, brain volume reduction, and predict the risk of disability progression (Disanto et al., 2017; Kuhle et al., 2019; Bjornevik et al., 2020; Siller et al., 2019). In addition, there is a correlation between sNFL levels and drug therapy, such as fingolimod in the treatment of relapse and remission MS, and Mitoxantrone, Rituximab, and natalizumab in the treatment of progressive MS, the content of sNFL in patients after treatment is significantly reduced, suggesting that sNFL can be used as an indicator for drug efficacy evaluation (Kuhle et al., 2015; Novakova et al., 2017). It is worth noting that the NFL is not a specific diagnostic indicator. Other neurological diseases, such as trauma, stroke, and aging, may also cause the increase of NFL content in CSF or serum (Chen et al., 2021).

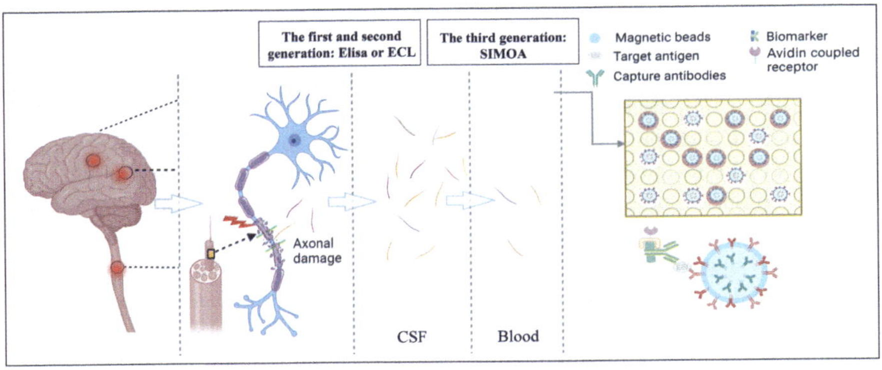

Fig. 5 Progress of measurement of NFL concentration. After neuronal axonal injury, NFL is released into the CSF and peripheral blood. The level of NFL in CSF is higher than in blood, which can be detected using Elisa or electrochemiluminescence (ECL) techniques. However, due to the low levels of NFL in peripheral blood, the application of Elisa and ECL assays is limited. The ultra-sensitive single molecule array (Simoa) can improve the detection sensitivity of low abundance proteins largely through reading signals from one magnetic bead with antigen-antibody complex. A study has reported the limit of detection for NFL by the three methods. (SIMOA: 0.62 pg/ml, ECL: 15.6 pg/mL and ELISA: 78.0 pg/mL). (Siller et al., 2019)

Glial Fibrillary Acid Protein (GFAP) GFAP is an intermediate filament protein that forms the astroglial cytoskeleton. It is released into CSF and serum after astrocyte injury and is used to indicate astrocyte injury. Studies have shown that serum levels of GFAP are significantly higher in MS patients compared to non-inflammatory neurological diseases and correlate with clinical symptoms and the number of lesions on MRI imaging, which is particularly pronounced in the progressive phase of MS (Axelsson et al., 2011; Abdelhak et al., 2019). In addition, increased serum GFAP is also seen in the acute stage of NMOSD relapse and correlates with the degree of disability and drug therapy of the patient (Chang et al., 2021). These studies suggest that the detection of serum levels of GFAP could be a potential marker for assessing disease prognosis.

Assay of Nerve Injury Markers The detection principle of NFL and GFAP protein is based on the double antibody sandwich method. NFL and GFAP levels in CSF are high and can be detected by applying traditional Elisa and electrochemiluminescence, but GFAP and NFL protein levels in serum are extremely low (fg to pg levels) and difficult to detect using traditional methods. With the development of ultrasensitive immunoassay platforms, such as single-molecule assay (Simoa) application, low levels of plasma proteins can be tested (see Fig. 5). A strong correlation between serum and CSF NFL levels was found (Disanto et al., 2017). Therefore, examination of neurological damage markers in blood may serve as a feasible, non-invasive approach to evaluate prognosis of MS or NMOSD.

5.4 *Immune Cell and Immune Molecular Assays*

The occurrence and development of inflammatory demyelinating diseases of the CNS are closely related to the immune response in the body. Therefore, the detection of immune cells or related factors can assess the degree of immune response and be used to monitor disease activity and predict relapse, etc. In addition, clinically exploring the strategy of new drug treatment, such as the application of monoclonal antibodies to deplete B cells (anti-CD19, CD20, etc.) for the treatment of MS or NMO, is required to monitor B cell subsets and function. These indicators play an important role in personalized medication management.

With the in-depth study of diseases, new immune molecular markers including adhesion molecules, cytokines, cytokine receptors, chemokines, etc., are emerging, which can reflect to, some extent, the disease activity and the effectiveness of disease-modifying therapeutics (DMT). CXCL13 is a B-cell chemokine that is increased in the CSF of MS patients and can reflect the degree of disease activity (Kuhle et al., 2015). Moreover, the CXCL13 levels in the CSF decreased when MS patients were treated with medication switched from first-line DMT drugs such as IFN-β, glatiramer acetate, or teriflunomide to fingolimod or natalizumab (Chen et al., 2021; Meier et al., 2023). In addition, decreased levels of polysaccharidase 3-like protein 1 (CHI3L1) in CSF have also been reported in MS patients when treated with fingolimod or mitoxantrone (Abdelhak et al., 2019; Meier et al., 2023). Therefore, CXCL13 and CHI3L1 molecules can serve as markers of drug treatment responsiveness. Additionally, soluble CD27 (sCD27) is elevated in the CSF of patients with progressive MS (Komori et al., 2015) and sCD27 also as an inflammatory marker is a potentially sensitive molecule for drug response to natalizumab and methylprednisolone (Romme Christensen et al., 2019). In contrast, kappa free light chain (KFLC) has important applications in early diagnosis of MS and prediction of CIS to MS outcome (Khademi et al., 2011). However, these molecular markers are in the clinical research stage, and their sensitivity and specificity are to be further validated in large clinical trials. In conclusion, abnormal immune response is a key factor in the development and progression of inflammatory demyelinating diseases of CNS, and mapping the neurological and immune profiles of CSF to find potential therapeutic targets and biomarkers can truly facilitate clinical translation.

6 Conclusion

In the past 30 years, our understanding of the pathological mechanisms of inflammatory demyelinating diseases of the CNS and the clinical management of these patients have been largely improved. Especially, the development and clinical application of DMTs have greatly enriched clinicians' drug selection and improved patients' clinical outcomes. However, as patients with CNS demyelinating diseases are highly heterogeneous and exhibit diverse symptoms, there is no clinical treatment suitable for all disease subtypes. FDA-approved drugs (first- and second-line) fail to adequately control the disease progression of patients with severe forms of relapsing MS. It is noteworthy that the application of new targeted drugs for the treatment of MS or NMOSD patients is still in the exploratory stage, most of the current drugs vary in efficacy, safety, and cost, there is no definitive answer on how to hold the balance between drug dose, duration of drug intervention, and prevention of disease relapse and safety. In addition, the long-term efficacy and side effects of the drugs as well as the tolerance of patients to the drugs need to be further investigated. Importantly, key scientific questions remain to be answered, such as the tissue specificity of immune cells in different tissues and organs of MS or NMOSD patients, such as the bone marrow, peripheral blood, CNS, and thymus tissue, and how immune cells in these organs regulate and interact. What are the specificities of the immunogenesis mechanisms in refractory NMOSD and patients with AQP4-Ab^{-}? Is it possible to find specific molecular markers to predict disease relapse or progression? In the face of these challenges, only through the full cooperation of clinicians, scientific and technical personnel can make significant breakthroughs in the diagnosis and treatment of central nervous system demyelinating diseases.

Acknowledgments This work was supported in part by National Science Foundation of China (82171284).

Conflict of Interest The authors declare no competing interests.

References

Abdelhak, A., Hottenrott, T., Morenas-Rodríguez, E., Suárez-Calvet, M., Zettl, U. K., Haass, C., Meuth, S. G., Rauer, S., Otto, M., Tumani, H., & Huss, A. (2019). Glial activation markers in CSF and serum from patients with primary progressive multiple sclerosis: Potential of serum GFAP as disease severity marker? *Frontiers in Neurology, 10*, 280.

Acharjee, S., Gordon, P. M. K., Lee, B. H., Read, J., Workentine, M. L., Sharkey, K. A., & Pittman, Q. J. (2021). Characterization of microglial transcriptomes in the brain and spinal cord of mice in early and late experimental autoimmune encephalomyelitis using a RiboTag strategy. *Scientific Reports, 11*, 14319.

Ajami, B., Samusik, N., Wieghofer, P., Ho, P. P., Crotti, A., Bjornson, Z., Prinz, M., Fantl, W. J., Nolan, G. P., & Steinman, L. (2018). Single-cell mass cytometry reveals distinct populations of brain myeloid cells in mouse neuroinflammation and neurodegeneration models. *Nature Neuroscience, 21*, 541–551.

Araki, M., Matsuoka, T., Miyamoto, K., Kusunoki, S., Okamoto, T., Murata, M., Miyake, S., Aranami, T., & Yamamura, T. (2014). Efficacy of the anti-IL-6 receptor antibody tocilizumab in neuromyelitis optica: A pilot study. *Neurology, 82*, 1302–1306.

Axelsson, M., Malmeström, C., Nilsson, S., Haghighi, S., Rosengren, L., & Lycke, J. (2011). Glial fibrillary acidic protein: A potential biomarker for progression in multiple sclerosis. *Journal of Neurology, 258*, 882–888.

Baecher-Allan, C., Kaskow, B. J., & Weiner, H. L. (2018). Multiple sclerosis: Mechanisms and immunotherapy. *Neuron, 97*, 742–768.

Bjornevik, K., Munger, K. L., Cortese, M., Barro, C., Healy, B. C., Niebuhr, D. W., Scher, A. I., Kuhle, J., & Ascherio, A. (2020). Serum neurofilament light chain levels in patients with presymptomatic multiple sclerosis. *JAMA Neurology, 77*, 58–64.

Boffa, G., Massacesi, L., Inglese, M., Mariottini, A., Capobianco, M., Lucia, M., Amato, M. P., Cottone, S., Gualandi, F., De Gobbi, M., Greco, R., Scimè, R., Frau, J., Zimatore, G. B., Bertolotto, A., Comi, G., Uccelli, A., Signori, A., Angelucci, E., Innocenti, C., Ciceri, F., Repice, A. M., Sormani, M. P., Saccardi, R., & Mancardi, G. (2021). Long-term clinical outcomes of hematopoietic stem cell transplantation in multiple sclerosis. *Neurology, 96*(8), e1215–e1226.

Bourahoui, A., De Seze, J., Guttierez, R., Onraed, B., Hennache, B., Ferriby, D., Stojkovic, T., & Vermersch, P. (2004). CSF isoelectrofocusing in a large cohort of MS and other neurological diseases. *European Journal of Neurology, 11*, 525–529.

Brunner, C., Müller, B., & Wirth, T. (2005). Bruton's tyrosine kinase is involved in innate and adaptive immunity. *Histology and Histopathology, 20*, 945–955.

Burman, J., Zetterberg, H., Fransson, M., Loskog, A. S., Raininko, R., & Fagius, J. (2014). Assessing tissue damage in multiple sclerosis: A biomarker approach. *Acta Neurologica Scandinavica, 130*, 81–89.

Burt, R. K., Balabanov, R., Han, X., Burns, C., Gastala, J., Jovanovic, B., Helenowski, I., Jitprapaikulsan, J., Fryer, J. P., & Pittock, S. J. (2019). Autologous nonmyeloablative hematopoietic stem cell transplantation for neuromyelitis optica. *Neurology, 93*, e1732–e1e41.

Cabre, P., Olindo, S., Marignier, R., Jeannin, S., Merle, H., & Smadja, D. (2013). Efficacy of mitoxantrone in neuromyelitis optica spectrum: Clinical and neuroradiological study. *Journal of Neurology, Neurosurgery, and Psychiatry, 84*, 511–516.

Cadavid, D., Balcer, L., Galetta, S., Aktas, O., Ziemssen, T., Vanopdenbosch, L., Frederiksen, J., Skeen, M., Jaffe, G. J., Butzkueven, H., Ziemssen, F., Massacesi, L., Chai, Y., Xu, L., & Freeman, S. (2017). Safety and efficacy of opicinumab in acute optic neuritis (RENEW): A randomised, placebo-controlled, phase 2 trial. *Lancet Neurology, 16*, 189–199.

Carnero Contentti, E., & Correale, J. (2021). Neuromyelitis optica spectrum disorders: From pathophysiology to therapeutic strategies. *Journal of Neuroinflammation, 18*, 208.

Carnero Contentti, E., Lopez, P. A., Pettinicchi, J. P., Pappolla, A., Miguez, J., Patrucco, L., Cristiano, E., Vrech, C., Tkachuk, V., Liwacki, S., Correale, J., Marrodan, M., Gaitán, M. I., Fiol, M., Negrotto, L., Ysrraelit, M. C., Burgos, M., Leguizamon, F., Tavolini, D., Deri, N., Balbuena, M. E., Mainella, C., Luetic, G., Blaya, P., Hryb, J., Menichini, M. L., Alvez Pinheiro, A., Nofal, P., Zanga, G., Barboza, A., Martos, I., Lazaro, L., Alonso, R., Silva, E., Bestoso, S., Fracaro, M. E., Carrá, A., Garcea, O., Fernandez Liguori, N., Alonso Serena, M., Caride, A., & Rojas, J. I. (2021). What percentage of AQP4-ab-negative NMOSD patients are MOG-ab positive? A study from the Argentinean multiple sclerosis registry (RelevarEM). *Multiple Sclerosis and Related Disorders, 49*, 102742.

Cencioni, M. T., Mattoscio, M., Magliozzi, R., Bar-Or, A., & Muraro, P. A. (2021). B cells in multiple sclerosis – From targeted depletion to immune reconstitution therapies. *Nature Reviews. Neurology, 17*, 399–414.

Cepok, S., Rosche, B., Grummel, V., Vogel, F., Zhou, D., Sayn, J., Sommer, N., Hartung, H. P., & Hemmer, B. (2005). Short-lived plasma blasts are the main B cell effector subset during the course of multiple sclerosis. *Brain, 128*, 1667–1676.

Chan, K. H., & Lee, C. Y. (2021). Treatment of neuromyelitis optica spectrum disorders. *International Journal of Molecular Sciences, 22*, 8638.

Chang, X., Huang, W., Wang, L., ZhangBao, J., Zhou, L., Lu, C., Wang, M., Yu, J., Li, H., Li, Y., Zhao, C., Lu, J., & Quan, C. (2021). Serum neurofilament light and GFAP are associated with disease severity in inflammatory disorders with aquaporin-4 or myelin oligodendrocyte glycoprotein antibodies. *Frontiers in Immunology, 12*, 647618.

Charabati, M., Grasmuck, C., Ghannam, S., Bourbonnière, L., Fournier, A. P., Lécuyer, M. A., Tastet, O., Kebir, H., Rébillard, R. M., Hoornaert, C., Gowing, E., Larouche, S., Fortin, O., Pittet, C., Filali-Mouhim, A., Lahav, B., Moumdjian, R., Bouthillier, A., Girard, M., Duquette, P., Cayrol, R., Peelen, E., Quintana, F. J., Antel, J. P., Flügel, A., Larochelle, C., Arbour, N.,

Zandee, S., & Prat, A. (2022). DICAM promotes T(H)17 lymphocyte trafficking across the blood-brain barrier during autoimmune neuroinflammation. *Science Translational Medicine, 14*, eabj0473.

Chen, B., Wu, Q., Ke, G., & Bu, B. (2017). Efficacy and safety of tacrolimus treatment for neuromyelitis optica spectrum disorder. *Scientific Reports, 7*, 831.

Chen, T., Lennon, V. A., Liu, Y. U., Bosco, D. B., Li, Y., Yi, M. H., Zhu, J., Wei, S., & Wu, L. J. (2020). Astrocyte-microglia interaction drives evolving neuromyelitis optica lesion. *The Journal of Clinical Investigation, 130*, 4025–4038.

Chen, J., Yang, X., Zhang, Y., Zheng, P., Wei, C., Mao, Z., He, W., & Jin, W. N. (2021). Reference values for plasma neurofilament light chain (NfL) in healthy Chinese. *Clinical Chemistry and Laboratory Medicine, 59*, e153–e1e6.

Chihara, N., Aranami, T., Sato, W., Miyazaki, Y., Miyake, S., Okamoto, T., Ogawa, M., Toda, T., & Yamamura, T. (2011). Interleukin 6 signaling promotes anti-aquaporin 4 autoantibody production from plasmablasts in neuromyelitis optica. *Proceedings of the National Academy of Sciences of the United States of America, 108*, 3701–3706.

Chin, H. J., Chae, D. W., Kim, Y. C., An, W. S., Ihm, C., Jin, D. C., Kim, S. G., Kim, Y. L., Kim, Y. S., Kim, Y. G., Koo, H. S., Lee, J. E., Lee, K. W., Oh, J., Park, J. H., Jiang, H., Lee, H., & Lee, S. K. (2021). Comparison of the efficacy and safety of tacrolimus and low-dose corticosteroid with high-dose corticosteroid for minimal change nephrotic syndrome in adults. *Journal of the American Society of Nephrology, 32*, 199–210.

Choi, S. R., Howell, O. W., Carassiti, D., Magliozzi, R., Gveric, D., Muraro, P. A., Nicholas, R., Roncaroli, F., & Reynolds, R. (2012). Meningeal inflammation plays a role in the pathology of primary progressive multiple sclerosis. *Brain, 135*, 2925–2937.

Collongues, N., & de Seze, J. (2011). Current and future treatment approaches for neuromyelitis optica. *Therapeutic Advances in Neurological Disorders, 4*, 111–121.

Cree, B. A. C., Bennett, J. L., Kim, H. J., Weinshenker, B. G., Pittock, S. J., Wingerchuk, D. M., Fujihara, K., Paul, F., Cutter, G. R., Marignier, R., Green, A. J., Aktas, O., Hartung, H. P., Lublin, F. D., Drappa, J., Barron, G., Madani, S., Ratchford, J. N., She, D., Cimbora, D., & Katz, E. (2019). Inebilizumab for the treatment of neuromyelitis optica spectrum disorder (N-MOmentum): A double-blind, randomised placebo-controlled phase 2/3 trial. *Lancet, 394*, 1352–1363.

Croxford, A. L., Lanzinger, M., Hartmann, F. J., Schreiner, B., Mair, F., Pelczar, P., Clausen, B. E., Jung, S., Greter, M., & Becher, B. (2015). The cytokine GM-CSF drives the inflammatory signature of CCR2+ monocytes and licenses autoimmunity. *Immunity, 43*, 502–514.

Cugurra, A., Mamuladze, T., Rustenhoven, J., Dykstra, T., Beroshvili, G., Greenberg, Z. J., Baker, W., Papadopoulos, Z., Drieu, A., Blackburn, S., Kanamori, M., Brioschi, S., Herz, J., Schuettpelz, L. G., Colonna, M., Smirnov, I., & Kipnis, J. (2021). Skull and vertebral bone marrow are myeloid cell reservoirs for the meninges and CNS parenchyma. *Science, 373*, eabf7844.

Dal-Bianco, A., Grabner, G., Kronnerwetter, C., Weber, M., Kornek, B., Kasprian, G., Berger, T., Leutmezer, F., Rommer, P. S., Trattnig, S., Lassmann, H., & Hametner, S. (2021). Long-term evolution of multiple sclerosis iron rim lesions in 7 T MRI. *Brain, 144*, 833–847.

Deenick, E. K., Chan, A., Ma, C. S., Gatto, D., Schwartzberg, P. L., Brink, R., & Tangye, S. G. (2010). Follicular helper T cell differentiation requires continuous antigen presentation that is independent of unique B cell signaling. *Immunity, 33*, 241–253.

Dejanovic, B., Huntley, M. A., De Mazière, A., Meilandt, W. J., Wu, T., Srinivasan, K., Jiang, Z., Gandham, V., Friedman, B. A., Ngu, H., Foreman, O., Carano, R. A. D., Chih, B., Klumperman, J., Bakalarski, C., Hanson, J. E., & Sheng, M. (2018). Changes in the synaptic proteome in tauopathy and rescue of tau-induced synapse loss by C1q antibodies. *Neuron, 100*, 1322–1336.e7.

Di Pauli, F., Mader, S., Rostasy, K., Schanda, K., Bajer-Kornek, B., Ehling, R., Deisenhammer, F., Reindl, M., & Berger, T. (2011). Temporal dynamics of anti-MOG antibodies in CNS demyelinating diseases. *Clinical Immunology, 138*, 247–254.

DiSano, K. D., Gilli, F., & Pachner, A. R. (2021). Memory B cells in multiple sclerosis: Emerging players in disease pathogenesis. *Frontiers in Immunology, 12*, 676686.

Disanto, G., Barro, C., Benkert, P., Naegelin, Y., Schädelin, S., Giardiello, A., Zecca, C., Blennow, K., Zetterberg, H., Leppert, D., Kappos, L., Gobbi, C., & Kuhle, J. (2017). Serum neurofilament light: A biomarker of neuronal damage in multiple sclerosis. *Annals of Neurology, 81*, 857–870.

Dobson, R., Ramagopalan, S., Davis, A., & Giovannoni, G. (2013). Cerebrospinal fluid oligoclonal bands in multiple sclerosis and clinically isolated syndromes: A meta-analysis of prevalence, prognosis and effect of latitude. *Journal of Neurology, Neurosurgery, and Psychiatry, 84*, 909–914.

Dombrowski, Y., O'Hagan, T., Dittmer, M., Penalva, R., Mayoral, S. R., Bankhead, P., Fleville, S., Eleftheriadis, G., Zhao, C., Naughton, M., Hassan, R., Moffat, J., Falconer, J., Boyd, A., Hamilton, P., Allen, I. V., Kissenpfennig, A., Moynagh, P. N., Evergren, E., Perbal, B., Williams, A. C., Ingram, R. J., Chan, J. R., Franklin, R. J. M., & Fitzgerald, D. C. (2017). Regulatory T cells promote myelin regeneration in the central nervous system. *Nature Neuroscience, 20*, 674–680.

Domingues, H. S., Portugal, C. C., Socodato, R., & Relvas, J. B. (2016). Oligodendrocyte, astrocyte, and microglia crosstalk in myelin development, damage, and repair. *Frontiers in Cell and Development Biology, 4*, 71.

Dos Passos, G. R., Oliveira, L. M., da Costa, B. K., Apostolos-Pereira, S. L., Callegaro, D., Fujihara, K., & Sato, D. K. (2018). MOG-IgG-associated optic neuritis, encephalitis, and myelitis: Lessons learned from neuromyelitis optica spectrum disorder. *Frontiers in Neurology, 9*, 217.

Duan, T., Smith, A. J., & Verkman, A. S. (2018). Complement-dependent bystander injury to neurons in AQP4-IgG seropositive neuromyelitis optica. *Journal of Neuroinflammation, 15*, 294.

Elsone, L., Panicker, J., Mutch, K., Boggild, M., Appleton, R., & Jacob, A. (2014). Role of intravenous immunoglobulin in the treatment of acute relapses of neuromyelitis optica: Experience in 10 patients. *Multiple Sclerosis, 20*, 501–504.

Enriquez, C. A. G., Espiritu, A. I., & Pasco, P. M. D. (2019). Efficacy and tolerability of mitoxantrone for neuromyelitis optica spectrum disorder: A systematic review. *Journal of Neuroimmunology, 332*, 126–134.

Faissner, S., Plemel, J. R., Gold, R., & Yong, V. W. (2019). Progressive multiple sclerosis: From pathophysiology to therapeutic strategies. *Nature Reviews. Drug Discovery, 18*, 905–922.

Farina, G., Magliozzi, R., Pitteri, M., Reynolds, R., Rossi, S., Gajofatto, A., Benedetti, M. D., Facchiano, F., Monaco, S., & Calabrese, M. (2017). Increased cortical lesion load and intrathecal inflammation is associated with oligoclonal bands in multiple sclerosis patients: A combined CSF and MRI study. *Journal of Neuroinflammation, 14*, 40.

Fransen, N. L., Hsiao, C. C., van der Poel, M., Engelenburg, H. J., Verdaasdonk, K., Vincenten, M. C. J., Remmerswaal, E. B. M., Kuhlmann, T., Mason, M. R. J., Hamann, J., Smolders, J., & Huitinga, I. (2020). Tissue-resident memory T cells invade the brain parenchyma in multiple sclerosis white matter lesions. *Brain, 143*, 1714–1730.

Frieser, D., Pignata, A., Khajavi, L., Shlesinger, D., Gonzalez-Fierro, C., Nguyen, X. H., Yermanos, A., Merkler, D., Höftberger, R., Desestret, V., Mair, K. M., Bauer, J., Masson, F., & Liblau, R. S. (2022). Tissue-resident CD8(+) T cells drive compartmentalized and chronic autoimmune damage against CNS neurons. *Science Translational Medicine, 14*, eabl6157.

Frischer, J. M., Bramow, S., Dal-Bianco, A., Lucchinetti, C. F., Rauschka, H., Schmidbauer, M., Laursen, H., Sorensen, P. S., & Lassmann, H. (2009). The relation between inflammation and neurodegeneration in multiple sclerosis brains. *Brain, 132*, 1175–1189.

Frischer, J. M., Weigand, S. D., Guo, Y., Kale, N., Parisi, J. E., Pirko, I., Mandrekar, J., Bramow, S., Metz, I., Brück, W., Lassmann, H., & Lucchinetti, C. F. (2015). Clinical and pathological insights into the dynamic nature of the white matter multiple sclerosis plaque. *Annals of Neurology, 78*, 710–721.

Fujihara, K., Bennett, J. L., de Seze, J., Haramura, M., Kleiter, I., Weinshenker, B. G., Kang, D., Mughal, T., & Yamamura, T. (2020). Interleukin-6 in neuromyelitis optica spectrum disorder pathophysiology. *Neurology Neuroimmunology & Neuroinflammation, 7*, e841.

Gagliani, N., Amezcua Vesely, M. C., Iseppon, A., Brockmann, L., Xu, H., Palm, N. W., de Zoete, M. R., Licona-Limón, P., Paiva, R. S., Ching, T., Weaver, C., Zi, X., Pan, X., Fan, R., Garmire, L. X., Cotton, M. J., Drier, Y., Bernstein, B., Geginat, J., Stockinger, B., Esplugues, E., Huber, S., & Flavell, R. A. (2015). Th17 cells transdifferentiate into regulatory T cells during resolution of inflammation. *Nature, 523*, 221–225.

Gal, R. L., Vedula, S. S., & Beck, R. (2012). Corticosteroids for treating optic neuritis. *Cochrane Database of Systematic Reviews, 4*, Cd001430.

Garg, N., Padron, E. J., Rammohan, K. W., & Goodman, C. F. (2022). Bruton's tyrosine kinase inhibitors: The next frontier of B-cell-targeted therapies for cancer, autoimmune disorders, and multiple sclerosis. *Journal of Clinical Medicine, 11*, 6139.

Genchi, A., Brambilla, E., Sangalli, F., Radaelli, M., Bacigaluppi, M., Furlan, R., Andolfo, A., Drago, D., Magagnotti, C., Scotti, G. M., Greco, R., Vezzulli, P., Ottoboni, L., Bonopane, M., Capilupo, D., Ruffini, F., Belotti, D., Cabiati, B., Cesana, S., Matera, G., Leocani, L., Martinelli, V., Moiola, L., Vago, L., Panina-Bordignon, P., Falini, A., Ciceri, F., Uglietti, A., Sormani, M. P., Comi, G., Battaglia, M. A., Rocca, M. A., Storelli, L., Pagani, E., Gaipa, G., & Martino, G. (2023). Neural stem cell transplantation in patients with progressive multiple sclerosis: An open-label, phase 1 study. *Nature Medicine, 29*, 75–85.

Gillen, K. M., Mubarak, M., Nguyen, T. D., & Pitt, D. (2018). Significance and in vivo detection of iron-laden microglia in white matter multiple sclerosis lesions. *Frontiers in Immunology, 9*, 255.

Giovannoni, F., & Quintana, F. J. (2020). The role of astrocytes in CNS inflammation. *Trends in Immunology, 41*, 805–819.

Gran, B., Hemmer, B., Vergelli, M., McFarland, H. F., & Martin, R. (1999). Molecular mimicry and multiple sclerosis: Degenerate T-cell recognition and the induction of autoimmunity. *Annals of Neurology, 45*, 559–567.

Graves, J. S., Krysko, K. M., Hua, L. H., Absinta, M., Franklin, R. J. M., & Segal, B. M. (2023). Ageing and multiple sclerosis. *Lancet Neurology, 22*, 66–77.

Green, A. J., Gelfand, J. M., Cree, B. A., Bevan, C., Boscardin, W. J., Mei, F., Inman, J., Arnow, S., Devereux, M., Abounasr, A., Nobuta, H., Zhu, A., Friessen, M., Gerona, R., von Büdingen, H. C., Henry, R. G., Hauser, S. L., & Chan, J. R. (2017). Clemastine fumarate as a remyelinating therapy for multiple sclerosis (ReBUILD): A randomised, controlled, double-blind, crossover trial. *Lancet, 390*, 2481–2489.

Hauser, S. L., & Cree, B. A. C. (2020). Treatment of multiple sclerosis: A review. *The American Journal of Medicine, 133*, 1380–1390.e2.

Hauser, S. L., Bar-Or, A., Cohen, J. A., Comi, G., Correale, J., Coyle, P. K., Cross, A. H., de Seze, J., Leppert, D., Montalban, X., Selmaj, K., Wiendl, H., Kerloeguen, C., Willi, R., Li, B., Kakarieka, A., Tomic, D., Goodyear, A., Pingili, R., Häring, D. A., Ramanathan, K., Merschhemke, M., & Kappos, L. (2020). Ofatumumab versus teriflunomide in multiple sclerosis. *The New England Journal of Medicine, 383*, 546–557.

Herisson, F., Frodermann, V., Courties, G., Rohde, D., Sun, Y., Vandoorne, K., Wojtkiewicz, G. R., Masson, G. S., Vinegoni, C., Kim, J., Kim, D. E., Weissleder, R., Swirski, F. K., Moskowitz, M. A., & Nahrendorf, M. (2018). Direct vascular channels connect skull bone marrow and the brain surface enabling myeloid cell migration. *Nature Neuroscience, 21*, 1209–1217.

Howe, C. L., Kaptzan, T., Magaña, S. M., Ayers-Ringler, J. R., LaFrance-Corey, R. G., & Lucchinetti, C. F. (2014). Neuromyelitis optica IgG stimulates an immunological response in rat astrocyte cultures. *Glia, 62*, 692–708.

Huh, S. Y., Kim, S. H., Hyun, J. W., Joung, A. R., Park, M. S., Kim, B. J., & Kim, H. J. (2014). Mycophenolate mofetil in the treatment of neuromyelitis optica spectrum disorder. *JAMA Neurology, 71*, 1372–1378.

Jacob, A., Matiello, M., Weinshenker, B. G., Wingerchuk, D. M., Lucchinetti, C., Shuster, E., Carter, J., Keegan, B. M., Kantarci, O. H., & Pittock, S. J. (2009). Treatment of neuromyelitis optica with mycophenolate mofetil: Retrospective analysis of 24 patients. *Archives of Neurology, 66*, 1128–1133.

Jarius, S., Paul, F., Aktas, O., Asgari, N., Dale, R. C., de Seze, J., Franciotta, D., Fujihara, K., Jacob, A., Kim, H. J., Kleiter, I., Kümpfel, T., Levy, M., Palace, J., Ruprecht, K., Saiz, A., Trebst, C., Weinshenker, B. G., & Wildemann, B. (2018). MOG encephalomyelitis: International recommendations on diagnosis and antibody testing. *Journal of Neuroinflammation, 15*, 134.

Jarius, S., Paul, F., Weinshenker, B. G., Levy, M., Kim, H. J., & Wildemann, B. (2020). Neuromyelitis optica. *Nature Reviews Disease Primers, 6*, 85.

Kandjani, O. J., Yaqoubi, S., Vahdati, S. S., Borhannejad, B., Dastmalchi, S., & Alizadeh, A. A. (2023). S1PR1 modulators in multiple sclerosis: Efficacy, safety, comparison, and chemical structure insights. *European Journal of Medicinal Chemistry, 250*, 115182.

Kaneko, K., Sato, D. K., Nakashima, I., Ogawa, R., Akaishi, T., Takai, Y., Nishiyama, S., Takahashi, T., Misu, T., Kuroda, H., Tanaka, S., Nomura, K., Hashimoto, Y., Callegaro, D., Steinman, L., Fujihara, K., & Aoki, M. (2018). CSF cytokine profile in MOG-IgG+ neurological disease is similar to AQP4-IgG+ NMOSD but distinct from MS: A cross-sectional study and potential therapeutic implications. *Journal of Neurology, Neurosurgery, and Psychiatry, 89*, 927–936.

Kebir, H., Kreymborg, K., Ifergan, I., Dodelet-Devillers, A., Cayrol, R., Bernard, M., Giuliani, F., Arbour, N., Becher, B., & Prat, A. (2007). Human TH17 lymphocytes promote blood-brain barrier disruption and central nervous system inflammation. *Nature Medicine, 13*, 1173–1175.

Khademi, M., Kockum, I., Andersson, M. L., Iacobaeus, E., Brundin, L., Sellebjerg, F., Hillert, J., Piehl, F., & Olsson, T. (2011). Cerebrospinal fluid CXCL13 in multiple sclerosis: A suggestive prognostic marker for the disease course. *Multiple Sclerosis, 17*, 335–343.

Khoy, K., Mariotte, D., Defer, G., Petit, G., Toutirais, O., & Le Mauff, B. (2020). Natalizumab in multiple sclerosis treatment: From biological effects to immune monitoring. *Frontiers in Immunology, 11*, 549842.

Kieseier, B. C. (2011). The mechanism of action of interferon-β in relapsing multiple sclerosis. *CNS Drugs, 25*, 491–502.

Kim, S. H., Huh, S. Y., Lee, S. J., Joung, A., & Kim, H. J. (2013). A 5-year follow-up of rituximab treatment in patients with neuromyelitis optica spectrum disorder. *JAMA Neurology, 70*, 1110–1117.

Komori, M., Blake, A., Greenwood, M., Lin, Y. C., Kosa, P., Ghazali, D., Winokur, P., Natrajan, M., Wuest, S. C., Romm, E., Panackal, A. A., Williamson, P. R., Wu, T., & Bielekova, B. (2015). Cerebrospinal fluid markers reveal intrathecal inflammation in progressive multiple sclerosis. *Annals of Neurology, 78*, 3–20.

Korn, T., Mitsdoerffer, M., Croxford, A. L., Awasthi, A., Dardalhon, V. A., Galileos, G., Vollmar, P., Stritesky, G. L., Kaplan, M. H., Waisman, A., Kuchroo, V. K., & Oukka, M. (2008). IL-6 controls Th17 immunity in vivo by inhibiting the conversion of conventional T cells into Foxp3+ regulatory T cells. *Proceedings of the National Academy of Sciences of the United States of America, 105*, 18460–18465.

Kuhle, J., Disanto, G., Lorscheider, J., Stites, T., Chen, Y., Dahlke, F., Francis, G., Shrinivasan, A., Radue, E. W., Giovannoni, G., & Kappos, L. (2015). Fingolimod and CSF neurofilament light chain levels in relapsing-remitting multiple sclerosis. *Neurology, 84*, 1639–1643.

Kuhle, J., Kropshofer, H., Haering, D. A., Kundu, U., Meinert, R., Barro, C., Dahlke, F., Tomic, D., Leppert, D., & Kappos, L. (2019). Blood neurofilament light chain as a biomarker of MS disease activity and treatment response. *Neurology, 92*, e1007–e1e15.

Kuhlmann, T., Ludwin, S., Prat, A., Antel, J., Brück, W., & Lassmann, H. (2017). An updated histological classification system for multiple sclerosis lesions. *Acta Neuropathologica, 133*, 13–24.

Lamb, Y. N. (2022). Ocrelizumab: A review in multiple sclerosis. *Drugs, 82*, 323–334.

Lécuyer, M. A., Saint-Laurent, O., Bourbonnière, L., Larouche, S., Larochelle, C., Michel, L., Charabati, M., Abadier, M., Zandee, S., Haghayegh Jahromi, N., Gowing, E., Pittet, C., Lyck, R., Engelhardt, B., & Prat, A. (2017). Dual role of ALCAM in neuroinflammation and blood-brain barrier homeostasis. *Proceedings of the National Academy of Sciences of the United States of America, 114*, E524–E533.

Levy, M., & Mealy, M. A. (2014). Purified human C1-esterase inhibitor is safe in acute relapses of neuromyelitis optica. *Neurology Neuroimmunology & Neuroinflammation, 1*, e5.

Li, R., Wang, J., Li, C., Liu, X., Chu, M., Chang, Y., Wang, Y., Wang, X., Yu, B., Ling, L., Yang, H., Yang, H., Hu, X., & Qiu, W. (2021). Rescue immunoadsorption treatment for neuromyelitis optica spectrum disorder attacks unresponsive to intravenous methylprednisolone. *Journal of Neuroimmunology, 356*, 577604.

Li, J., Zaslavsky, M., Su, Y., Guo, J., Sikora, M. J., van Unen, V., Christophersen, A., Chiou, S. H., Chen, L., Li, J., Ji, X., Wilhelmy, J., McSween, A. M., Palanski, B. A., Mallajosyula, V. V. A., Bracey, N. A., Dhondalay, G. K. R., Bhamidipati, K., Pai, J., Kipp, L. B., Dunn, J. E., Hauser, S. L., Oksenberg, J. R., Satpathy, A. T., Robinson, W. H., Dekker, C. L., Steinmetz, L. M., Khosla, C., Utz, P. J., Sollid, L. M., Chien, Y. H., Heath, J. R., Fernandez-Becker, N. Q., Nadeau, K. C., Saligrama, N., & Davis, M. M. (2022a). KIR(+)CD8(+) T cells suppress pathogenic T cells and are active in autoimmune diseases and COVID-19. *Science, 376*, eabi9591.

Li, R., Tang, H., Burns, J. C., Hopkins, B. T., Le Coz, C., Zhang, B., de Barcelos, I. P., Romberg, N., Goldstein, A. C., Banwell, B. L., Luning Prak, E. T., Mingueneau, M., & Bar-Or, A. (2022b). BTK inhibition limits B-cell-T-cell interaction through modulation of B-cell metabolism: Implications for multiple sclerosis therapy. *Acta Neuropathologica, 143*, 505–521.

Liddelow, S. A., Guttenplan, K. A., Clarke, L. E., Bennett, F. C., Bohlen, C. J., Schirmer, L., Bennett, M. L., Münch, A. E., Chung, W. S., Peterson, T. C., Wilton, D. K., Frouin, A., Napier, B. A., Panicker, N., Kumar, M., Buckwalter, M. S., Rowitch, D. H., Dawson, V. L., Dawson, T. M., Stevens, B., & Barres, B. A. (2017). Neurotoxic reactive astrocytes are induced by activated microglia. *Nature, 541*, 481–487.

Linhares, U. C., Schiavoni, P. B., Barros, P. O., Kasahara, T. M., Teixeira, B., Ferreira, T. B., Alvarenga, R., Hygino, J., Vieira, M. M., Bittencourt, V. C., Andrade, R. M., Andrade, A. F., & Bento, C. A. (2013). The ex vivo production of IL-6 and IL-21 by CD4+ T cells is directly associated with neurological disability in neuromyelitis optica patients. *Journal of Clinical Immunology, 33*, 179–189.

Liu, J., Tan, G., Li, B., Zhang, J., Gao, Y., Cao, Y., Jia, Z., & Sugimoto, K. (2021). Serum aquaporin 4-immunoglobulin G titer and neuromyelitis optica spectrum disorder activity and severity: A systematic review and meta-analysis. *Frontiers in Neurology, 12*, 746959.

López-Chiriboga, A. S., Majed, M., Fryer, J., Dubey, D., McKeon, A., Flanagan, E. P., Jitprapaikulsan, J., Kothapalli, N., Tillema, J. M., Chen, J., Weinshenker, B., Wingerchuk, D., Sagen, J., Gadoth, A., Lennon, V. A., Keegan, B. M., Lucchinetti, C., & Pittock, S. J. (2018). Association of MOG-IgG serostatus with relapse after acute disseminated encephalomyelitis and proposed diagnostic criteria for MOG-IgG-associated disorders. *JAMA Neurology, 75*, 1355–1363.

Lotan, I., McGowan, R., & Levy, M. (2021). Anti-IL-6 therapies for neuromyelitis optica spectrum disorders: A systematic review of safety and efficacy. *Current Neuropharmacology, 19*, 220–232.

Lubetzki, C., Zalc, B., Williams, A., Stadelmann, C., & Stankoff, B. (2020). Remyelination in multiple sclerosis: From basic science to clinical translation. *Lancet Neurology, 19*, 678–688.

Lucchinetti, C. F., Mandler, R. N., McGavern, D., Bruck, W., Gleich, G., Ransohoff, R. M., Trebst, C., Weinshenker, B., Wingerchuk, D., Parisi, J. E., & Lassmann, H. (2002). A role for humoral mechanisms in the pathogenesis of Devic's neuromyelitis optica. *Brain, 125*, 1450–1461.

Lucchinetti, C. F., Guo, Y., Popescu, B. F., Fujihara, K., Itoyama, Y., & Misu, T. (2014). The pathology of an autoimmune astrocytopathy: Lessons learned from neuromyelitis optica. *Brain Pathology, 24*, 83–97.

Magliozzi, R., Howell, O. W., Reeves, C., Roncaroli, F., Nicholas, R., Serafini, B., Aloisi, F., & Reynolds, R. (2010). A gradient of neuronal loss and meningeal inflammation in multiple sclerosis. *Annals of Neurology, 68*, 477–493.

Margoni, M., Preziosa, P., Filippi, M., & Rocca, M. A. (2022). Anti-CD20 therapies for multiple sclerosis: Current status and future perspectives. *Journal of Neurology, 269*, 1316–1334.

Mascanfroni, I. D., Yeste, A., Vieira, S. M., Burns, E. J., Patel, B., Sloma, I., Wu, Y., Mayo, L., Ben-Hamo, R., Efroni, S., Kuchroo, V. K., Robson, S. C., & Quintana, F. J. (2013). IL-27 acts on DCs to suppress the T cell response and autoimmunity by inducing expression of the immunoregulatory molecule CD39. *Nature Immunology, 14*, 1054–1063.

Mayer, M. C., & Meinl, E. (2012). Glycoproteins as targets of autoantibodies in CNS inflammation: MOG and more. *Therapeutic Advances in Neurological Disorders, 5*, 147–159.

Mazzitelli, J. A., Smyth, L. C. D., Cross, K. A., Dykstra, T., Sun, J., Du, S., Mamuladze, T., Smirnov, I., Rustenhoven, J., & Kipnis, J. (2022). Cerebrospinal fluid regulates skull bone marrow niches via direct access through Dural channels. *Nature Neuroscience, 25*, 555–560.

Mealy, M. A., & Levy, M. (2019). A pilot safety study of ublituximab, a monoclonal antibody against CD20, in acute relapses of neuromyelitis optica spectrum disorder. *Medicine (Baltimore), 98*, e15944.

Mealy, M. A., Shin, K., John, G., & Levy, M. (2015). Bevacizumab is safe in acute relapses of neuromyelitis optica. *Clinical and Experimental Neuroimmunology, 6*, 413–418.

Meier, S., Willemse, E. A. J., Schaedelin, S., Oechtering, J., Lorscheider, J., Melie-Garcia, L., Cagol, A., Barakovic, M., Galbusera, R., Subramaniam, S., Barro, C., Abdelhak, A., Thebault, S., Achtnichts, L., Lalive, P., Müller, S., Pot, C., Salmen, A., Disanto, G., Zecca, C., D'Souza, M., Orleth, A., Khalil, M., Buchmann, A., Du Pasquier, R., Yaldizli, Ö., Derfuss, T., Berger, K., Hermesdorf, M., Wiendl, H., Piehl, F., Battaglini, M., Fischer, U., Kappos, L., Gobbi, C., Granziera, C., Bridel, C., Leppert, D., Maleska Maceski, A., Benkert, P., & Kuhle, J. (2023). Serum glial fibrillary acidic protein compared with Neurofilament light chain as a biomarker for disease progression in multiple sclerosis. *JAMA Neurology, 80*, 287–297.

Merlini, A., Haberl, M., Strauß, J., Hildebrand, L., Genc, N., Franz, J., Chilov, D., Alitalo, K., Flügel-Koch, C., Stadelmann, C., Flügel, A., & Odoardi, F. (2022). Distinct roles of the meningeal layers in CNS autoimmunity. *Nature Neuroscience, 25*, 887–899.

Misu, T., Höftberger, R., Fujihara, K., Wimmer, I., Takai, Y., Nishiyama, S., Nakashima, I., Konno, H., Bradl, M., Garzuly, F., Itoyama, Y., Aoki, M., & Lassmann, H. (2013). Presence of six different lesion types suggests diverse mechanisms of tissue injury in neuromyelitis optica. *Acta Neuropathologica, 125*, 815–827.

Møllgård, K., Beinlich, F. R. M., Kusk, P., Miyakoshi, L. M., Delle, C., Plá, V., Hauglund, N. L., Esmail, T., Rasmussen, M. K., Gomolka, R. S., Mori, Y., & Nedergaard, M. (2023). A mesothelium divides the subarachnoid space into functional compartments. *Science, 379*, 84–88.

Molnarfi, N., Schulze-Topphoff, U., Weber, M. S., Patarroyo, J. C., Prod'homme, T., Varrin-Doyer, M., Shetty, A., Linington, C., Slavin, A. J., Hidalgo, J., Jenne, D. E., Wekerle, H., Sobel, R. A., Bernard, C. C., Shlomchik, M. J., & Zamvil, S. S. (2013). MHC class II-dependent B cell APC function is required for induction of CNS autoimmunity independent of myelin-specific antibodies. *The Journal of Experimental Medicine, 210*, 2921–2937.

Montalban, X., Arnold, D. L., Weber, M. S., Staikov, I., Piasecka-Stryczynska, K., Willmer, J., Martin, E. C., Dangond, F., Syed, S., & Wolinsky, J. S. (2019). Placebo-controlled trial of an oral BTK inhibitor in multiple sclerosis. *The New England Journal of Medicine, 380*, 2406–2417.

Muehler, A., Peelen, E., Kohlhof, H., Gröppel, M., & Vitt, D. (2020). Vidofludimus calcium, a next generation DHODH inhibitor for the treatment of relapsing-remitting multiple sclerosis. *Multiple Sclerosis and Related Disorders, 43*, 102129.

Nakamura, M., Nakazawa, T., Doi, H., Hariya, T., Omodaka, K., Misu, T., Takahashi, T., Fujihara, K., & Nishida, K. (2010). Early high-dose intravenous methylprednisolone is effective in preserving retinal nerve fiber layer thickness in patients with neuromyelitis optica. *Graefe's Archive for Clinical and Experimental Ophthalmology, 248*, 1777–1785.

Narayan, R., Simpson, A., Fritsche, K., Salama, S., Pardo, S., Mealy, M., Paul, F., & Levy, M. (2018). MOG antibody disease: A review of MOG antibody seropositive neuromyelitis optica spectrum disorder. *Multiple Sclerosis and Related Disorders, 25*, 66–72.

Novakova, L., Zetterberg, H., Sundström, P., Axelsson, M., Khademi, M., Gunnarsson, M., Malmeström, C., Svenningsson, A., Olsson, T., Piehl, F., Blennow, K., & Lycke, J. (2017). Monitoring disease activity in multiple sclerosis using serum neurofilament light protein. *Neurology, 89*, 2230–2237.

Okamoto, T., Ogawa, M., Lin, Y., Murata, M., Miyake, S., & Yamamura, T. (2008). Treatment of neuromyelitis optica: Current debate. *Therapeutic Advances in Neurological Disorders, 1*, 5–12.

Palace, J., Lin, D. Y., Zeng, D., Majed, M., Elsone, L., Hamid, S., Messina, S., Misu, T., Sagen, J., Whittam, D., Takai, Y., Leite, M. I., Weinshenker, B., Cabre, P., Jacob, A., Nakashima, I., Fujihara, K., & Pittock, S. J. (2019). Outcome prediction models in AQP4-IgG positive neuromyelitis optica spectrum disorders. *Brain, 142*, 1310–1323.

Papadopoulos, M. C., & Verkman, A. S. (2012). Aquaporin 4 and neuromyelitis optica. *Lancet Neurology, 11*, 535–544.

Papp, V., Magyari, M., Aktas, O., Berger, T., Broadley, S. A., Cabre, P., Jacob, A., Kira, J. I., Leite, M. I., Marignier, R., Miyamoto, K., Palace, J., Saiz, A., Sepulveda, M., Sveinsson, O., & Illes, Z. (2021). Worldwide incidence and prevalence of neuromyelitis optica: A systematic review. *Neurology, 96*, 59–77.

Pittock, S. J., Lennon, V. A., McKeon, A., Mandrekar, J., Weinshenker, B. G., Lucchinetti, C. F., O'Toole, O., & Wingerchuk, D. M. (2013). Eculizumab in AQP4-IgG-positive relapsing neuromyelitis optica spectrum disorders: An open-label pilot study. *Lancet Neurology, 12*, 554–562.

Pittock, S. J., Berthele, A., Fujihara, K., Kim, H. J., Levy, M., Palace, J., Nakashima, I., Terzi, M., Totolyan, N., Viswanathan, S., Wang, K. C., Pace, A., Fujita, K. P., Armstrong, R., & Wingerchuk, D. M. (2019). Eculizumab in aquaporin-4-positive neuromyelitis optica spectrum disorder. *The New England Journal of Medicine, 381*, 614–625.

Pittock, S. J., Fujihara, K., Palace, J., Berthele, A., Kim, H. J., Oreja-Guevara, C., Nakashima, I., Levy, M., Shang, S., Yountz, M., Miller, L., Armstrong, R., & Wingerchuk, D. M. (2022). Eculizumab monotherapy for NMOSD: Data from PREVENT and its open-label extension. *Multiple Sclerosis, 28*, 480–486.

Pittock, S. J., Barnett, M., Bennett, J. L., Berthele, A., de Sèze, J., Levy, M., Nakashima, I., Oreja-Guevara, C., Palace, J., Paul, F., Pozzilli, C., Yountz, M., Allen, K., Mashhoon, Y., & Kim, H. J. (2023). Ravulizumab in aquaporin-4-positive neuromyelitis optica spectrum disorder. *Annals of Neurology, 93*, 1053–1068.

Popescu, B. F., & Lucchinetti, C. F. (2012). Pathology of demyelinating diseases. *Annual Review of Pathology, 7*, 185–217.

Popescu, B. F., Frischer, J. M., Webb, S. M., Tham, M., Adiele, R. C., Robinson, C. A., Fitz-Gibbon, P. D., Weigand, S. D., Metz, I., Nehzati, S., George, G. N., Pickering, I. J., Brück, W., Hametner, S., Lassmann, H., Parisi, J. E., Yong, G., & Lucchinetti, C. F. (2017). Pathogenic implications of distinct patterns of iron and zinc in chronic MS lesions. *Acta Neuropathologica, 134*, 45–64.

Pröbstel, A. K., Dornmair, K., Bittner, R., Sperl, P., Jenne, D., Magalhaes, S., Villalobos, A., Breithaupt, C., Weissert, R., Jacob, U., Krumbholz, M., Kuempfel, T., Blaschek, A., Stark, W., Gärtner, J., Pohl, D., Rostasy, K., Weber, F., Forne, I., Khademi, M., Olsson, T., Brilot, F., Tantsis, E., Dale, R. C., Wekerle, H., Hohlfeld, R., Banwell, B., Bar-Or, A., Meinl, E., & Derfuss, T. (2011). Antibodies to MOG are transient in childhood acute disseminated encephalomyelitis. *Neurology, 77*, 580–588.

Qin, C., Tian, D. S., Zhou, L. Q., Shang, K., Huang, L., Dong, M. H., You, Y. F., Xiao, J., Xiong, Y., Wang, W., Pang, H., Guo, J. J., Cai, S. B., Wang, D., Li, C. R., Zhang, M., Bu, B. T., & Wang, W. (2023). Anti-BCMA CAR T-cell therapy CT103A in relapsed or refractory AQP4-IgG seropositive neuromyelitis optica spectrum disorders: Phase 1 trial interim results. *Signal Transduction and Targeted Therapy, 8*, 5.

Ransohoff, R. M., & Engelhardt, B. (2012). The anatomical and cellular basis of immune surveillance in the central nervous system. *Nature Reviews. Immunology, 12*, 623–635.

Rasouli, J., Casella, G., Yoshimura, S., Zhang, W., Xiao, D., Garifallou, J., Gonzalez, M. V., Wiedeman, A., Kus, A., Mari, E. R., Fortina, P., Hakonarson, H., Long, S. A., Zhang, G. X., Ciric, B., & Rostami, A. (2020). A distinct GM-CSF(+) T helper cell subset requires T-bet to adopt a T(H)1 phenotype and promote neuroinflammation. *Science Immunology, 5*, eaba9953.

Ratelade, J., Zhang, H., Saadoun, S., Bennett, J. L., Papadopoulos, M. C., & Verkman, A. S. (2012). Neuromyelitis optica IgG and natural killer cells produce NMO lesions in mice without myelin loss. *Acta Neuropathologica, 123*, 861–872.

Rawji, K. S., Mishra, M. K., & Yong, V. W. (2016). Regenerative capacity of macrophages for remyelination. *Frontiers in Cell and Development Biology, 4*, 47.

Rawji, K. S., Gonzalez Martinez, G. A., Sharma, A., & Franklin, R. J. M. (2020). The role of astrocytes in remyelination. *Trends in Neurosciences, 43*, 596–607.

Reich, D. S., Lucchinetti, C. F., & Calabresi, P. A. (2018). Multiple sclerosis. *The New England Journal of Medicine, 378*, 169–180.

Reich, D. S., Arnold, D. L., Vermersch, P., Bar-Or, A., Fox, R. J., Matta, A., Turner, T., Wallström, E., Zhang, X., Mareš, M., Khabirov, F. A., & Traboulsee, A. (2021). Safety and efficacy of tolebrutinib, an oral brain-penetrant BTK inhibitor, in relapsing multiple sclerosis: A phase 2b, randomised, double-blind, placebo-controlled trial. *Lancet Neurology, 20*, 729–738.

Riddoch, D., & Thompson, R. A. (1970). Immunoglobulin levels in the cerebrospinal fluid. *British Medical Journal, 1*, 396–399.

Robinson, W. H., & Steinman, L. (2022). Epstein-Barr virus and multiple sclerosis. *Science, 375*, 264–265.

Rodríguez-Lorenzo, S., Konings, J., van der Pol, S., Kamermans, A., Amor, S., van Horssen, J., Witte, M. E., Kooij, G., & de Vries, H. E. (2020). Inflammation of the choroid plexus in progressive multiple sclerosis: Accumulation of granulocytes and T cells. *Acta Neuropathologica Communications, 8*, 9.

Roed, H. G., Langkilde, A., Sellebjerg, F., Lauritzen, M., Bang, P., Mørup, A., & Frederiksen, J. L. (2005). A double-blind, randomized trial of IV immunoglobulin treatment in acute optic neuritis. *Neurology, 64*, 804–810.

Romme Christensen, J., Komori, M., von Essen, M. R., Ratzer, R., Börnsen, L., Bielekova, B., & Sellebjerg, F. (2019). CSF inflammatory biomarkers responsive to treatment in progressive multiple sclerosis capture residual inflammation associated with axonal damage. *Multiple Sclerosis, 25*, 937–946.

Rustenhoven, J., Drieu, A., Mamuladze, T., de Lima, K. A., Dykstra, T., Wall, M., Papadopoulos, Z., Kanamori, M., Salvador, A. F., Baker, W., Lemieux, M., Da Mesquita, S., Cugurra, A., Fitzpatrick, J., Sviben, S., Kossina, R., Bayguinov, P., Townsend, R. R., Zhang, Q., Erdmann-Gilmore, P., Smirnov, I., Lopes, M. B., Herz, J., & Kipnis, J. (2021). Functional characterization of the dural sinuses as a neuroimmune interface. *Cell, 184*, 1000–1016.e27.

Salzer, J., Svenningsson, R., Alping, P., Novakova, L., Björck, A., Fink, K., Islam-Jakobsson, P., Malmeström, C., Axelsson, M., Vågberg, M., Sundström, P., Lycke, J., Piehl, F., & Svenningsson, A. (2016). Rituximab in multiple sclerosis: A retrospective observational study on safety and efficacy. *Neurology, 87*, 2074–2081.

Sato, D. K., Lana-Peixoto, M. A., Fujihara, K., & de Seze, J. (2013). Clinical spectrum and treatment of neuromyelitis optica spectrum disorders: Evolution and current status. *Brain Pathology, 23*, 647–660.

Sato, D. K., Callegaro, D., Lana-Peixoto, M. A., Waters, P. J., de Haidar Jorge, F. M., Takahashi, T., Nakashima, I., Apostolos-Pereira, S. L., Talim, N., Simm, R. F., Lino, A. M., Misu, T., Leite, M. I., Aoki, M., & Fujihara, K. (2014). Distinction between MOG antibody-positive and AQP4 antibody-positive NMO spectrum disorders. *Neurology, 82*, 474–481.

Schafflick, D., Xu, C. A., Hartlehnert, M., Cole, M., Schulte-Mecklenbeck, A., Lautwein, T., Wolbert, J., Heming, M., Meuth, S. G., Kuhlmann, T., Gross, C. C., Wiendl, H., Yosef, N., & Meyer Zu Horste, G. (2020). Integrated single cell analysis of blood and cerebrospinal fluid leukocytes in multiple sclerosis. *Nature Communications, 11*, 247.

Schnell, A., Huang, L., Singer, M., Singaraju, A., Barilla, R. M., Regan, B. M. L., Bollhagen, A., Thakore, P. I., Dionne, D., Delorey, T. M., Pawlak, M., Meyer Zu Horste, G., Rozenblatt-Rosen, O., Irizarry, R. A., Regev, A., & Kuchroo, V. K. (2021). Stem-like intestinal Th17 cells give rise to pathogenic effector T cells during autoimmunity. *Cell, 184*, 6281–98.e23.

Schrempf, W., & Ziemssen, T. (2007). Glatiramer acetate: Mechanisms of action in multiple sclerosis. *Autoimmunity Reviews, 6*, 469–475.

Shi, K., Li, H., Chang, T., He, W., Kong, Y., Qi, C., Li, R., Huang, H., Zhu, Z., Zheng, P., Ruan, Z., Zhou, J., Shi, F. D., & Liu, Q. (2022a). Bone marrow hematopoiesis drives multiple sclerosis progression. *Cell, 185*, 2234–2247.e17.

Shi, M., Chu, F., Jin, T., & Zhu, J. (2022b). Progress in treatment of neuromyelitis optica spectrum disorders (NMOSD): Novel insights into therapeutic possibilities in NMOSD. *CNS Neuroscience & Therapeutics, 28*, 981–991.

Siller, N., Kuhle, J., Muthuraman, M., Barro, C., Uphaus, T., Groppa, S., Kappos, L., Zipp, F., & Bittner, S. (2019). Serum neurofilament light chain is a biomarker of acute and chronic neuronal damage in early multiple sclerosis. *Multiple Sclerosis, 25*, 678–686.

Sospedra, M., & Martin, R. (2005). Immunology of multiple sclerosis. *Annual Review of Immunology, 23*, 683–747.

Stangel, M., Fredrikson, S., Meinl, E., Petzold, A., Stüve, O., & Tumani, H. (2013). The utility of cerebrospinal fluid analysis in patients with multiple sclerosis. *Nature Reviews. Neurology, 9*, 267–276.

Steinman, L. (2014). Immunology of relapse and remission in multiple sclerosis. *Annual Review of Immunology, 32*, 257–281.

Steinman, L., Fox, E., Hartung, H. P., Alvarez, E., Qian, P., Wray, S., Robertson, D., Huang, D., Selmaj, K., Wynn, D., Cutter, G., Mok, K., Hsu, Y., Xu, Y., Weiss, M. S., Bosco, J. A., Power, S. A., Lee, L., Miskin, H. P., & Cree, B. A. C. (2022). Ublituximab versus teriflunomide in relapsing multiple sclerosis. *The New England Journal of Medicine, 387*, 704–714.

Syed, Y. Y. (2021). Alemtuzumab: A review in relapsing remitting multiple sclerosis. *Drugs, 81*, 157–168.

Tahara, M., Oeda, T., Okada, K., Kiriyama, T., Ochi, K., Maruyama, H., Fukaura, H., Nomura, K., Shimizu, Y., Mori, M., Nakashima, I., Misu, T., Umemura, A., Yamamoto, K., & Sawada, H. (2020). Safety and efficacy of rituximab in neuromyelitis optica spectrum disorders (RIN-1 study): A multicentre, randomised, double-blind, placebo-controlled trial. *Lancet Neurology, 19*, 298–306.

Takai, Y., Misu, T., Suzuki, H., Takahashi, T., Okada, H., Tanaka, S., Okita, K., Sasou, S., Watanabe, M., Namatame, C., Matsumoto, Y., Ono, H., Kaneko, K., Nishiyama, S., Kuroda, H., Nakashima, I., Lassmann, H., Fujihara, K., Itoyama, Y., & Aoki, M. (2021). Staging of astrocytopathy and complement activation in neuromyelitis optica spectrum disorders. *Brain, 144*, 2401–2415.

Tanaka, M., Kinoshita, M., & Tanaka, K. (2015). Corticosteroid and tacrolimus treatment in neuromyelitis optica related disorders. *Multiple Sclerosis, 21*, 669.

Ten, V. S., Yao, J., Ratner, V., Sosunov, S., Fraser, D. A., Botto, M., Sivasankar, B., Morgan, B. P., Silverstein, S., Stark, R., Polin, R., Vannucci, S. J., Pinsky, D., & Starkov, A. A. (2010). Complement component c1q mediates mitochondria-driven oxidative stress in neonatal hypoxic-ischemic brain injury. *The Journal of Neuroscience, 30*, 2077–2087.

Tian, D. C., Li, Z., Yuan, M., Zhang, C., Gu, H., Wang, Y., & Shi, F. D. (2020). Incidence of neuromyelitis optica spectrum disorder (NMOSD) in China: A national population-based study. *The Lancet Regional Health Western Pacific, 2*, 100021.

Torke, S., Pretzsch, R., Häusler, D., Haselmayer, P., Grenningloh, R., Boschert, U., Brück, W., & Weber, M. S. (2020). Inhibition of Bruton's tyrosine kinase interferes with pathogenic B-cell development in inflammatory CNS demyelinating disease. *Acta Neuropathologica, 140*, 535–548.

Touil, H., Kobert, A., Lebeurrier, N., Rieger, A., Saikali, P., Lambert, C., Fawaz, L., Moore, C. S., Prat, A., Gommerman, J., Antel, J. P., Itoyama, Y., Nakashima, I., & Bar-Or, A. (2018). Human central nervous system astrocytes support survival and activation of B cells: Implications for MS pathogenesis. *Journal of Neuroinflammation, 15*, 114.

Traboulsee, A., Greenberg, B. M., Bennett, J. L., Szczechowski, L., Fox, E., Shkrobot, S., Yamamura, T., Terada, Y., Kawata, Y., Wright, P., Gianella-Borradori, A., Garren, H., & Weinshenker, B. G. (2020). Safety and efficacy of satralizumab monotherapy in neuromyelitis optica spectrum disorder: A randomised, double-blind, multicentre, placebo-controlled phase 3 trial. *Lancet Neurology, 19*, 402–412.

Tradtrantip, L., Yao, X., Su, T., Smith, A. J., & Verkman, A. S. (2017). Bystander mechanism for complement-initiated early oligodendrocyte injury in neuromyelitis optica. *Acta Neuropathologica, 134*, 35–44.

Vaknin-Dembinsky, A., Brill, L., Orpaz, N., Abramsky, O., & Karussis, D. (2010). Preferential increase of B-cell activating factor in the cerebrospinal fluid of neuromyelitis optica in a white population. *Multiple Sclerosis, 16*, 1453–1457.

Varrin-Doyer, M., Spencer, C. M., Schulze-Topphoff, U., Nelson, P. A., Stroud, R. M., Cree, B. A., & Zamvil, S. S. (2012). Aquaporin 4-specific T cells in neuromyelitis optica exhibit a Th17 bias and recognize clostridium ABC transporter. *Annals of Neurology, 72*, 53–64.

Velasco, M., Zarco, L. A., Agudelo-Arrieta, M., Torres-Camacho, I., Garcia-Cifuentes, E., & Muñoz, O. (2021). Effectiveness of treatments in neuromyelitis optica to modify the course of disease in adult patients. Systematic review of literature. *Multiple Sclerosis and Related Disorders, 50*, 102869.

Venken, K., Hellings, N., Thewissen, M., Somers, V., Hensen, K., Rummens, J. L., Medaer, R., Hupperts, R., & Stinissen, P. (2008). Compromised CD4+ CD25(high) regulatory T-cell function in patients with relapsing-remitting multiple sclerosis is correlated with a reduced frequency of FOXP3-positive cells and reduced FOXP3 expression at the single-cell level. *Immunology, 123*, 79–89.

Vincent, T., Saikali, P., Cayrol, R., Roth, A. D., Bar-Or, A., Prat, A., & Antel, J. P. (2008). Functional consequences of neuromyelitis optica-IgG astrocyte interactions on blood-brain barrier permeability and granulocyte recruitment. *Journal of Immunology, 181*, 5730–5737.

Walton, C., King, R., Rechtman, L., Kaye, W., Leray, E., Marrie, R. A., Robertson, N., La Rocca, N., Uitdehaag, B., van der Mei, I., Wallin, M., Helme, A., Angood Napier, C., Rijke, N., & Baneke, P. (2020). Rising prevalence of multiple sclerosis worldwide: Insights from the Atlas of MS, third edition. *Multiple Sclerosis, 26*, 1816–1821.

Wang, L., Liu, K., Tan, X., Zhou, L., Zhang, Y., Liu, X., Fu, Y., Qiu, W., & Yang, H. (2020). Remedial effect of intravenous cyclophosphamide in corticosteroid-refractory patients in the acute phase of neuromyelitis optica spectrum disorder-related optic neuritis. *Frontiers in Neurology, 11*, 612097.

Wang, Y., Zhong, X., Wang, H., Peng, Y., Shi, F., Jia, D., Yang, H., Zeng, Q., Quan, C., ZhangBao, J., Lee, M., Qi, J., Chen, X., & Qiu, W. (2023). Batoclimab as an add-on therapy in neuromyelitis optica spectrum disorder patients with acute attacks. *European Journal of Neurology, 30*, 195–203.

Wheeler, M. A., Jaronen, M., Covacu, R., Zandee, S. E. J., Scalisi, G., Rothhammer, V., Tjon, E. C., Chao, C. C., Kenison, J. E., Blain, M., Rao, V. T. S., Hewson, P., Barroso, A., Gutiérrez-Vázquez, C., Prat, A., Antel, J. P., Hauser, R., & Quintana, F. J. (2019). Environmental control of astrocyte pathogenic activities in CNS inflammation. *Cell, 176*, 581–96.e18.

Wingerchuk, D. M., Lennon, V. A., Lucchinetti, C. F., Pittock, S. J., & Weinshenker, B. G. (2007). The spectrum of neuromyelitis optica. *Lancet Neurology, 6*, 805–815.

Wingerchuk, D. M., Banwell, B., Bennett, J. L., Cabre, P., Carroll, W., Chitnis, T., de Seze, J., Fujihara, K., Greenberg, B., Jacob, A., Jarius, S., Lana-Peixoto, M., Levy, M., Simon, J. H., Tenembaum, S., Traboulsee, A. L., Waters, P., Wellik, K. E., & Weinshenker, B. G. (2015). International consensus diagnostic criteria for neuromyelitis optica spectrum disorders. *Neurology, 85*, 177–189.

Wu, Y., Zhong, L., & Geng, J. (2019). Neuromyelitis optica spectrum disorder: Pathogenesis, treatment, and experimental models. *Multiple Sclerosis and Related Disorders, 27*, 412–418.

Yadav, S. K., Soin, D., Ito, K., & Dhib-Jalbut, S. (2019). Insight into the mechanism of action of dimethyl fumarate in multiple sclerosis. *Journal of Molecular Medicine (Berlin, Germany), 97*, 463–472.

Yamamura, T., Kleiter, I., Fujihara, K., Palace, J., Greenberg, B., Zakrzewska-Pniewska, B., Patti, F., Tsai, C. P., Saiz, A., Yamazaki, H., Kawata, Y., Wright, P., & De Seze, J. (2019). Trial of satralizumab in neuromyelitis optica spectrum disorder. *The New England Journal of Medicine, 381*, 2114–2124.

Yong, H. Y. F., & Yong, V. W. (2022). Mechanism-based criteria to improve therapeutic outcomes in progressive multiple sclerosis. *Nature Reviews. Neurology, 18*, 40–55.

Zhang, C., Tian, D. C., Yang, C. S., Han, B., Wang, J., Yang, L., & Shi, F. D. (2017). Safety and efficacy of bortezomib in patients with highly relapsing Neuromyelitis Optica Spectrum disorder. *JAMA Neurology, 74*, 1010–1012.

Zhang, C., Zhang, M., Qiu, W., Ma, H., Zhang, X., Zhu, Z., Yang, C. S., Jia, D., Zhang, T. X., Yuan, M., Feng, Y., Yang, L., Lu, W., Yu, C., Bennett, J. L., & Shi, F. D. (2020). Safety and efficacy of tocilizumab versus azathioprine in highly relapsing neuromyelitis optica spectrum disorder (TANGO): An open-label, multicentre, randomised, phase 2 trial. *Lancet Neurology, 19*, 391–401.

Zhang, C., Zhang, T. X., Liu, Y., Jia, D., Zeng, P., Du, C., Yuan, M., Liu, Q., Wang, Y., & Shi, F. D. (2021). B-cell compartmental features and molecular basis for therapy in autoimmune disease. *Neurology Neuroimmunology & Neuroinflammation, 8*, e1070.

Zipfel, P. F., & Skerka, C. (2009). Complement regulators and inhibitory proteins. *Nature Reviews. Immunology, 9*, 729–740.

Multilayer Analysis of RNA Sequencing Data in Alzheimer's Disease to Unravel Molecular Mysteries

Dilara Uzuner, Atılay İlgün, Elif Düz, Fatma Betül Bozkurt, and Tunahan Çakır

Abstract Alzheimer's disease (AD) is a complex disease, and numerous cellular events may be involved in etiology. RNAseq-based transcriptome data hold multi-layer information content, which could be crucial in unraveling molecular mysteries of AD. It enables quantification of gene expression levels, identification of genomic variants, and elucidation of splicing anomalies such as exon skipping and intron retention. Additional integration of this information into protein-protein interaction networks and genome-scale metabolic models from the literature has potential to decipher functional modules and affected mechanisms for complex scenarios such as AD. In this chapter, we review the application areas of the multilayer content of RNAseq and associated integrative approaches available, with a special focus on AD.

Keywords Alzheimer's disease · RNAseq · Splicing events · Genomic variants · Gene expression · Protein-protein interactions · Genome-scale metabolic models

1 Introduction

Different factors contribute to progression of Alzheimer's disease (AD), but the exact mechanism of the disease is still a mystery. A systems-oriented approach is necessary to elucidate molecular players behind the complex behavior of the disease. RNAseq technology is a widely used method to measure gene expression levels in cells at genome scale. Compared to other genome-scale data collection approaches, RNAseq data are easy to reach, cost-effective and cover almost all the genes in the organism. Besides, it stores multiple layers of information not limited to gene expression levels but also covering genomic variants and splicing anomalies

D. Uzuner · A. İlgün · E. Düz · F. B. Bozkurt · T. Çakır (✉)
Department of Bioengineering, Gebze Technical University, Gebze, Kocaeli, Turkey
e-mail: tcakir@gtu.edu.tr

A. C.-H. Yu et al. (eds.), *Systems Neuroscience*, Advances in Neurobiology 41,
https://doi.org/10.1007/978-3-031-69188-1_9

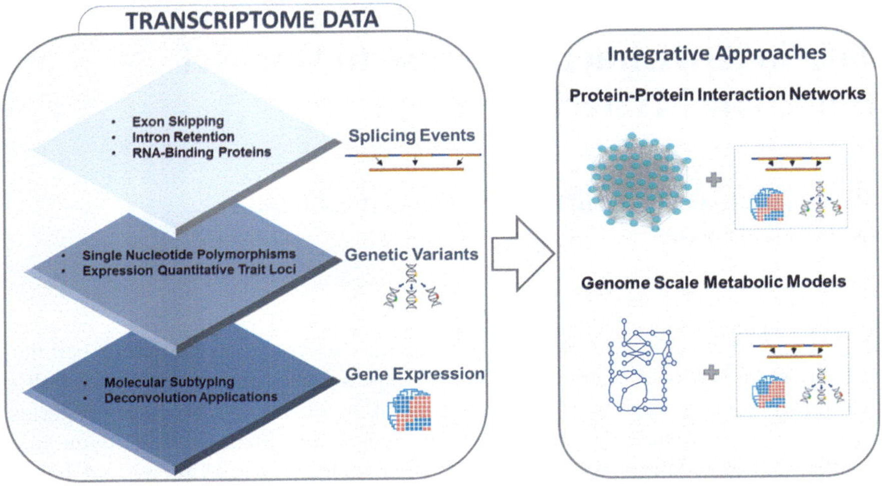

Fig. 1 Different layers of information stored in RNAseq data with potential applications in Alzheimer's disease research

(Fig. 1). The information-rich nature of RNAseq data makes it crucial to enlighten molecular mechanisms of AD.

In this chapter, we will give an overview of novel aspects of RNAseq data analysis in AD, focusing on its multilayer content from the perspective of AD (Fig. 1). The first layer, gene expression levels, will be reviewed in terms of their use for molecular subtyping of AD and their deconvolution to predict cell type composition and cell-type-specific expression. The second layer, genomic variants, can be used to identify genomic variants pointing to the risk for AD, and relevant literature will be provided. Then, the third layer, abnormal splicing events, will be discussed in terms of their importance to decode molecular mechanisms of AD. Finally, we will cover the examples from literature that utilize information-rich content of RNAseq data by mapping it onto molecular interaction networks, including protein-protein interaction networks and genome-scale metabolic networks.

2 AD-Specific RNAseq Datasets in Literature

Transcriptomics studies play a crucial role in investigating causes and mechanisms of various diseases. Also, they have undebatable importance in the development of disease-specific medicine applications and in revealing secrets of multifactorial and polygenic neurodegenerative diseases including Alzheimer's disease. The number of RNAseq datasets in the literature is increasing day by day, and RNAseq technology is replacing microarrays. One of the most important reasons for this change is that RNAseq data are more informative.

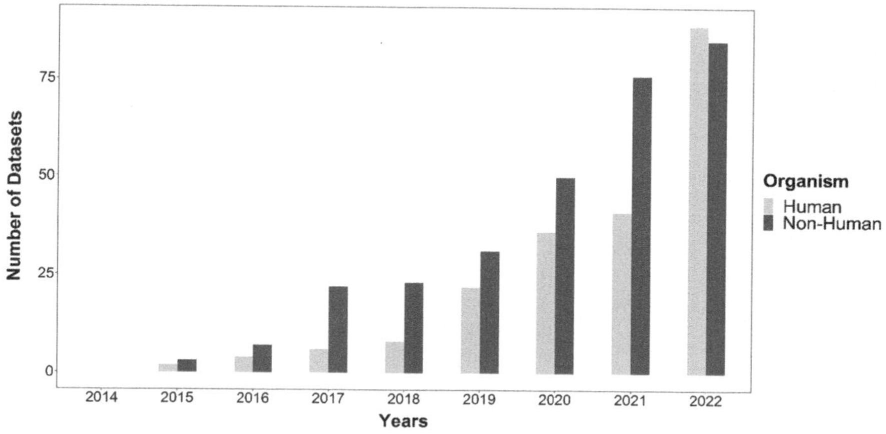

Fig. 2 Number of published human RNAseq datasets per year identified with a search of "Alzheimer's disease" keyword in GEO Database. Light gray and dark gray bars correspond to human and non-human datasets, respectively

2.1 *Bulk RNAseq Datasets for AD*

By February 2023, in Gene Expression Omnibus (GEO) (https://www.ncbi.nlm.nih.gov/geo) (Clough & Barrett, 2016), a comprehensive transcriptome database under National Center for Biotechnology Information, the search with "Alzheimer's disease" keyword leads to 523 RNAseq datasets, and 215 of them are human-derived (Fig. 2). There are also databases that store AD omics data and require special access. To illustrate, Synapse (https://www.synapse.org) is a research platform that contains large AD patient cohorts from multiple distinct studies conducted by Rush Alzheimer's disease Center (Religious Orders Study/Memory and Aging Project, ROSMAP), Mayo Clinic brain bank (Allen et al., 2016), and Mount Sinai Brain Bank (MSBB) (Wang et al., 2018). They include ~1000 RNAseq samples in total from AD patients spanning multiple brain regions (Table 1). Those large cohorts are commonly used in AD-related studies and have revealed many important findings regarding AD pathology. One advantage of Synapse database is the availability of long-term clinic data of the patients. Long-term clinical follow-up provides valuable information about the progression of disease. Since AD is a heterogeneous disease, focusing on dataset variability can reveal molecular mysteries of the disease.

2.2 *Single-Cell RNAseq Datasets for AD*

Transcriptome data types can be divided into two main groups: bulk and single cell. While bulk RNAseq studies provide a holistic view of gene expression levels

Table 1 Three large AD datasets deposited in Synapse platform

Dataset	Accession ID	Brain region	# of AD samples	# of control samples
ROSMAP	syn17024112	Dorsolateral prefrontal cortex (DLPFC)	403	164
Mayo Clinic Brain Bank	syn9738945	Temporal cortex (TCX)	82	78
Mayo Clinic Brain Bank	syn9738945	Cerebellum (CER)	82	77
Mount Sinai Brain Bank	syn7416949	Anterior prefrontal cortex (aPFC)	139	93
Mount Sinai Brain Bank	syn7416949	Superior temporal gyrus (STG)	148	86
Mount Sinai Brain Bank	syn7416949	Parahippocampal gyrus (PHG)	146	83
Mount Sinai Brain Bank	syn7416949	Inferior frontal gyrus (IFG)	130	91

for a specific tissue, single-cell RNAseq (scRNAseq) approaches provide whole transcriptome profiles of thousands of single cells in a given tissue. scRNAseq can capture the differences in the affected molecular mechanisms by AD across different cell types, and it can also be used to detect subtypes within the cell types. Focusing on certain cell types and their subtypes might provide a better understanding of disease pathogenesis. In literature, the number of AD-related scRNAseq datasets is growing day by day. Distribution of the number of publicly available scRNAseq datasets in GEO database for AD is given in Fig. 3. Several scRNAseq datasets were recently made available for AD; (Del-Aguila et al., 2019; Gerrits et al., 2021; Grubman et al., 2019; Leng et al., 2021; Mathys et al., 2019; Nguyen et al., 2020; Zhou et al., 2020).

In Mathys et al. (2019), scRNAseq was performed on the 80,660 nuclei obtained from the prefrontal cortex of 48 individuals with AD. As a huge advantage of single-cell transcriptomic studies, they were able to investigate cell subpopulations. They found that the major disease-associated changes were occurring in early phase of the disease, and those changes were cell-type-specific. Moreover, myelination-related gene impairments were shown to be not specific to oligodendrocytes; in fact, other major cell types had also impairments on their myelin-related genes.

Another study published in the same year, (Grubman et al., 2019) reported gene expression data of 13,214 nuclei from entorhinal cortex of individuals with the AD and control. They also focused on cell subpopulations and cell-type-specific transcriptional changes during AD. They found that APOE gene, a known risk factor gene for Alzheimer's disease, is upregulated in AD microglial subpopulations and repressed in AD oligodendrocyte progenitor cells and astrocyte subpopulations.

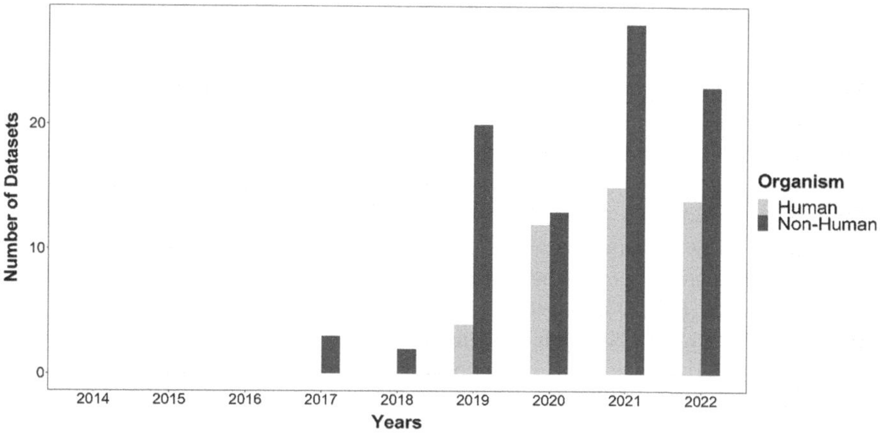

Fig. 3 Distribution of the number of Alzheimer's Disease scRNAseq datasets in GEO database by year. Light gray and dark gray bars correspond to human and non-human datasets respectively

Investigation of the cell type and cell-subtype-specific transcriptional changes during AD is informative and useful to unravel the molecular mysteries of the disease. However, the number of captured cells or nuclei per cell type is important while working with cell subtypes. Astrocytes, for example, despite their high abundance in the brain, are mostly captured with a very low yield in these studies; only about 3–18% of the captured cells are astrocytes. For this reason, Sadick and colleagues (Sadick et al., 2022) implemented astrocyte enrichment strategy in their single-cell collection approach to increase the number of captured astrocytes per donor. Their scRNAseq data for AD were highly enriched for astrocytes such that ~41,000 of the ~80,000 nuclei were astrocytes. The authors used the astrocyte data and identified nine astrocyte subpopulations in the prefrontal cortex. Identified astrocyte subpopulations had unique transcriptomic signatures, and they exhibited upregulation/downregulation of specific GO terms such as angiogenesis, blood-brain barrier maintenance, lipid storage oxidative stress during AD.

3 First Layer: Gene Expression Levels

This section reviews two recent developments regarding the gene expression layer of RNAseq data (Fig. 1): (i) the use of RNAseq data to identify molecular subtypes of AD and (ii) the deconvolution of RNAseq data to identify cell type composition and cell-type-specific expression levels.

3.1 Molecular Subtyping of AD

Alzheimer's disease shows high multiformity, and the course of disease can differ among patients. This implies availability of different subtypes within AD. The subtypes may display different clinical phenotypes in terms of disease progression or different age onsets. Omics data from a sufficient number of AD patients provide an opportunity to identify molecular subtypes of the disease and might pave the way for developing new therapeutic targets or drugs that can be more efficient in disease treatment. There are different studies focusing on disease subtype determination for AD. Murray et al. (2011) identified three different AD subtypes using postmortem neurofibrillary tangles (NFT) density as a criterion. Those three subtypes were named as typical AD, hippocampal-sparing AD, and limbic-predominant AD. In addition to previous disease subtype determination methods conducted by in vivo magnetic resonance imaging technologies, it became possible to determine disease subtypes from RNAseq data (Zheng & Xu, 2021). Transcriptomic data from ROSMAP cohort were recently analyzed with the non-negative matrix factorization method, and two AD molecular subtypes were determined (Zheng & Xu, 2021). Those molecular subtypes were called synaptic type and inflammatory type. The determined subtypes were shown to have different molecular signatures. While the synaptic type was characterized by disrupted synaptic plasticity, the inflammatory type was characterized by the disruption of certain inflammatory pathways like IL2. Moreover, those molecular subtypes were found to be associated with APOE genotypes and sex. Synaptic type was observed commonly in *APOE* $\varepsilon3\varepsilon4$ genotype and in males, while inflammatory type was commonly observed in *APOE* $\varepsilon3\varepsilon3$ genotype and in females. Those findings were also validated using two publicly available datasets in GEO.

Another RNAseq-based molecular subtyping study for AD was conducted by Ryan Neff and colleagues (2021). RNAseq data from ROSMAP and MSBB datasets (Table 1), which are among the largest AD cohorts in literature, were used for molecular subtyping of AD in that study. Since PHG brain region from the MSBB dataset was identified to be the most vulnerable brain region compared to other brain regions based on differentially expressed genes, the authors used RNAseq from this region for the molecular subtyping of AD. Three major classes and five molecular subtypes were determined using a method called Weighted Sample Gene Network Analysis. They also showed the validity of the same subtypes for the ROSMAP dataset. Additionally, certain genes were found to be closely associated with the classification of those subtypes such as *GABRB2, MSN, PLP1, LRP10*, and *ATP6V1A*. Molecular subtyping based on RNAseq data might reveal unknown mechanisms playing roles in AD, and novel biomarkers and drugs specific for the subtypes can be identified.

3.2 Deconvolution-Based Approaches to Determine Cell Type Composition and Cell-Type-Specific Gene Expressions from Transcriptome Data

Single-cell transcriptome data can provide critical molecular information that cannot be captured from bulk-RNAseq data. However, scRNAseq data are not as common as bulk data in the literature. In addition, the collection and analysis of scRNAseq data are costlier and time-consuming than bulk data. For this reason, deconvolution methods that can predict cell type composition or cell-type-specific gene expression values from bulk transcriptomic data have emerged.

Deconvolution methods can be divided into two: complete deconvolution and partial (supervised) deconvolution. Complete deconvolution methods make prediction of both cell type proportion and cell-type-specific expression (Li & Wu, 2019; Zaitsev et al., 2019). swCAM (Chen et al., 2021) and bMIND (Wang et al., 2021b) are examples of complete deconvolution algorithms. In contrast, partial deconvolution only predicts cell type proportion. For partial deconvolution, the list of signature genes for each cell type is crucial, and it has significant impacts on the accuracy of the deconvolution (Sutton et al., 2022). Partial deconvolution methods that perform cell type composition predictions from transcriptome data include DeconRNASeq (Gong & Szustakowski, 2013), CIBERSORT (Newman et al., 2015), dtangle (Hunt et al., 2018), and MuSiC (Wang et al., 2019).

It is crucial to select the appropriate algorithm to apply for AD transcriptome data because the disease has strong connections with different cell types in the brain such as astrocytes, oligodendrocytes, or microglia. Certain algorithms might be better candidates for deconvoluting specific tissues. For instance, MIND and swCAM can be used for deconvoluting human brain cells. Microglia are important in AD pathology. In a study (Olah et al., 2018), expression of a set of microglial genes was shown to be increased by aging. This pattern was also detected by using MIND deconvolution algorithm, thereby revealing the power of this algorithm to detect important biological gene expression patterns in bulk data (Wang et al., 2021b). swCAM (Chen et al., 2021) is a complete deconvolution algorithm that was applied on the ROSMAP (De Jager et al., 2018) dataset, and eight different cell groups were identified. swCAM algorithm was also compared with bMIND (Wang et al., 2021b) algorithm by replicating the cell group expressions for ROSMAP dataset. For each identified cell group, Spearman correlation coefficients were found to be 0.74 ± 0.04 with p-value <0.05, revealing high concordance between the results of the two independent deconvolution algorithms. bMIND is the newer version of the MIND algorithm, and it relies on the integration of bulk and single-cell data to estimate cell-type-specific expression (Wang et al., 2021b). Therefore, bMIND takes the advantage of using both data types. bMIND algorithm is also capable of performing cell-type-specific differential expression, which makes it quite powerful compared to other algorithms (Wang et al., 2021b). It was used to predict AD-related differentially expressed genes in excitatory neurons based on three major AD datasets (ROSMAP, Mayo Clinic and MSBB), and the results were compared with the genes identified in Mathys et al. using scRNAseq data. A significant overlap was found between the results, validating the prediction performance of the bMIND algorithm (Wang et al., 2021b).

Another algorithm developed for brain cells is called Enrichment Score Homogenate RNA Deconvolution (ESHRD) (Piras et al., 2020), which allows for conducting cell-type-specific differential expression as well as pathway analysis. ESHRD method was also applied to major AD cohorts like ROSMAP, Mayo Clinic, and MSBB studies. By the usage of ESHRD algorithm, novel gene expression changes in oligodendrocytes and endothelial cells during AD were detected. To illustrate, endothelial cells-related genes were found to be upregulated in TCX, STG, PHG, CER, and DLPFC brain regions during AD (Piras et al., 2020). Moreover, vascular development and angiogenesis-related pathways were found to be enriched (Piras et al., 2020).

CIBERSORT, another deconvolution algorithm, can be used for detecting cell type proportion (Newman et al., 2015). Despite the fact that it is not initially optimized for deconvolution of brain cells, recent studies (Sutton et al., 2022) revealed that it produces better outputs compared to complete deconvolution algorithms on human brain transcriptome data. Additionally, CIBEXSORTx is the newer version of the CIBERSORT that allows for predicting cell-type-specific gene expression in addition to cell type abundance (Newman et al., 2019).

The benchmarking studies would be a guiding light to select appropriate deconvolution algorithms (Sutton et al., 2022). According to G. J. Sutton et al., best algorithms for deconvoluting brain transcriptome data are CIBERSORT, dtangle, and MuSic. It is not possible to use every deconvolution algorithm in perfect compatibility and perfect integrity with every dataset in the literature. For this reason, there is a need to develop more sophisticated algorithms to get optimum results for every data.

4 Second Layer: Genetic Variants in AD from RNAseq Data

Identification and analysis of genetic variants are mostly based on DNA sequencing in the form of whole genome sequencing or whole exome sequencing data. However, RNAseq is also based on sequencing of exonic regions, and it can be used to identify candidate pathogenic variants. Genome Analysis Toolkit (GATK), one of the most known variant discovery toolkits for genome sequences, has also a specific pipeline for the variant discovery from RNAseq (Brouard & Bissonnette, 2022). In literature, there are several studies that apply variant discovery from RNAseq successfully for different conditions (Bakhtiarizadeh & Alamouti, 2020; Jehl et al., 2021; Lezmi & Benvenisty, 2021).

It is clear that genomic variants play key roles in the etiology of AD. Most known genetic mutations associated with AD are pathogenic mutations on *APP*, *PSEN1*, and *PSEN2* genes. Also, many disease-associated variants were reported on AD susceptibility genes including *APOE* (Liu et al., 2021; Neuner et al., 2020;

Sims et al., 2020). In the literature, there are plenty of variant discovery studies for AD based on whole exome sequencing (WES) and whole genome sequencing (WGS) data (Li et al., 2018b; Prokopenko et al., 2021). In the specific case of AD, Akila Parvathy Dharshini et al. (2019) have performed variant discovery analysis and differentially expressed gene analysis from the same RNAseq dataset, and they compared the detected variants with the genome-wide association studies (GWAS) data. As a result of this analysis, they found 297 common variants with the GWAS, and 20 of the variants were novel variants. They also reported that the variants interrupted the binding of transcription factors and miRNAs, which entailed essential biological functions such as myelination, neuronal death, and mitochondrial energy metabolism.

Still there are some challenges for variant detection from RNAseq. First, the technical errors during mRNA-to-cDNA translation may cause false variants. Also read depths show high variability between genes and within genes in RNAseq caused by the high difference in the number of reads in exon and intron regions due to alternative splicing (Han et al., 2015; Jehl et al., 2021). Nevertheless, with the advancement of the approach, these bottlenecks will be overcome. As it allows to determine both gene expression levels and genetic variants from the same sample, RNAseq analysis has a great potential to shed light on the understanding of genetic mechanisms.

GWAS datasets provide substantial results to the discovery of genetics of complex diseases. They aim to elucidate genetic variation loci that are associated with a phenotype. The most studied genetic variations are single-nucleotide polymorphisms (SNPs). Interestingly, majority of disease-associated variants are found in intronic regions according to GWAS. This makes it hard to understand disease susceptibility of GWAS loci. One of the ways to understand disease susceptibility of intronic regions is expression quantitative trait loci (eQTL). The goal of eQTL analysis is the association of genetic variants with changes in gene expression levels (Wang et al., 2021b). Transcriptome-wide association studies (TWASs) aim to identify an association between gene expression regulation determined by eQTL and phenotype. In the literature, there is a growing number of TWAS studies on AD. Hu et al. (2018) analyzed two GWAS and five eQTL datasets together and targeted seven AD-associated genes with 27 SNPs. Gockley et al. (2021) used three large cohorts, Mayo Clinic, Mount Sinai Brain Bank, and ROSMAP, to perform TWAS analysis and predicted AD-associated genes. TWAS has a further strength in that it can be applied to various physical traits rather than disease phenotypes. Zhao et al. (2021) applied TWAS to associate brain structural variation in 211 samples and found 29 statistically significant candidate genes. Further studies on eQTL analysis and TWAS will be new steps in linking intronic variants with AD phenotype.

5 Third Layer: Splicing Anomalies in AD from RNAseq Data

The alterations in splicing patterns are also important for disease pathology. Precursor messenger RNA (pre-mRNA) splicing is an essential process in eukaryotic

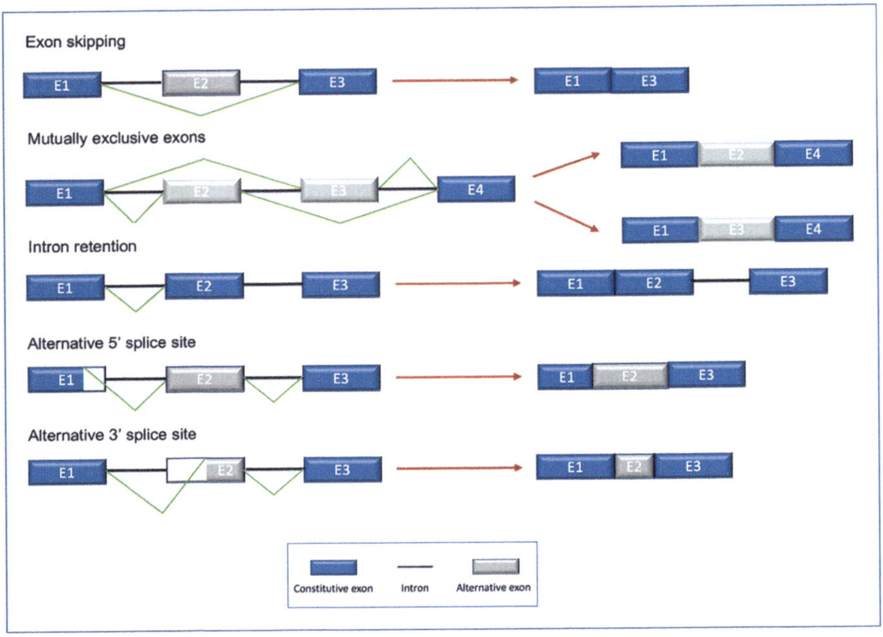

Fig. 4 Five major types of alternative splicing events

organisms that removes introns and joins exons together to produce mature mRNA. The strict regulation and high flexibility of this posttranscriptional process by a mechanism called alternative splicing (AS) enable the generation of several transcripts encoding for protein isoforms having varying characteristics in terms of structure, function, stability, and cellular localization. Alternative splicing may arise due to several mechanisms such as exon skipping, intron retention, mutually exclusive exon usage, or changes in acceptor/donor sites (Fig. 4). This mechanism is especially common in neuronal cells where the splicing patterns are ever-changing for the maintenance of cellular homeostasis and promotion of neurogenesis, migration, and synaptic transmission (Douglas & Wood, 2011; Li et al., 2021a; Ward & Cooper, 2010). These continuous alterations in the splicing patterns and regulation of alternative splicing predispose neuronal tissues to splicing defects resulting in neurodegenerative diseases including AD (Deschênes & Chabot, 2017; Su et al., 2018). Particularly, recent studies have shown that regulatory mutations and changes in subcellular localization of splicing factors and spliceosome components are related to aberrant splicing, which may have an impact on the onset and progression of AD (Bagyinszky et al., 2020; Biamonti et al., 2021; Deschênes & Chabot, 2017; Su et al., 2018).

As yet, numerous computational tools (exon-based, isoform-based, and event-based methods) have been developed for the identification and quantification of alternative splicing from RNAseq data. Isoform-based approaches first aim at full-length transcript reconstruction and relative abundance estimation of these transcripts in each sample on the basis of sequence reads. Statistical tests are then performed to identify significant alterations in relative transcripts across different experimental groups. The accuracy of the transcript quantification determines the performance of these methods. For these methods, Cufflinks/cuffdiff2 (Trapnell et al., 2010) and DiffSplice (Hu et al., 2012) tools that are commonly used in differential splicing studies can be given as examples. Exon-based and event-based approaches are called count-based approaches. DEXSeq (Anders et al., 2012), edgeR (Robinson et al., 2009), JunctionSeq (Hartley & Mullikin, 2016), and limma tools (Ritchie et al., 2015) use the exon-based approach, which assigns different features, for instance, exons or junctions, to read counts. The disadvantage of this methodology is that the splicing event type cannot be inferred, but only differential expression of exons/junctions could be identified between different conditions. On the other hand, in event-based approaches, represented by tools such as dSpliceType (Zhu et al., 2015), MAJIQ (Vaquero-Garcia et al., 2016), rMATS (Shen et al., 2014), and SUPPA/SUPPA2 (Trincado et al., 2018), the percent spliced in (PSI) values are calculated for each splicing event, which is the proportion of mRNAs expressed from a gene having a specific alternative splicing event, to quantify splicing events (Alamancos et al., 2015). Systematic evaluation of these 10 differential splicing tools using human and mouse RNAseq datasets showed that all of the exon-based approaches and two event-based approaches (MAJIQ and rMATS) had better performances, and the performance of exon-based approaches is generally greater than that of the isoform-based and event-based approaches. Also, it was stated that the usage of different datasets or sample sizes significantly affected the performance of these computational tools (Mehmood et al., 2020).

Recent advances in RNAseq technologies together with the development of computational tools have allowed transcriptome-wide studies of alternative splicing at an exceptional scale. The targeting of AS due to the involvement of disrupted splicing in human disorders resulted in development of novel therapeutics (Zhao, 2019). Currently, the gold standard for disease diagnosis by clinical genetic tests is the detection, annotation, and categorization of sequence variants. However, this approach is not capable of linking AS and other regulatory events to diseases. The use of RNAseq in clinical applications is essential for the interpretation and identification of splicing events and splicing variants together with others so that the diagnostic rates could be significantly improved (Marco-Puche et al., 2019).

5.1 Exon Skipping in AD from RNAseq Data

Exon skipping (ES) is the most prevalent alternative splicing event that contributes to several human disorders. As previously stated, mounting evidence from the analyses of transcriptomic data of aging brains suggests that aberrant splicing is a common hallmark of several neurodegenerative disorders including AD. ExonSkipAD (Yang et al., 2021) is an ES annotation database, which provides a resource for numerous annotation categories of ES events in AD. The database was created by analyzing three large transcriptomic datasets: ROSMAP, MSSB, and Mayo Clinic. The analysis identified 16,414 genes having approximately 156,000, 69,000, and 231,000 exon skipping events for each dataset, respectively (Yang et al., 2021).

Integrative approaches used to investigate the associations between sequence variants and ES events in AD have greatly contributed to our understanding of AD pathogenesis. *TREM2* is one of the most widely known AD risk genes. This gene encodes an immunoglobulin receptor that is involved in immune response and inflammatory pathways associated with AD development. Besides, *TREM2* is linked to beta-amyloid deposition and neurofibrillary tangles, which are the two defining characteristics of AD. The presence of low-frequency genetic variants in *TREM2* is associated with three to four times increased risk of AD. An integrative study of RNAseq and WGS data utilized a splicing-guided aggregation method where several *TREM2* variants with low frequency were gathered to explore the functional effect of those variants on alternative splicing in AD (Seonggyun Han et al., 2021). They detected an exon skipping event in the exon 2 of *TREM2*, and the frequency of this ES event was higher in individuals carrying at least one low-frequency variant causing loss of function for a splicing regulatory element. Besides, differentially expressed genes identified by comparing AD patients with high and low exon 2 skipping were enriched in immune-associated pathways (Han et al., 2021). Further studies are required to investigate several other AD risk genes having low-frequency variants to gain more insights into the functional effects of these variants on AS in AD.

5.2 Intron Retention in AD from RNAseq Data

Intron retention (IR) is one of the several modes of alternative splicing mechanisms. Its function in the regulation of gene expression has been recently acknowledged. Usually, the nonsense-mediated decay (NMD) pathway is triggered by premature termination codons located in most of the retained introns and causes degradation of intron-retained mature mRNAs. As a result, there is a low chance that most intron-retained mRNAs would be translated into proteins. However, NMD may be bypassed in some cases allowing modified proteins to be formed. Several studies have reported the association between intron retention events and AD. Integrative analysis of RNAseq and WGS data for AD revealed a significant association between the variant rs7982 in Clusterin gene and IR of the fifth exon in the gene. The decrease in the IR rates was found to be significantly greater in females compared to males, suggesting a gender-related function of rs7982 in AD pathogenesis through the regulation of splicing (Han et al., 2020). A genome-wide integrative analysis of IR using genetic, transcriptomic, and proteomic data of patients with AD and AD mouse models detected thousands of IR events in 2173 human and in 1736 mouse genes and identified differentially expressed genes that could not be identified via traditional exon-level methods. In addition, intron expression in genes involved in innate immunity such as *HMBOX1* was found to be significantly correlated with AD in humans. The study identified peptides that are most likely to be translated from intron-retained mRNAs by using mass spectrometry and determined genes associated with splicing that might regulate intron retention (Li et al., 2021b).

Phosphorylation and truncation of Tau are a fundamental step in the pathogenesis of AD. In a recent study, a new human-specific truncated form of Tau produced by the intron 12 retention was detected in human neuroblastoma cells and human brain RNA samples by qPCR, and the results were confirmed on a large human RNAseq data. The levels of this truncated isoform were lower in individuals with AD compared to non-demented controls, indicating that the absence of this isoform might have a significant role in the AD pathology. Intriguingly, this new Tau isoform has a less tendency to aggregate compared to other isoforms of Tau. The evidence provided by the study suggests the association of this new Tau isoform with the *GSK3* beta inhibition, which would mediate the retention of 12th intron by the serine/arginine-rich splicing factor 2 modulation (Garcia-Escudero et al., 2021). Although these studies have shown increased levels of intron retention in AD patients in especially protein-coding genes, future studies should also consider their functional roles in non-coding RNAs to better understand AD development and progression.

5.3 RNA-Binding Proteins

RNA-binding proteins (RBPs) are recruited by splicing regulatory sequences to assist or inhibit the spliceosome assembly on an AS exon located nearby. The serine-arginine (SR) factors and heterogeneous ribonucleoproteins (hnRNPs) are the two RNA-binding protein families that promote or inhibit exon inclusion (Giuseppe Biamonti et al., 2021). A hnRNP family member, the long isoform of hnRNP D-like (L-DL) particularly localizes to nuclear speckles, and its levels are lower in the hippocampus of aged mice brains. The downregulation of L-DL results in cognitive impairment in mice. The analysis of L-DL interactome and alternative RNA splicing measured by RNAseq revealed that the role of L-DL is to recruit other speckle proteins and regulate gene expression associated with cytoskeleton and synapse via modifying RNA splicing. These splicing alterations are regulated by L-DL-mediated interaction of SF3B3, an essential component of U2 small nuclear ribonucleoproteins (snRNP), and U2AF65, a U2 spliceosome protein that directs the binding of U2 snRNP to RNA (Zhang et al., 2021). The definition of a protein that connects alternative splicing and AD-related cognitive decline is significant for the development of potential therapeutics that can correct splicing abnormalities.

6 Mapping RNAseq Layers on Molecular Interaction Networks in AD

The three layers of RNAseq data were reviewed in the previous sections (gene expression, genetic variants, and splicing events). The data associated with each layer can be mapped on molecular interaction networks to uncover hidden information in these layers and to unravel molecular mysteries of AD (Fig. 1). Such approaches are reviewed in this section.

6.1 Genome-Scale Metabolic Networks

To understand the biology of complex diseases such as AD, traditional molecular biology techniques are not enough. A systems biology approach is needed that considers the organism as a complete system. Genome-scale metabolic models/networks (GEMs) are useful tools to identify metabolic alterations and predict drug target candidates and biomarker genes or metabolites. GEMs are models that include all known enzymatic reactions and their gene associations for an organism. For the reactions controlled by the enzymes, the genes encoding the enzymes are associated with that reaction in the form of GPR (Gene-Protein-Reaction) rules in these models. The most recent version of human GEM includes 13,024 reactions controlled by 2920 genes (Robinson et al., 2020). Thanks to the gene associations, gene expression data can be mapped to GEMs. By using GEMs and RNAseq data, a great extent of information about biological systems can be obtained (Fig. 5).

Transcriptome data mapping onto GEMs allows to extract context-specific metabolic models. There are several bioinformatics tools to reconstruct context-specific metabolic models using RNAseq data and GEMs. Some of these tools, including iMAT (Zur et al., 2010) and tINIT (Agren et al., 2014) algorithms, were successfully applied for AD research. These algorithms are commonly used in investigating diseases because they do not require the definition of a biological cellular objective to construct optimum condition-specific metabolic models.

Recently, Varma et al. (2021) reconstructed the AD and healthy state metabolic models from the RNAseq data of a large cohort using the iMAT algorithm, and they showed that cholesterol biosynthesis and catabolism reactions were differentially activated between AD and control models. Also, Baloni et al. (2020) used ROSMAP, MSBB, and Mayo Clinic RNAseq data and constructed brain region-specific metabolic models for seven different brain regions and individual-specific metabolic models for AD and healthy brains by the iMAT algorithm. They indicated that taurine transporters and cholesterol metabolism were mainly affected in the DLPFC, TCX and PHG regions. Also, they reported that cholesterol clearance pathways and bile acid pathways have different active reaction sets in AD and control brain according to individual-specific metabolic models. Bayraktar et al. used ROSMAP and Mayo Clinic RNAseq data and the tINIT algorithm to reconstruct brain region-

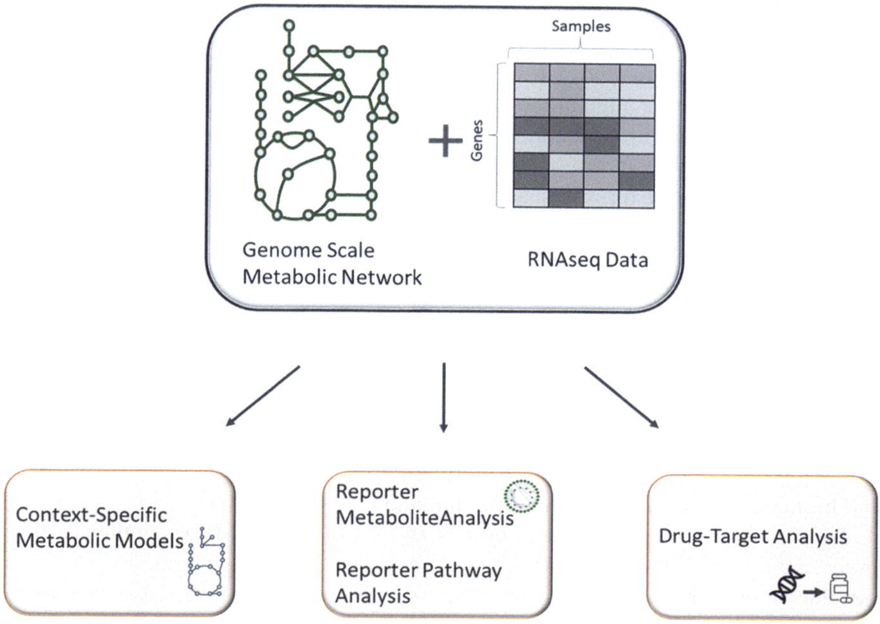

Fig. 5 Different applications of RNAseq data integration to genome-scale metabolic networks

specific AD metabolic models and examined differences in reaction, gene, and metabolite levels (Bayraktar et al., 2021). Similarly, Lam and colleagues firstly performed a clustering on RNAseq data from different sources and then applied tINIT algorithm on a brain-specific metabolic model to reconstruct cluster-specific metabolic models for AD and Parkinson's disease (PD) and performed flux balance analysis using these metabolic models (Lam et al., 2021). Then, they compared flux states of reactions in each cluster and observed different flux distributions in sex-hormone-related pathways between AD-PD and control clusters. In another study, researchers constructed genome-scale metabolic models for model organisms (Wang et al., 2021a). They also integrated RNAseq data from the brains of different experimental transgenic AD mice with mouse GEM to create AD-specific mouse metabolic models. Using these genetic strain-specific mouse metabolic models, they documented metabolic differences between genetic models, such as APP overexpression and Trem2 knockout mice (Wang et al., 2021a).

Reporter pathway analyses and reporter metabolite analyses are other approaches that are based on the integrative analysis of transcriptome data and GEMs. Reporter metabolite analysis (RMA) is a graph-theoretical strategy of data analysis to uncover the transcriptional regulatory architecture of metabolic networks. It uses topological information of the gene-metabolite network based on gene-reaction relationships in genome-scale metabolic models. Using transcriptome data, it identifies the metabolites around which the most important transcriptional changes occur in the gene-metabolite network (Patil & Nielsen, 2005). Turkez et al. performed RMA using RNAseq data of postmortem brain samples of AD and healthy subjects and the brain-specific metabolic model *iBrain2845* and revealed energy metabolism-related metabolites as reporter metabolites (Turkez et al., 2023). Moreover, Wang et al. (2021a) applied RMA using the mouse genome-scale metabolic model and RNAseq data of each transgenic mouse model. They found that the lysosomal pathway was affected in APP overexpression animal models and identified some metabolites in this pathway as reporter metabolites. Lam et al. (2021) also performed RMA for each cluster they identified from AD-PD transcriptomic datasets by using cluster-specific metabolic models. Similar to their flux balance analysis results, they found sex-hormone-related metabolites such as androsterone and pregnanediol as reporter metabolites common to all clusters. Reporter pathway analysis (RPA) is an algorithm developed based on RMA. Here, a pathway-metabolite network is created, and this network is used to identify pathways associated with the most important transcriptional changes based on RMA results (Çakır, 2015). Moolamalla and Vinod used both RPA and RMA in three different neuropsychiatric disorders to find biomarkers for these diseases (Moolamalla & Vinod, 2020).

Another research area that uses GEMs and transcriptome data together is drug target prediction algorithms. Metabolic Transformation Algorithm (MTA) was introduced in 2013 to predict genes whose deletion converts disease metabolic state to healthy metabolic state using transcriptome data and GEMs (Yizhak et al., 2013). The MTA algorithm uses the iMAT algorithm first to create a disease state metabolic model and calculate the flux values of the reactions in this model. Second, the changes in the expressions of genes between healthy and disease states are calculated from transcriptome data to be used as input to the MTA algorithm. Then, by applying systematic gene knockout simulations for each gene in the GEM, the gene knockouts that bring the disease-state flux values closer to the healthy state are identified (Yizhak et al., 2013). Stempler et al. (2014) used the MTA and detected candidate drug target pathways for AD. They found vitamin D, nucleotides, and steroid metabolism as the top-scoring candidate pathways. Later, Valcarcel et al. (2019) have proposed an upgraded version of MTA, called robust MTA. They applied robust MTA to the same AD transcriptome data analyzed in Stempler et al. (2014) and reported that Solute Carrier Family member genes could be target genes to reverse AD metabolism to healthy state.

6.2 Protein-Protein Interaction (PPI) Networks

PPI networks are structures that consist of multiple proteins and their physical interactions. There are many databases that store PPI interactions, and most of them include experimentally validated or computationally predicted PPIs (Kuzmanov & Emili, 2013). The interactions between proteins can be detected with many experimental methods. For example, yeast two-hybrid detects binary protein interactions, affinity-purification mass spectrometry detects co-membership in protein complexes

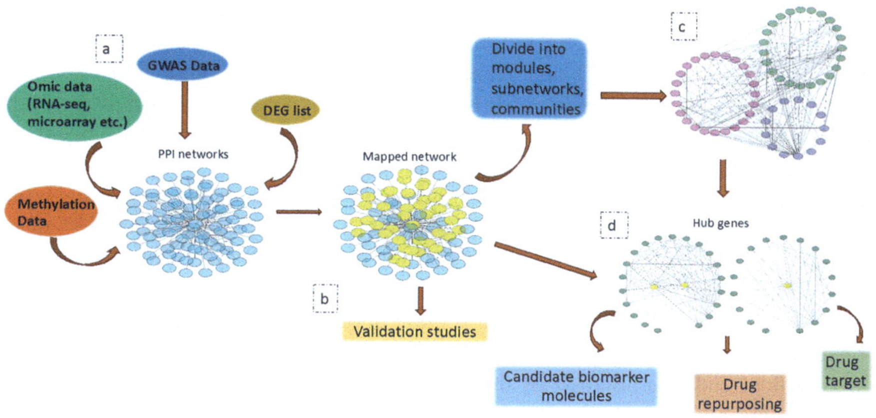

Fig. 6 Different usages of RNAseq data and PPI networks together. (**a**) Different types of data can be mapped to the PPI networks. (**b**) Mapped network can be used to validate co-expression interactions. (**c**) Mapped network can be divided into subgroups with algorithms. (**d**) Hub genes from subgroups of mapped network can be used to identify candidate biomarker molecules, drug repurposing, and drug target studies

(Nibbe et al., 2011), NMR spectroscopy and affinity chromatography detect weak interactions between proteins, and coimmunoprecipitation detects the interaction of complexes by targeting one of them (Rao et al., 2014).

PPI network-based analyses are important because they provide information about physical and biochemical interactions within the cell (Fig. 6) (Vrahatis et al., 2020). The most recent version of human PPI network based on BioGrid Database (version 4.4.221) (Oughtred et al., 2021) includes 27,611 proteins and 928,279 physical interactions between them. Disruption in PPI networks such as changes in interactions, or emergence/disappearance of interactions, can create changes in functionality, which can trigger diseases (Kuzmanov & Emili, 2013).

By integrating omics data with PPI networks, subnetworks, communities or modules can be created via subnetwork discovery algorithms to discover disease-altered parts of the PPI network (Vrahatis et al., 2020). These subnetworks facilitate research by reducing the set of disease-affected proteins and their interactions from the large PPI interaction network to smaller, focused, and meaningful subsets. In addition, disease-specific subnetworks bring a new perspective to molecules whose function is not known (guilt by association), as well as to molecules whose functions are known but have not yet been found to be related to the condition or disease of interest (Nibbe et al., 2011).

There are many different usages of PPI network in the literature that combines AD RNAseq data with organism-specific human PPI network, and one common approach is mapping differentially expressed genes (DEGs) on PPI network. Song and colleagues performed deconvolution analysis to specify cell types from the blood RNAseq data and estimated DEGs by comparing each of them with control groups. They built a PPI network for DEGs, and this PPI was then divided into modules, which were shown to be enriched with AD-related terms (Song et al., 2022). In another study, AD and Huntington's disease were compared to find a relation between their trigger mechanisms. DEG analysis was performed for both of them separately, and a PPI network of the common DEGs was created. They then divided the DEG-specific PPI network into modules to integrate them with miRNA interactions. Two drug molecules were associated with one of the hub proteins in this module based on DrugBank and can be used as a treatment for AD (Mahbub et al., 2022). In another study, CIBERSORT (Chen et al., 2018) algorithm was used to define cell types from blood RNAseq data. DEGs from 12 cell types were mapped on a PPI network, and the hub genes in these subnetworks were identified as AD biomarkers (Shigemizu et al., 2020). Beckmann and colleagues used proteomic and genomic data and combination of them, and they created co-expression networks and modules. DEGs were combined with the network information from KEGG and human PPI, and hub genes were identified to see if they regulate known AD genes. *VGF* was identified in the networks as a key regulator in AD and validated experimentally (Beckmann et al., 2020). Mapping DEGs to PPI network was also used by Rooij and colleagues (van Rooij et al., 2019). They divided DEG-PPI network into modules by Markov Clustering Algorithm, and hub genes in these clusters were compared with known AD genes to find AD-related genes and pathways (van Rooij et al., 2019). Miyashita and colleagues divided AD RNAseq data into four Braak stages and compared them to obtain DEGs. The DEGs were mapped to a PPI network, and their interactions were identified as candidate genes for AD (Miyashita et al., 2014).

GWAS data are also commonly used in PPI integration. Rosenthal and colleagues collected AD-related gene list from GWAS studies and created a PPI network for these candidate genes. Network propagation and network proximity approaches were used to increase the connectivity of this network, and Louvain (Blondel et al., 2008) and CDAPS algorithms (Singhal et al., 2020) were used to divide the network into more connected communities. This method was repeated for RNAseq AD data (Rosenthal et al., 2022). DEGs were calculated and mapped to the PPI network, and communities were detected using the Louvain algorithm (Rosenthal et al., 2022). In another study, RNAseq and GWAS data were used to find rank-based biomarkers by SCUDO (Lauria et al., 2015) algorithm. These biomarkers were then mapped to the PPI network, and subnetworks were created with the NetWalker algorithm (Komurov et al., 2012) to detect AD-related SNPs (Caberlotto et al., 2016). In another study, Luo and colleagues created a disease prediction algorithm that combines PPI data, clinical RNAseq data for AD, and OMIM data. They created differential co-expression network by comparing control and disease expression data and combined the results with OMIM data and PPI network. Genes were scored with logistic regression and a model was trained. They predicted candidate genes for AD (Luo et al., 2019).

Some studies use tissue-specific PPI networks. Canchi and colleagues calculated DEGs from RNAseq data and combined them with brain-specific interactome and divided this network into communities with Louvain algorithm and focused on common transcription factors to understand the disease mechanism better (Canchi et al., 2019).

The epigenetic mechanisms alter gene activity without changing DNA sequence but regulate gene expression, and most of them trigger disease mechanisms (Aboud et al., 2020). In Kim and colleagues' study, they created co-methylation networks from methylation data. Hub genes in these networks and differentially expressed genes from RNAseq data were combined and mapped into PPI network, and the mapped network was used for enrichment analysis. Enrichment analysis showed that genes are related to the AD mechanisms (Kim et al., 2023). In another study, RNAseq data, methylation data, and PPI data were used to create co-functional brain network called ADBrainNexus. This brain network was made specific for 147 known AD-related genes based on the literature. A machine learning algorithm was used for the prediction of probable AD genes by using this network. Enrichment analysis proved that the genes in this AD-brain specific network were related to AD mechanisms (Lin et al., 2021). Another study used methylation and RNAseq data to see the effect of 5-hydroxymethylcytosine modification on AD. Differentially methylated regions were identified, and 140 of them showed change in expression levels. AD-related genes, differentially methylated regions and differentially expressed genes were given as an input to DAPPLE (Trost et al., 2013) tool, and direct and indirect PPI networks were created for these genes. Hub genes were identified and known AD-related genes were expanded with these genes, and epigenetic modifiers of AD mechanism were enlightened (Bernstein et al., 2016). Another study used AD with psychiatric disorder-related data, and first, differentially methylated positions and genes were identified between genders, and SNPs information was obtained. Then, expression data from RNAseq analysis for these genes were combined with PPI data and divided into modules with functional epigenetic modules algorithm (Xia et al., 2021). In another study, LeafCutter algorithm (Li et al., 2018a) was used to estimate the "percentage spliced-in" values (PSI) from RNAseq data. This analysis reveals spliced intron loci of 21 genes. The PPI network for these 21 genes was created and divided into the modules (Raj et al., 2018).

There are also multiomics AD datasets in the literature, and network-based analyses were performed for these datasets. AlzCode is one of them, and it includes different types of omics data (single cell, proteome, transcriptome) and network information from different sources (co-expression, miRNA, PPI etc.) together with some extra information (neuropathological and clinical traits). This information was combined within the tool, and user can search the relation of a gene or sets of gene with AD (Lin et al., 2022). AlzGps is another database based on multiomics datasets from the literature, and the developers curated and combined them with different types of networks and created disease-specific subnetworks and modules (Zhou et al., 2021). Users can search for any genes in this database and can obtain PPI network for these genes and their neighbors, drug targets, the number of studies that reported that gene as significantly changed for AD and general information for that gene (Zhou et al., 2021).

AD omics data were used for drug repurposing and drug candidate analysis. For example, RNAseq, microarray, and scRNAseq datasets were used together to find drug candidates for AD (Zhou et al., 2022). In that study, a co-expression network was first created from healthy brain samples and divided into modules with the WGCNA algorithm (Langfelder & Horvath, 2008). Two of these co-expression modules were found to be enriched in terms of AD-related genes from GWAS studies in the literature. Then, they used scRNAseq data and showed that one of the modules was enriched with inhibitory neurons while the other was enriched with excitatory neurons. They used Connectivity Map drug transcriptome database (Lamb et al., 2006) to identify drug candidates that could reverse the expression of the genes in these modules, and Gly-His-Lys acetate salt (GHK) was the top candidate. Later, they extracted protein interactions of the genes in these modules from the human PPI network to identify hub genes. They experimentally showed that GHK provides a neuroprotective effect in the cell line models of AD, and it restores the dysregulation of the first 10 hub genes (Zhou et al., 2022).

Some studies use PPI network for validation. In Zhang and Wong's study, they used different correlation methods to create a co-expression network from AD RNAseq data and divided these networks into modules. Then, a PPI network was created by using STRING (Szklarczyk et al., 2018) database and hub genes were identified. The interaction of these hub genes in the largest co-expression module was checked for validation, and this module was found to be enriched with known AD-related genes and mechanisms (Zhang & Wong, 2022). In another study, a human PPI network was used for validating AD variants and their interactions that were obtained via network propagation approach (Scelsi et al., 2021). Bayraktar and colleagues used human PPI network to support their findings. They created co-expression networks with WGCNA algorithm by using AD RNAseq data and performed metabolic network analysis with same data separately. The important genes in these networks and their interactions were verified by using the PPI network. (Bayraktar et al., 2021). Tasaki and colleagues used RNAseq data and measured expression changes and protein abundance with a machine learning algorithm, and they compared predicted and real protein data by using co-expression modules. The PPI network was used to validate the co-expression interactions in their study (Tasaki et al., 2022).

7 Future Remarks

Since the completion of the human genome project, the transcriptome data have been increasingly used in the literature. RNAseq data especially have a wide range of uses since they are based on gene sequences, allowing more diverse and deeper analyses ranging from expression alterations to genetic variants and abnormal splicing events. In this chapter, we reviewed different applications of RNAseq data in AD studies. RNAseq data still have some technical problems (such as library preparation, short read lengths) and practical confusion (such as posttranslational modifications). However, with the application of artificial intelligence techniques and the widespread use of machines that can read longer sequences, these technical and practical problems will be reduced in the future. Since RNAseq data allow the identification of disease subtypes and cell types, analysis of variants, splicing events, affected protein-protein interactions, and metabolic pathways from a single sample, it will shed light on future studies to understand the complexity of AD.

Acknowledgments This work was financially supported by TUBITAK, The Scientific and Technological Research Council of Turkey (Project Code: 120S824). Also, Dilara Uzuner is funded by TUBITAK 2211-A PhD Student Scholarship and Council of Higher Education (CoHE) 100/2000 PhD Scholarship Programmes.

Conflicts of Interest The authors declare that they have no competing interests.

References

Aboud, N. M. A., Simpson, B. N., & Jialal, I. (2020). *Genetics, epigenetic mechanism*. StatPearls.

Agren, R., Mardinoglu, A., Asplund, A., Kampf, C., Uhlen, M., & Nielsen, J. (2014). Identification of anticancer drugs for hepatocellular carcinoma through personalized genome-scale metabolic modeling. *Molecular Systems Biology, 10*(3), 721.

Akila Parvathy Dharshini, S., Taguchi, Y. H., & Michael Gromiha, M. (2019). Exploring the selective vulnerability in Alzheimer disease using tissue specific variant analysis. *Genomics, 111*(4), 936–949. https://doi.org/10.1016/j.ygeno.2018.05.024

Alamancos, G. P., Pagès, A., Trincado, J. L., Bellora, N., & Eyras, E. (2015). Leveraging transcript quantification for fast computation of alternative splicing profiles. *RNA, 21*(9), 1521–1531. https://doi.org/10.1261/rna.051557.115

Allen, M., Carrasquillo, M. M., Funk, C., Heavner, B. D., Zou, F., Younkin, C. S., et al. (2016). Human whole genome genotype and transcriptome data for Alzheimer's and other neurodegenerative diseases. *Scientific Data, 3*, 160089. https://doi.org/10.1038/sdata.2016.89

Anders, S., Reyes, A., & Huber, W. (2012). Detecting differential usage of exons from RNA-seq data. *Genome Research, 22*(10), 2008–2017. https://doi.org/10.1101/gr.133744.111

Bagyinszky, E., Giau, V. V., & An, S. A. (2020). Transcriptomics in Alzheimer's disease: Aspects and challenges. *International Journal of Molecular Sciences, 21*(10). https://doi.org/10.3390/ijms21103517

Bakhtiarizadeh, M. R., & Alamouti, A. A. (2020). RNA-Seq based genetic variant discovery provides new insights into controlling fat deposition in the tail of sheep. *Scientific Reports, 10*(1), 1–13.

Baloni, P., Funk, C. C., Yan, J., Yurkovich, J. T., Kueider-Paisley, A., Nho, K., et al. (2020). Metabolic network analysis reveals altered bile acid synthesis and metabolism in Alzheimer's disease. *Cell Reports Medicine, 1*(8), 100138. https://doi.org/10.1016/j.xcrm.2020.100138

Bayraktar, A., Lam, S., Altay, O., Li, X., Yuan, M., Zhang, C., et al. (2021). Revealing the molecular mechanisms of Alzheimer's disease based on network analysis. *International Journal of Molecular Sciences, 22*(21), 11556. Retrieved from https://www.mdpi.com/1422-0067/22/21/11556

Beckmann, N. D., Lin, W.-J., Wang, M., Cohain, A. T., Charney, A. W., Wang, P., et al. (2020). Multiscale causal networks identify VGF as a key regulator of Alzheimer's disease. *Nature Communications, 11*(1), 3942. https://doi.org/10.1038/s41467-020-17405-z

Bernstein, A. I., Lin, Y., Street, R. C., Lin, L., Dai, Q., Yu, L., et al. (2016). 5-Hydroxymethylation-associated epigenetic modifiers of Alzheimer's disease modulate Tau-induced neurotoxicity. *Human Molecular Genetics, 25*(12), 2437–2450. https://doi.org/10.1093/hmg/ddw109

Biamonti, G., Amato, A., Belloni, E., Di Matteo, A., Infantino, L., Pradella, D., & Ghigna, C. (2021). Alternative splicing in Alzheimer's disease. *Aging Clinical and Experimental Research, 33*(4), 747–758. https://doi.org/10.1007/s40520-019-01360-x

Blondel, V. D., Guillaume, J.-L., Lambiotte, R., & Lefebvre, E. (2008). Fast unfolding of communities in large networks. *Journal of Statistical Mechanics: Theory and Experiment, 2008*(10), P10008. https://doi.org/10.1088/1742-5468/2008/10/P10008

Brouard, J.-S., & Bissonnette, N. (2022). Variant calling from RNA-seq data using the GATK joint genotyping workflow. In C. Ng & S. Piscuoglio (Eds.), *Variant calling: Methods and protocols* (pp. 205–233). Springer.

Caberlotto, L., Marchetti, L., Lauria, M., Scotti, M., & Parolo, S. (2016). Integration of transcriptomic and genomic data suggests candidate mechanisms for APOE4-mediated pathogenic action in Alzheimer's disease. *Scientific Reports, 6*(1), 32583. https://doi.org/10.1038/srep32583

Çakır, T. (2015). Reporter pathway analysis from transcriptome data: Metabolite-centric versus reaction-centric approach. *Scientific Reports, 5*(1), 14563. https://doi.org/10.1038/srep14563

Canchi, S., Raao, B., Masliah, D., Rosenthal, S. B., Sasik, R., Fisch, K. M., et al. (2019). Integrating gene and protein expression reveals perturbed functional networks in Alzheimer's disease. *Cell Reports, 28*(4), 1103–1116.e1104. https://doi.org/10.1016/j.celrep.2019.06.073

Chen, B., Khodadoust, M. S., Liu, C. L., Newman, A. M., & Alizadeh, A. A. (2018). Profiling tumor infiltrating immune cells with CIBERSORT. In L. von Stechow (Ed.), Cancer systems biology: Methods and protocols (pp. 243-259). : Springer New York.

Chen, L., Wu, C. T., Lin, C. H., Dai, R., Liu, C., Clarke, R., et al. (2021). swCAM: Estimation of subtype-specific expressions in individual samples with unsupervised sample-wise deconvolution. *Bioinformatics*. https://doi.org/10.1093/bioinformatics/btab839

Clough, E., & Barrett, T. (2016). The gene expression omnibus database. In *Statistical genomics* (pp. 93–110). Springer.

De Jager, P. L., Ma, Y., McCabe, C., Xu, J., Vardarajan, B. N., Felsky, D., et al. (2018). A multi-omic atlas of the human frontal cortex for aging and Alzheimer's disease research. *Scientific Data, 5*, 180142. https://doi.org/10.1038/sdata.2018.142

Del-Aguila, J. L., Li, Z., Dube, U., Mihindukulasuriya, K. A., Budde, J. P., Fernandez, M. V., et al. (2019). A single-nuclei RNA sequencing study of Mendelian and sporadic AD in the human brain. *Alzheimer's Research & Therapy, 11*(1), 71. https://doi.org/10.1186/s13195-019-0524-x

Deschênes, M., & Chabot, B. (2017). The emerging role of alternative splicing in senescence and aging. *Aging Cell, 16*(5), 918–933. https://doi.org/10.1111/acel.12646

Douglas, A. G., & Wood, M. J. (2011). RNA splicing: Disease and therapy. *Briefings in Functional Genomics, 10*(3), 151–164. https://doi.org/10.1093/bfgp/elr020

Garcia-Escudero, V., Ruiz-Gabarre, D., Gargini, R., Perez, M., Garcia, E., Cuadros, R., et al. (2021). A new non-aggregative splicing isoform of human Tau is decreased in Alzheimer's disease. *Acta Neuropathologica, 142*(1), 159–177. https://doi.org/10.1007/s00401-021-02317-z

Gerrits, E., Brouwer, N., Kooistra, S. M., Woodbury, M. E., Vermeiren, Y., Lambourne, M., et al. (2021). Distinct amyloid-β and tau-associated microglia profiles in Alzheimer's disease. *Acta Neuropathologica, 141*(5), 681–696. https://doi.org/10.1007/s00401-021-02263-w

Gockley, J., Montgomery, K. S., Poehlman, W. L., Wiley, J. C., Liu, Y., Gerasimov, E., et al. (2021). Multi-tissue neocortical transcriptome-wide association study implicates 8 genes across 6 genomic loci in Alzheimer's disease. *Genome Medicine, 13*(1), 76. https://doi.org/10.1186/s13073-021-00890-2

Gong, T., & Szustakowski, J. D. (2013). DeconRNASeq: A statistical framework for deconvolution of heterogeneous tissue samples based on mRNA-seq data. *Bioinformatics, 29*(8), 1083–1085. https://doi.org/10.1093/bioinformatics/btt090

Grubman, A., Chew, G., Ouyang, J. F., Sun, G., Choo, X. Y., McLean, C., et al. (2019). A single-cell atlas of entorhinal cortex from individuals with Alzheimer's disease reveals cell-type-specific gene expression regulation. *Nature Neuroscience, 22*(12), 2087–2097. https://doi.org/10.1038/s41593-019-0539-4

Han, Y., Gao, S., Muegge, K., Zhang, W., & Zhou, B. (2015). Advanced applications of RNA sequencing and challenges. *Bioinformatics and Biology Insights, 9s1*, BBI.S28991. https://doi.org/10.4137/BBI.S28991

Han, S., Nho, K., & Lee, Y. (2020). Alternative splicing regulation of an Alzheimer's risk variant in CLU. *International Journal of Molecular Sciences, 21*(19). https://doi.org/10.3390/ijms21197079

Han, S., Na, Y., Koh, I., Nho, K., & Lee, Y. (2021). Alternative splicing regulation of low-frequency genetic variants in exon 2 of TREM2 in Alzheimer's disease by splicing-based aggregation. *International Journal of Molecular Sciences, 22*(18). https://doi.org/10.3390/ijms22189865

Hartley, S. W., & Mullikin, J. C. (2016). Detection and visualization of differential splicing in RNA-seq data with JunctionSeq. *Nucleic Acids Research, 44*(15), e127–e127. https://doi.org/10.1093/nar/gkw501

Hu, Y., Huang, Y., Du, Y., Orellana, C. F., Singh, D., Johnson, A. R., et al. (2012). DiffSplice: the genome-wide detection of differential splicing events with RNA-seq. *Nucleic Acids Research, 41*(2), e39–e39. https://doi.org/10.1093/nar/gks1026

Hu, Y., Zhao, T., Zang, T., Zhang, Y., & Cheng, L. (2018). Identification of Alzheimer's disease-related genes based on data integration method. *Frontiers in Genetics, 9*, 703. https://doi.org/10.3389/fgene.2018.00703

Hunt, G. J., Freytag, S., Bahlo, M., & Gagnon-Bartsch, J. A. (2018). dtangle: Accurate and robust cell type deconvolution. *Bioinformatics, 35*(12), 2093–2099. https://doi.org/10.1093/bioinformatics/bty926

Jehl, F., Degalez, F., Bernard, M., Lecerf, F., Lagoutte, L., Desert, C., et al. (2021). RNA-seq data for reliable SNP detection and genotype calling: Interest for coding variant characterization and cis-regulation analysis by allele-specific expression in livestock species. *Frontiers in Genetics, 12*, 655707. https://doi.org/10.3389/fgene.2021.655707

Kim, J. P., Kim, B.-H., Bice, P. J., Seo, S. W., Bennett, D. A., Saykin, A. J., & Nho, K. (2023). Integrative co-methylation network analysis identifies novel DNA methylation signatures and their target genes in Alzheimer's disease. *Biological Psychiatry*. https://doi.org/10.1016/j.biopsych.2022.06.020

Komurov, K., Dursun, S., Erdin, S., & Ram, P. T. (2012). NetWalker: A contextual network analysis tool for functional genomics. *BMC Genomics, 13*(1), 282. https://doi.org/10.1186/1471-2164-13-282

Kuzmanov, U., & Emili, A. (2013). Protein-protein interaction networks: Probing disease mechanisms using model systems. *Genome Medicine, 5*(4), 37. https://doi.org/10.1186/gm441

Lam, S., Hartmann, N., Benfeitas, R., Zhang, C., Arif, M., Turkez, H., et al. (2021). Systems analysis reveals ageing-related perturbations in retinoids and sex hormones in Alzheimer's and Parkinson's diseases. *Biomedicine, 9*(10). https://doi.org/10.3390/biomedicines9101310

Lamb, J., Crawford, E. D., Peck, D., Modell, J. W., Blat, I. C., Wrobel, M. J., et al. (2006). The connectivity map: Using gene-expression signatures to connect small molecules, genes, and disease. *Science, 313*(5795), 1929–1935. https://doi.org/10.1126/science.1132939

Langfelder, P., & Horvath, S. (2008). WGCNA: An R package for weighted correlation network analysis. *BMC Bioinformatics, 9*(1), 559. https://doi.org/10.1186/1471-2105-9-559

Lauria, M., Moyseos, P., & Priami, C. (2015). SCUDO: A tool for signature-based clustering of expression profiles. *Nucleic Acids Research, 43*(W1), W188–W192. https://doi.org/10.1093/nar/gkv449

Leng, K., Li, E., Eser, R., Piergies, A., Sit, R., Tan, M., et al. (2021). Molecular characterization of selectively vulnerable neurons in Alzheimer's disease. *Nature Neuroscience, 24*(2), 276–287. https://doi.org/10.1038/s41593-020-00764-7

Lezmi, E., & Benvenisty, N. (2021). Identification of cancer-related mutations in human pluripotent stem cells using RNA-seq analysis. *Nature Protocols, 16*(9), 4522–4537.

Li, Z., & Wu, H. (2019). TOAST: Improving reference-free cell composition estimation by cross-cell type differential analysis. *Genome Biology, 20*(1), 190. https://doi.org/10.1186/s13059-019-1778-0

Li, Z., Del-Aguila, J. L., Dube, U., Budde, J., Martinez, R., Black, K., et al. (2018a). Genetic variants associated with Alzheimer's disease confer different cerebral cortex cell-type population structure. *Genome Medicine, 10*(1), 43. https://doi.org/10.1186/s13073-018-0551-4

Li, Y. I., Knowles, D. A., Humphrey, J., Barbeira, A. N., Dickinson, S. P., Im, H. K., & Pritchard, J. K. (2018b). Annotation-free quantification of RNA splicing using LeafCutter. *Nature Genetics, 50*(1), 151–158. https://doi.org/10.1038/s41588-017-0004-9

Li, H.-D., Funk, C. C., McFarland, K., Dammer, E. B., Allen, M., Carrasquillo, M. M., et al. (2021a). Integrative functional genomic analysis of intron retention in human and mouse brain with Alzheimer's disease. *Alzheimer's & Dementia, 17*(6), 984–1004. https://doi.org/10.1002/alz.12254

Li, D., McIntosh, C. S., Mastaglia, F. L., Wilton, S. D., & Aung-Htut, M. T. (2021b). Neurodegenerative diseases: A hotbed for splicing defects and the potential therapies. *Translational Neurodegeneration, 10*(1), 16. https://doi.org/10.1186/s40035-021-00240-7

Lin, C.-X., Li, H.-D., Deng, C., Liu, W., Erhardt, S., Wu, F.-X., et al. (2021). An integrated brain-specific network identifies genes associated with neuropathologic and clinical traits of Alzheimer's disease. *Briefings in Bioinformatics, 23*(1). https://doi.org/10.1093/bib/bbab522

Lin, C.-X., Li, H.-D., Deng, C., Erhardt, S., Wang, J., Peng, X., & Wang, J. (2022). AlzCode: A platform for multiview analysis of genes related to Alzheimer's disease. *Bioinformatics, 38*(7), 2030–2032. https://doi.org/10.1093/bioinformatics/btac033

Liu, C.-C., Murray, M. E., Li, X., Zhao, N., Wang, N., Heckman, M. G., et al. (2021). APOE3-Jacksonville (V236E) variant reduces self-aggregation and risk of dementia. *Science Translational Medicine, 13*(613), eabc9375. https://doi.org/10.1126/scitranslmed.abc9375

Luo, P., Tian, L. P., Ruan, J., & Wu, F. X. (2019). Disease gene prediction by integrating PPI networks, clinical RNA-Seq data and OMIM data. *IEEE/ACM Transactions on Computational Biology and Bioinformatics, 16*(1), 222–232. https://doi.org/10.1109/TCBB.2017.2770120

Mahbub, N. I., Hasan, M. I., Rahman, M. H., Naznin, F., Islam, M. Z., & Moni, M. A. (2022). Identifying molecular signatures and pathways shared between Alzheimer's and Huntington's disorders: A bioinformatics and systems biology approach. *Informatics in Medicine Unlocked, 30*, 100888. https://doi.org/10.1016/j.imu.2022.100888

Marco-Puche, G., Lois, S., Benítez, J., & Trivino, J. C. (2019). RNA-Seq perspectives to improve clinical diagnosis. *Frontiers in Genetics, 10*. https://doi.org/10.3389/fgene.2019.01152

Mathys, H., Davila-Velderrain, J., Peng, Z., Gao, F., Mohammadi, S., Young, J. Z., et al. (2019). Single-cell transcriptomic analysis of Alzheimer's disease. *Nature, 570*(7761), 332–337. https://doi.org/10.1038/s41586-019-1195-2

Mehmood, A., Laiho, A., Venäläinen, M. S., McGlinchey, A. J., Wang, N., & Elo, L. L. (2020). Systematic evaluation of differential splicing tools for RNA-seq studies. *Briefings in Bioinformatics, 21*(6), 2052–2065. https://doi.org/10.1093/bib/bbz126

Miyashita, A., Hatsuta, H., Kikuchi, M., Nakaya, A., Saito, Y., Tsukie, T., et al. (2014). Genes associated with the progression of neurofibrillary tangles in Alzheimer's disease. *Translational Psychiatry, 4*(6), e396–e396. https://doi.org/10.1038/tp.2014.35

Moolamalla, S. T. R., & Vinod, P. K. (2020). Genome-scale metabolic modelling predicts biomarkers and therapeutic targets for neuropsychiatric disorders. *Computers in Biology and Medicine, 125*, 103994. https://doi.org/10.1016/j.compbiomed.2020.103994

Murray, M. E., Graff-Radford, N. R., Ross, O. A., Petersen, R. C., Duara, R., & Dickson, D. W. (2011). Neuropathologically defined subtypes of Alzheimer's disease with distinct clinical characteristics: A retrospective study. *The Lancet Neurology, 10*(9), 785–796. https://doi.org/10.1016/S1474-4422(11)70156-9

Neff, R. A., Wang, M., Vatansever, S., Guo, L., Ming, C., Wang, Q., et al. (2021). Molecular subtyping of Alzheimer's disease using RNA sequencing data reveals novel mechanisms and targets. *Science Advances, 7*(2). https://doi.org/10.1126/sciadv.abb5398

Neuner, S. M., Tcw, J., & Goate, A. M. (2020). Genetic architecture of Alzheimer's disease. *Neurobiology of Disease, 143*, 104976. https://doi.org/10.1016/j.nbd.2020.104976

Newman, A. M., Liu, C. L., Green, M. R., Gentles, A. J., Feng, W., Xu, Y., et al. (2015). Robust enumeration of cell subsets from tissue expression profiles. *Nature Methods, 12*(5), 453–457. https://doi.org/10.1038/nmeth.3337

Newman, A. M., Steen, C. B., Liu, C. L., Gentles, A. J., Chaudhuri, A. A., Scherer, F., et al. (2019). Determining cell type abundance and expression from bulk tissues with digital cytometry. *Nature Biotechnology, 37*(7), 773–782. https://doi.org/10.1038/s41587-019-0114-2

Nguyen, A. T., Wang, K., Hu, G., Wang, X., Miao, Z., Azevedo, J. A., et al. (2020). APOE and TREM2 regulate amyloid-responsive microglia in Alzheimer's disease. *Acta Neuropathologica, 140*(4), 477–493. https://doi.org/10.1007/s00401-020-02200-3

Nibbe, R. K., Chowdhury, S. A., Koyutürk, M., Ewing, R., & Chance, M. R. (2011). Protein–protein interaction networks and subnetworks in the biology of disease. *WIREs Systems Biology and Medicine, 3*(3), 357–367. https://doi.org/10.1002/wsbm.121

Olah, M., Patrick, E., Villani, A.C., et al. (2018). A transcriptomic atlas of aged human microglia. *Nat Commun, 9*, 539. https://doi.org/10.1038/s41467-018-02926-5

Oughtred, R., Rust, J., Chang, C., Breitkreutz, B. J., Stark, C., Willems, A., et al. (2021). The BioGRID database: A comprehensive biomedical resource of curated protein, genetic, and chemical interactions. *Protein Science, 30*(1), 187–200. https://doi.org/10.1002/pro.3978

Patil, K. R., & Nielsen, J. (2005). Uncovering transcriptional regulation of metabolism by using metabolic network topology. *Proceedings of the national academy of sciences, 102*(8), 2685–2689. https://www.pnas.org/doi/full/10.1073/pnas.0406811102

Piras, I. S., Bleul, C., Talboom, J. S., De Both, M. D., Schrauwen, I., Halliday, G., et al. (2020). ESHRD: Deconvolution of brain homogenate RNA expression data to identify cell-type-specific alterations in Alzheimer's disease. *Aging (Albany NY), 12*(5), 4124–4162. https://doi.org/10.18632/aging.102840

Prokopenko, D., Morgan, S. L., Mullin, K., Hofmann, O., Chapman, B., Kirchner, R., et al. (2021). Whole-genome sequencing reveals new Alzheimer's disease–associated rare variants in loci related to synaptic function and neuronal development. *Alzheimer's & Dementia, 17*(9), 1509–1527. https://doi.org/10.1002/alz.12319

Raj, T., Li, Y. I., Wong, G., Humphrey, J., Wang, M., Ramdhani, S., et al. (2018). Integrative transcriptome analyses of the aging brain implicate altered splicing in Alzheimer's disease susceptibility. *Nature Genetics, 50*(11), 1584–1592. https://doi.org/10.1038/s41588-018-0238-1

Rao, V. S., Srinivas, K., Sujini, G. N., & Kumar, G. N. (2014). Protein-protein interaction detection: Methods and analysis. *International Journal of Proteomics, 2014*, 147648. https://doi.org/10.1155/2014/147648

Ritchie, M. E., Phipson, B., Wu, D., Hu, Y., Law, C. W., Shi, W., & Smyth, G. K. (2015). limma powers differential expression analyses for RNA-sequencing and microarray studies. *Nucleic Acids Research, 43*(7), e47–e47. https://doi.org/10.1093/nar/gkv007

Robinson, M. D., McCarthy, D. J., & Smyth, G. K. (2009). edgeR: A Bioconductor package for differential expression analysis of digital gene expression data. *Bioinformatics, 26*(1), 139–140. https://doi.org/10.1093/bioinformatics/btp616

Robinson, J. L., Kocabaş, P., Wang, H., Cholley, P.-E., Cook, D., Nilsson, A., et al. (2020). An atlas of human metabolism. *Science Signaling, 13*(624), eaaz1482. https://doi.org/10.1126/scisignal.aaz1482

Rosenthal, S. B., Wang, H., Shi, D., Liu, C., Abagyan, R., McEvoy, L. K., & Chen, C.-H. (2022). Mapping the gene network landscape of Alzheimer's disease through integrating genomics and transcriptomics. *PLoS Computational Biology, 18*(2), e1009903. https://doi.org/10.1371/journal.pcbi.1009903

Sadick, J. S., O'Dea, M. R., Hasel, P., Dykstra, T., Faustin, A., & Liddelow, S. A. (2022). Astrocytes and oligodendrocytes undergo subtype-specific transcriptional changes in Alzheimer's disease. *Neuron, 110*(11), 1788–1805.e1710. https://doi.org/10.1016/j.neuron.2022.03.008

Scelsi, M. A., Napolioni, V., Greicius, M. D., Altmann, A., & for the Alzheimer's Disease Neuroimaging Initiative Initiative, & the Alzheimer's Disease Sequencing Project. (2021). Network propagation of rare variants in Alzheimer's disease reveals tissue-specific hub genes

and communities. *PLoS Computational Biology, 17*(1), e1008517. https://doi.org/10.1371/journal.pcbi.1008517
Shen, S., Park, J. W., Lu, Z.-X., Lin, L., Henry, M. D., Wu, Y. N., et al. (2014). rMATS: Robust and flexible detection of differential alternative splicing from replicate RNA-seq data. *Proceedings of the National Academy of Sciences, 111*(51), E5593–E5601. https://doi.org/10.1073/pnas.1419161111
Shigemizu, D., Mori, T., Akiyama, S., Higaki, S., Watanabe, H., Sakurai, T., et al. (2020). Identification of potential blood biomarkers for early diagnosis of Alzheimer's disease through RNA sequencing analysis. *Alzheimer's Research & Therapy, 12*(1), 87. https://doi.org/10.1186/s13195-020-00654-x
Sims, R., Hill, M., & Williams, J. (2020). The multiplex model of the genetics of Alzheimer's disease. *Nature Neuroscience, 23*(3), 311–322. https://doi.org/10.1038/s41593-020-0599-5
Singhal, A., Cao, S., Churas, C., Pratt, D., Fortunato, S., Zheng, F., & Ideker, T. (2020). Multiscale community detection in cytoscape. *PLoS Computational Biology, 16*(10), e1008239. https://doi.org/10.1371/journal.pcbi.1008239
Song, L., Yang, Y. T., Guo, Q., Zhao, X.-M., & ZIB Consortium. (2022). Cellular transcriptional alterations of peripheral blood in Alzheimer's disease. *BMC Medicine, 20*(1), 266. https://doi.org/10.1186/s12916-022-02472-4
Stempler, S., Yizhak, K., & Ruppin, E. (2014). Integrating transcriptomics with metabolic modeling predicts biomarkers and drug targets for Alzheimer's disease. *PLoS One, 9*(8), e105383. https://doi.org/10.1371/journal.pone.0105383
Su, C. H., Dhananjaya, D., & Tarn, W. Y. (2018). Alternative splicing in neurogenesis and brain development. *Frontiers in Molecular Biosciences, 5*, 12. https://doi.org/10.3389/fmolb.2018.00012
Sutton, G. J., Poppe, D., Simmons, R. K., Walsh, K., Nawaz, U., Lister, R., et al. (2022). Comprehensive evaluation of deconvolution methods for human brain gene expression. *Nature Communications, 13*(1), 1358. https://doi.org/10.1038/s41467-022-28655-4
Szklarczyk, D., Gable, A. L., Lyon, D., Junge, A., Wyder, S., Huerta-Cepas, J., et al. (2018). STRING v11: Protein–protein association networks with increased coverage, supporting functional discovery in genome-wide experimental datasets. *Nucleic Acids Research, 47*(D1), D607–D613. https://doi.org/10.1093/nar/gky1131
Tasaki, S., Xu, J., Avey, D. R., Johnson, L., Petyuk, V. A., Dawe, R. J., et al. (2022). Inferring protein expression changes from mRNA in Alzheimer's dementia using deep neural networks. *Nature Communications, 13*(1), 655. https://doi.org/10.1038/s41467-022-28280-1
Trapnell, C., Williams, B. A., Pertea, G., Mortazavi, A., Kwan, G., van Baren, M. J., et al. (2010). Transcript assembly and quantification by RNA-Seq reveals unannotated transcripts and isoform switching during cell differentiation. *Nature Biotechnology, 28*(5), 511–515. https://doi.org/10.1038/nbt.1621
Trincado, J. L., Entizne, J. C., Hysenaj, G., Singh, B., Skalic, M., Elliott, D. J., & Eyras, E. (2018). SUPPA2: Fast, accurate, and uncertainty-aware differential splicing analysis across multiple conditions. *Genome Biology, 19*(1), 40. https://doi.org/10.1186/s13059-018-1417-1
Trost, B., Arsenault, R., Griebel, P., Napper, S., & Kusalik, A. (2013). DAPPLE: A pipeline for the homology-based prediction of phosphorylation sites. *Bioinformatics, 29*(13), 1693–1695. https://doi.org/10.1093/bioinformatics/btt265
Turkez, H., Altay, O., Yildirim, S., Li, X., Yang, H., Bayram, C., et al. (2023). Combined metabolic activators improve metabolic functions in the animal models of neurodegenerative diseases. *Life Sciences, 314*, 121325. https://doi.org/10.1016/j.lfs.2022.121325
Valcarcel, L. V., Torrano, V., Tobalina, L., Carracedo, A., & Planes, F. J. (2019). rMTA: Robust metabolic transformation analysis. *Bioinformatics, 35*(21), 4350–4355. https://doi.org/10.1093/bioinformatics/btz231
van Rooij, J. G. J., Meeter, L. H. H., Melhem, S., Nijholt, D. A. T., Wong, T. H., Rozemuller, A., et al. (2019). Hippocampal transcriptome profiling combined with protein-protein interaction analysis elucidates Alzheimer's disease pathways and genes. *Neurobiology of Aging, 74*, 225–233. https://doi.org/10.1016/j.neurobiolaging.2018.10.023

Vaquero-Garcia, J., Barrera, A., Gazzara, M. R., González-Vallinas, J., Lahens, N. F., Hogenesch, J. B., et al. (2016). A new view of transcriptome complexity and regulation through the lens of local splicing variations. *eLife, 5*, e11752. https://doi.org/10.7554/eLife.11752

Varma, V. R., Büşra Lüleci, H., Oommen, A. M., Varma, S., Blackshear, C. T., Griswold, M. E., et al. (2021). Abnormal brain cholesterol homeostasis in Alzheimer's disease—A targeted metabolomic and transcriptomic study. *npj Aging and Mechanisms of Disease, 7*(1), 11. https://doi.org/10.1038/s41514-021-00064-9

Vrahatis, A. G., Vlamos, P., Gonidi, M., Sagiadinou, M., & Avramouli, A. (2020 September, 25-27). *Network biomarkers for Alzheimer's disease via a graph-based approach.* Paper presented at the 2020 5th South-East Europe design automation, computer engineering, computer networks and social media conference (SEEDA-CECNSM).

Wang, M., Beckmann, N. D., Roussos, P., Wang, E., Zhou, X., Wang, Q., et al. (2018). The Mount Sinai cohort of large-scale genomic, transcriptomic and proteomic data in Alzheimer's disease. *Scientific Data, 5*, 180185. https://doi.org/10.1038/sdata.2018.185

Wang, X., Park, J., Susztak, K., Zhang, N. R., & Li, M. (2019). Bulk tissue cell type deconvolution with multi-subject single-cell expression reference. *Nature Communications, 10*(1), 380. https://doi.org/10.1038/s41467-018-08023-x

Wang, H., Robinson, J. L., Kocabas, P., Gustafsson, J., Anton, M., Cholley, P.-E., et al. (2021a). Genome-scale metabolic network reconstruction of model animals as a platform for translational research. *Proceedings of the National Academy of Sciences, 118*(30), e2102344118. https://doi.org/10.1073/pnas.2102344118

Wang, J., Roeder, K., & Devlin, B. (2021b). Bayesian estimation of cell type-specific gene expression with prior derived from single-cell data. *Genome Research, 31*(10), 1807–1818. https://doi.org/10.1101/gr.268722.120

Ward, A. J., & Cooper, T. A. (2010). The pathobiology of splicing. *The Journal of Pathology, 220*(2), 152–163. https://doi.org/10.1002/path.2649

Xia, Y., Dai, R., Wang, K., Jiao, C., Zhang, C., Xu, Y., et al. (2021). Sex-differential DNA methylation and associated regulation networks in human brain implicated in the sex-biased risks of psychiatric disorders. *Molecular Psychiatry, 26*(3), 835–848. https://doi.org/10.1038/s41380-019-0416-2

Yang, M., Ke, Y., Kim, P., & Zhou, X. (2021). ExonSkipAD provides the functional genomic landscape of exon skipping events in Alzheimer's disease. *Briefings in Bioinformatics, 22*(5). https://doi.org/10.1093/bib/bbaa438

Yizhak, K., Gabay, O., Cohen, H., & Ruppin, E. (2013). Model-based identification of drug targets that revert disrupted metabolism and its application to ageing. *Nature Communications, 4*, 2632. https://doi.org/10.1038/ncomms3632

Zaitsev, K., Bambouskova, M., Swain, A., & Artyomov, M. N. (2019). Complete deconvolution of cellular mixtures based on linearity of transcriptional signatures. *Nature Communications, 10*(1), 2209. https://doi.org/10.1038/s41467-019-09990-5

Zhang, T., & Wong, G. (2022). Gene expression data analysis using Hellinger correlation in weighted gene co-expression networks (WGCNA). *Computational and Structural Biotechnology Journal, 20*, 3851–3863. https://doi.org/10.1016/j.csbj.2022.07.018

Zhang, Q., Zhang, J., Ye, J., Li, X., Liu, H., Ma, X., et al. (2021). Nuclear speckle specific hnRNP D-like prevents age- and AD-related cognitive decline by modulating RNA splicing. *Molecular Neurodegeneration, 16*(1), 66. https://doi.org/10.1186/s13024-021-00485-w

Zhao, S. (2019). Alternative splicing, RNA-seq and drug discovery. *Drug Discovery Today, 24*(6), 1258–1267. https://doi.org/10.1016/j.drudis.2019.03.030

Zhao, B., Shan, Y., Yang, Y., Yu, Z., Li, T., Wang, X., et al. (2021). Transcriptome-wide association analysis of brain structures yields insights into pleiotropy with complex neuropsychiatric traits. *Nature Communications, 12*(1), 2878. https://doi.org/10.1038/s41467-021-23130-y

Zheng, C., & Xu, R. (2021). Molecular subtyping of Alzheimer's disease with consensus non-negative matrix factorization. *PLoS One, 16*(5), e0250278. https://doi.org/10.1371/journal.pone.0250278

Zhou, Y., Song, W. M., Andhey, P. S., Swain, A., Levy, T., Miller, K. R., et al. (2020). Human and mouse single-nucleus transcriptomics reveal TREM2-dependent and TREM2-independent cellular responses in Alzheimer's disease. *Nature Medicine, 26*(1), 131–142. https://doi.org/10.1038/s41591-019-0695-9

Zhou, Y., Fang, J., Bekris, L. M., Kim, Y. H., Pieper, A. A., Leverenz, J. B., et al. (2021). AlzGPS: A genome-wide positioning systems platform to catalyze multi-omics for Alzheimer's drug discovery. *Alzheimer's Research & Therapy, 13*(1), 24. https://doi.org/10.1186/s13195-020-00760-w

Zhou, J., Li, Q., Wu, W., Zhang, X., Zuo, Z., Lu, Y., et al. (2022). Discovery of novel drug candidates for Alzheimer's disease by molecular network modeling. *Frontiers in Aging Neuroscience, 14*, 850217. https://doi.org/10.3389/fnagi.2022.850217

Zhu, D., Deng, N., & Bai, C. (2015). A generalized dSpliceType framework to detect differential splicing and differential expression events using RNA-Seq. *IEEE Transactions on Nanobioscience, 14*(2), 192–202. https://doi.org/10.1109/TNB.2015.2388593

Zur, H., Ruppin, E., & Shlomi, T. (2010). iMAT: An integrative metabolic analysis tool. *Bioinformatics, 26*(24), 3140–3142. https://doi.org/10.1093/bioinformatics/btq602

Opportunities for System Neuroscience

Russell W. Chan, Bradley Jay Edelman, Shui Ying Tsang, Kai Gao, and Albert Cheung-Hoi Yu

Abstract Systems neuroscience explores the intricate organization and dynamic function of neural circuits and networks within the brain. By elucidating how these complex networks integrate to execute mental operations, this field aims to deepen our understanding of the biological basis of cognition, behavior, and consciousness. In this chapter, we outline the promising future of systems neuroscience, highlighting the emerging opportunities afforded by powerful technological innovations and their applications. Cutting-edge tools such as awake functional MRI, ultrahigh field strength neuroimaging, functional ultrasound imaging, and optoacoustic techniques have revolutionized the field, enabling unprecedented observation and analysis of brain activity. The insights gleaned from these advanced methodologies have empowered the development of a suite of exciting applications across diverse domains. These include brain-machine interfaces (BMIs) for neural prosthetics, cognitive enhancement therapies, personalized mental health interventions, and precision medicine approaches. As our comprehension of neural systems continues

R. W. Chan
Hai Kang Life Corporation Ltd, Hong Kong, China

Hong Kong Centre for Cerebro-cardiovascular Health Engineering (COCHE), Hong Kong, China

B. J. Edelman
Brain-Wide Circuits for Behavior Research Group, Max Planck Institute of Biological Intelligence, Planegg, Germany

Emotion Research Department, Max Planck Institute of Psychiatry, Munich, Germany

S. Y. Tsang
Hai Kang Life Corporation Ltd, Hong Kong, China

K. Gao
Children's Medical Center, Peking University First Hospital, Beijing, China

A. C.-H. Yu (✉)
Hai Kang Life Corporation Ltd, Hong Kong, China

Neuroscience Research Institute, Peking University, Beijing, China
e-mail: achy@haikanglife.com; achy@bjmu.edu.cn

A. C.-H. Yu et al. (eds.), *Systems Neuroscience*, Advances in Neurobiology 41,
https://doi.org/10.1007/978-3-031-69188-1_10

to grow, it is envisioned that these and related applications will become increasingly refined and impactful in improving human health and well-being.

Keywords Systems neuroscience · Physiological mechanisms · Mental operation · Bioimaging applications · Technological advancements

Systems neuroscience studies how neurons work together in complex networks to give rise to integrative functions such as vision, audition, somatosensation, and motor activity, as well as higher-order functions such as the neural basis of behavior, self-awareness, and empathy. Where neuroscience broadly encompasses the investigation of the nervous system at multiple levels, ranging from molecular and cellular processes to the functioning of large-scale systems (including behavior and cognition), systems neuroscience specifically focuses on understanding the organization and function of neural circuits and networks in the brain. In other words, systems neuroscience integrates genetic, molecular and cellular approaches to understand neural cell interaction, brain structure and function, and to study high-level mental functions. The main goal of systems neuroscience is to explain the bridge between the physiological mechanisms and the corresponding mental operations.

In addition to playing a crucial role in advancing our understanding of the brain and its functions, systems neuroscience has been instrumental in the development of various scientific and technological applications. This field has witnessed significant advancements in recent years, leading to exciting applications across various domains:

Brain-Machine Interfaces (BMIs) Systems neuroscience has contributed to the development of advanced BMIs, enabling direct communication between the brain and external devices. These interfaces allow individuals with paralysis or diseases such as stroke and amyotrophic lateral sclerosis to control robotic limbs, prostheses, or computer systems using only their neural activity (Edelman et al., 2019; Chen et al., 2019). Systems neuroscience research has focused on decoding neural signals associated with movement, mental intention, and sensation to create more intuitive and accurate BMIs.

Neural Prosthetics Systems neuroscience has played a crucial role in the development of neural prosthetics that restore sensory feedback to individuals with limb loss. By integrating neural recordings with artificial sensors, researchers have enabled amputees to perceive touch, pressure, and temperature through their prosthetic limbs, improving the dexterity and sense of embodiment associated with these devices (Flesher et al., 2021; Tan et al., 2014).

Cognitive Enhancement By understanding the neural circuits involved in attention, memory, and decision-making, researchers can develop interventions to improve these cognitive functions. Techniques such as neurofeedback, brain stimulation, and cognitive training programs are being explored to enhance cognitive performance in healthy individuals and those with cognitive impairments (Anguera et al., 2013; Wessel et al., 2023).

Mental Health Interventions By investigating the neural mechanisms underlying psychiatric conditions such as depression, anxiety, and addiction, researchers can identify potential targets for therapeutic interventions. Techniques like deep brain stimulation, transcranial magnetic stimulation, and real-time fMRI neurofeedback are being explored as potential treatments for these disorders (Scangos et al., 2021; Diana et al., 2017).

Precision Medicine in Disease Systems neuroscience provides insights into the heterogeneity of neurological and psychiatric disorders, as well as individual variability in treatment responses. By investigating how alterations in neural circuits contribute to diseases such as Alzheimer's disease (AD), Parkinson's disease (PD), epilepsy, schizophrenia, and depression, systems neuroscience can pave the way for improved diagnosis, treatment, and prevention strategies. Moreover, by understanding the neural mechanisms underlying different subtypes of neurological disorders, researchers can develop personalized treatment approaches that target specific circuit abnormalities (Gratton et al., 2020; Zerbi et al., 2021). This approach holds promise for improving treatment outcomes and reducing side effects.

Understanding Consciousness By studying the interactions between different brain regions and their contribution to subjective experience, researchers aim to understand how consciousness emerges from neural activity. This research has implications for our understanding of self-awareness, perception, and altered states of consciousness.

Artificial Intelligence (AI) and Robotics Understanding the brain's computational principles can inspire the development of more efficient and robust AI algorithms and robotic systems. By drawing inspiration from neuronal networks and cognitive processes, researchers aim to improve machine learning algorithms, pattern recognition systems, and robotic control strategies (Schrimpf et al., 2021). Insights from systems neuroscience can help create more efficient and adaptable AI systems.

Improving Learning and Education By uncovering the neural mechanisms of learning and memory, systems neuroscience can be applied to help optimize teaching methods, design effective training programs, and develop educational technologies that enhance learning and retention (Gilbert et al., 2001; Reinert et al., 2021).

Promoting Brain Health and Well-Being Systems neuroscience contributes to our knowledge of how lifestyle factors, such as sleep, exercise, and stress, impact brain function and overall well-being (Tai et al., 2022). This knowledge can inform strategies to promote brain health, prevent cognitive decline, and design interventions for mental health disorders.

Guiding Neural Engineering Systems neuroscience provides valuable information for designing neural implants, neurostimulation techniques, and other therapeutic interventions. By understanding the interaction between neural circuits and external stimuli, more precise and targeted approaches for treating neurological disorders and enhancing brain function can be developed.

Many of these technologies are well on their way to successful human translation, some of which are already widespread. For example, deep brain stimulation has been approved for over 20 years by the US food and drug administration (FDA) for a variety of disorders including Parkinson's disease and more recently obsessive-compulsive disorder. Transcranial magnetic stimulation was also approved in 2008 as a noninvasive therapy for treatment-resistant depression. Finally, FDA has also recently approved feasibility studies for various BMI devices from companies such as Neuralink and Synchron that may eventually help paralyzed individuals regain independence. The clinical acceptance of these technologies represents a significant advance in innovative neurotechnologies that can be used to treat humans. Nevertheless, the effects of many of these approaches, as well as the conditions that they treat, are still poorly understood. To further advance these applications, ongoing bidirectional work between basic systems neuroscience research and clinical translation must continue. Within this realm, priority should be given to the investigation into the mechanisms of both neurological dysfunctions and therapeutic treatments.

As a step in this direction, a recent explosion of techniques for studying neural dynamics has revolutionized systems neuroscience and has enabled significant discoveries in our understanding of brain function. Indeed, these tools have provided unprecedented observations of neural activity and elegant means of manipulating specific components of neuronal circuits, enabling systems neuroscientists to go beyond the description of reflexive behaviors or the first stages of sensory processing. In fact, the use of these techniques facilitates the understanding of how cognitive processes such as memory are implemented at the level of large-scale circuits to support adaptive behavior.

The complex connections distributed throughout the brain are often compared to circuitry that carries information across various functionally disparate regions. While many individual circuits have been implicated in specific cognitive and bodily functions, studying how circuits integrate with large-scale networks can provide a more comprehensive overview of brain function. Such an encompassing approach can provide valuable insights into various neurological disorders, which are often thought of as circuit-based deficits. Neurodegenerative disorders, in particular, such as PD and AD, impose devastating effects on patient's quality-of-life and are often accompanied by significant personal and societal costs. The treatment of these and other disorders has rapidly developed over the past few decades; however, these treatments primarily target symptoms rather than the disease mechanism itself. In fact, these diseases have yet to be cured in large part due an incomplete systems-level understanding of healthy brain states, which must precede the identification of dysfunctional counterparts. While the ultimate goal of this work is to develop therapies for human applications, animal research can notably aid in advancing such strategies.

As neural circuits often span multiple anatomical regions and interact with various functional networks, measuring increasingly large numbers of neurons is becoming paramount to understanding these complex systems. The ever-increasing desire for more information has driven various neural recording modalities to push the boundaries of both brain coverage and spatiotemporal resolution. Ideally, whole-brain recordings at cellular resolution would provide the massive datasets to understand brain function, but no technique currently exists. Alternatively, various technologies offer diverse advantages that can be combined to achieve similar feats. With such large datasets, however, large-scale computational modeling and machine learning/AI algorithms must be appropriately applied to ensure proper data handling and interpretability. The reliability and standardization of these tools also require community engagement and participation (Grandjean et al., 2023)—open-access data repositories and open-source code banks greatly improve the collective knowledge base and reproducibility of results, and the field must continue to move in this direction.

With this being said, the landscape of systems neuroscience exhibits a vastly different outlook than it once did. Traditional approaches to studying neural circuitry involved lesion studies or local electrophysiological recordings in pre-selected regions of interest. However, these low-throughput techniques provide limited output relative to the physiological burden that they impose. Therefore, more advanced techniques that provide a higher yield are becoming increasingly popular. For example, two-photon microscopy can record from over ten thousands of neurons at high temporal resolution (Stringer et al., 2019). Similarly, silicon probes (e.g. neuropixels) exhibit thousands of recording sites and multiple probes can be inserted and recorded from in parallel (Steinmetz et al., 2021). Finally, modalities such as functional MRI (fMRI) and positron emission tomography capture whole-brain hemodynamic and metabolic information, respectively. fMRI in particular has seen significant advances for preclinical applications in the past two decades and provides a particularly useful opportunity to study neural dynamics at the whole-brain and network scale.

Furthermore, optogenetic tools that are still only readily abundant for rodent applications allow for the precise manipulation of specific cell types and circuits. Combining optogenetic stimulation with whole-brain fMRI (i.e., ofMRI) therefore provides inclusive measurements of both neural circuits and brain networks (Chan et al., 2017). This approach disrupted the field of systems neuroscience by providing more information than any previous approach could. However, ofMRI can be limited through the need for anesthesia during recording, which restricts access to behavioral responses and measurements. This boundary may, however, be softened through the use of awake fMRI and ultrahigh field strengths (now reaching 18 Tesla), both of which are becoming increasingly more popular (Gutierrez-Barragan et al., 2022). An adjacent modality, functional ultrasound (fUS) imaging, shares many signal characteristics as fMRI and has recently been employed when asking similar circuit-based questions. Conveniently, fUS utilizes a small and portable recording probe that facilitates pairing with diverse head-fixed behavior rigs. Therefore, it would appear that fUS offers a unique opportunity to combine whole-brain recordings, optogenetic circuit manipulations, and behavioral assays all at the same time (Edelman et al., 2023). Despite this, fUS is a relatively novel imaging modality, and there are still many unknowns regarding how the vascular signal acquired relates to the underlying neuronal responses, which should be accounted for when interpreting these results. Another novel imaging modality, optoacoustic imaging, may soon enable the rapid acquisition of whole-brain neural dynamics, opening more opportunities for system neuroscience discoveries (Gottschalk et al., 2019). These emerging technologies will promote the continuous development of systems neuroscience and provide great improvement in therapies for human applications.

Systems neuroscience is experiencing an exciting period supported by unprecedented technological breakthroughs. It is in essence an interdisciplinary field, which has recently given rise to the emergence of new subfields, such as "systems developmental neuroscience" and "systems social neuroscience." In the near future, systems neuroscientists throughout these sub-fields will certainly need to work hand in hand with data and computational neuroscientists to standardize their experiments and analysis, and bridge the gap between data and understanding.

The use of multi-modal approaches combining different technologies and advanced signal processing is poised to increase the understanding of neural mechanisms underlying neurological and cognitive functions. Intracranial recordings provide millisecond resolution time courses of neural activity, allowing the investigation of causal relationships between the activation of specific brain areas and the emergence of behavior. This should lead to a major intellectual step forward regarding the organization of the human motor system and the spatiotemporal dynamics that govern the contribution of motor outputs to cognitive functions.

Overall, systems neuroscience is essential for unraveling the mysteries of the brain and its functions. With recent breakthroughs in imaging technology, there are now many tools available to achieve this and which can lead to advancements in various scientific disciplines and practical applications to improve human health, and our understanding of the mind.

References

Anguera, J. A., et al. (2013). Video game training enhances cognitive control in older adults. *Nature, 501*, 97–101.

Chan, R. W., et al. (2017). Low-frequency hippocampal–cortical activity drives brain-wide resting-state functional MRI connectivity. *Proceedings of the National Academy of Sciences, 114*, E6972–E6981.

Chen, X., Zhao, B., Wang, Y., & Gao, X. (2019). Combination of high-frequency SSVEP-based BCI and computer vision for controlling a robotic arm. *Journal of Neural Engineering, 16*, 026012.

Diana, M., et al. (2017). Rehabilitating the addicted brain with transcranial magnetic stimulation. *Nature Reviews Neuroscience, 18*, 685–693.

Edelman, B. J., et al. (2019). Noninvasive neuroimaging enhances continuous neural tracking for robotic device control. *Science Robotics*. https://doi.org/10.1126/scirobotics.aaw6844
Edelman, B. J. et al. (2023) The COMBO window: A chronic cranial implant for multiscale circuit interrogation in mice. Preprint at https://doi.org/10.1101/2023.10.02.560470.
Flesher, S. N., et al. (2021). A brain-computer interface that evokes tactile sensations improves robotic arm control. *Science, 372*, 831–836.
Gilbert, C. D., Sigman, M., & Crist, R. E. (2001). The neural basis of perceptual learning. *Neuron, 31*, 681–697.
Gottschalk, S., et al. (2019). Rapid volumetric optoacoustic imaging of neural dynamics across the mouse brain. *Nature Biomedical Engineering, 3*, 392–401.
Grandjean, J., et al. (2023). A consensus protocol for functional connectivity analysis in the rat brain. *Nature Neuroscience, 26*, 673–681.
Gratton, C., et al. (2020). Defining individual-specific functional neuroanatomy for precision psychiatry. *Biological Psychiatry, 88*, 28–39.
Gutierrez-Barragan, D., et al. (2022). Unique spatiotemporal fMRI dynamics in the awake mouse brain. *Current Biology, 32*, 631–644.e6.
Reinert, S., Hübener, M., Bonhoeffer, T., & Goltstein, P. M. (2021). Mouse prefrontal cortex represents learned rules for categorization. *Nature, 593*, 411–417.
Scangos, K. W., et al. (2021). Closed-loop neuromodulation in an individual with treatment-resistant depression. *Nature Medicine, 27*, 1696–1700.
Schrimpf, M., et al. (2021). The neural architecture of language: Integrative modeling converges on predictive processing. *Proceedings of the National Academy of Sciences, 118*, e2105646118.
Steinmetz, N. A., et al. (2021). Neuropixels 2.0: A miniaturized high-density probe for stable, long-term brain recordings. *Science, 372*, eabf4588.
Stringer, C., et al. (2019). Spontaneous behaviors drive multidimensional, brainwide activity. *Science, 364*, 255.
Tai, X. Y., Chen, C., Manohar, S., & Husain, M. (2022). Impact of sleep duration on executive function and brain structure. *Communications Biology, 5*, 1–10.
Tan, D. W., et al. (2014). A neural interface provides long-term stable natural touch perception. *Science Translational Medicine, 6*, 257ra138.
Wessel, M. J., et al. (2023). Noninvasive theta-burst stimulation of the human striatum enhances striatal activity and motor skill learning. *Nature Neuroscience, 26*, 2005–2016.
Zerbi, V., et al. (2021). Brain mapping across 16 autism mouse models reveals a spectrum of functional connectivity subtypes. *Molecular Psychiatry, 26*, 7610–7620.

Index

A. C.-H. Yu et al. (eds.), *Systems Neuroscience*, Advances in Neurobiology 41,
https://doi.org/10.1007/978-3-031-69188-1

M

N

O

P

R

S

T

SPRINGER NATURE

GPSR Compliance

The European Union's (EU) General Product Safety Regulation (GPSR) is a set of rules that requires consumer products to be safe and our obligations to ensure this.

If you have any concerns about our products, you can contact us on ProductSafety@springernature.com

In case Publisher is established outside the EU, the EU authorized representative is:

Springer Nature Customer Service Center GmbH
Europaplatz 3
69115 Heidelberg, Germany

The manufacturer's authorised representative in the EU is Springer Nature Customer Service Centre GmbH, Europaplatz 3, 69115 Heidelberg, Germany. If you have any concerns regarding our products, please contact ProductSafety@springernature.com

Printed and bound by CPI Group (UK) Ltd, Croydon, CR0 4YY
01/10/2026
02225961-0010